CONTEMPORARY PHYSICAL GEOGRAPHY

About the Author

Nater Singh Raina, earned M.A. degree from CSRD/SSS, JNU N. Delhi and obtained Ph.D. degree in Geomorphology under guidance of Prof. H.S. Sharma from University of Rajasthan, Jaipur in 1987. Having worked as Assistant Professor in P.G. of Geography, University of Jammu, Associate Professor with NCERT New Delhi, heading Government Degree College, Kargil is presently serving as Principal of Government Degree College, Hiranagar (J & K).

He has contribution of publishing five books and more than 25 research papers.

Contemporary Physical Geography

Nater Singh Raina

CONCEPT PUBLISHING COMPANY PVT. LTD.,
NEW DELHI-110 059

ISBN-13: 978-81-8069-762-3 (P.B.)

First Published 2012

Published and Printed by

Concept
Since 1974

Concept Publishing Company Pvt. Ltd.
Regd. Office:
A/15-16, Commercial Block, Mohan Garden
New Delhi-110059 (India)
Phones : 25351460, 25351794, *Fax* : 091-11-25357109
Email : publishing@conceptpub.com
Website: www.conceptpub.com

Editorial Office:
H-13, Bali Nagar, New Delhi-110 015, India.

Cataloging in Publication Data--*Courtesy:* D.K. Agencies (P) Ltd. <docinfo@dkagencies.com>

Raina, Nater Singh.
Contemporary physical geography / Nater Singh Raina.
p. cm.
Includes bibliographical references (p.) and index.
ISBN 9788180697616

1. Physical geography. I. Title.

DDC 910.02 22

PREFACE

The students and teachers of Geography needed a textbook on Physical Geography, rich in content, up dated with latest research and data in the field of Geomorphology, Climatology, Oceanography and soil sciences, with Indian examples. The present work is an effort with these objectives in view.

Contemporary Physical Geography has been written and organized keeping in view the syllabi of physical geography in various Universities of India. There are 27 chapters in the present book. First five chapters cover the definition of physical Geography, geological time scale, the solar system, earth movements and models of landscape evolution. In chapter six the formation, classification and other properties of the rocks are discussed. Descriptive classification of mass wasting is described in chapter seven. A detailed study of landscapes is given in chapter eight.

Evolution and development of glacial, coastal and aeolian landscapes are covered in chapter nine, ten and eleven, respectively. Applied Geomorphology has come up as a very important branch of earth sciences, which has been covered in chapter twelve.

Climatology has now passed from a rather academic pursuit to a widely known area of study which is discussed in chapter thirteen. The author has made a conscious effort to retain both literary and scientific components like atmosphere, winds, solar radiation and earth's energy budget, air pressure, atmospheric moisture and classification of climate in chapter fourteen, fifteen, sixteen, seventeen, eighteen and nineteen respectively. A scientific study of oceans covering all the aspects of Oceanography is covered in chapter twenty. Chapter twenty-one deals with environmental pollution and its contents.

A fundamental knowledge of soil science is a prerequisite for tackling many natural resource challenges that humanity will face in the 21st century. Contemporary Physical Geography emphasises soil as a natural resource and highlights the many interactions between soil and other components of ecosystem in chapter twenty-two. The role and nature of time in Geomorphology is discussed in chapter twenty-three. Lithostratigraphic dealing of Geology is given in chapter twenty-four, while climate and global climate models are discussed in chapter twenty-five and twenty six respectively. Mapping is an essential tool for the geomorphologists. Mapping techniques have become more important with time as changing technology allows more complex mapping. In the last chapter a detailed description of mapping and topographic symbols is given with suitable diagrams. Latest bibliography is presented at the end of the book for further reading.

It is hoped that the *Contemporary Physical Geography* shall fully meet the requirements not only of geography students but also enrich the knowledge of those who wish to sit for various competitive examinations being conducted by the UPSC, State Public Service Commissions and other examinations bodies across India.

Nater Singh Raina

ACKNOWLEDGEMENTS

I express my sincere gratitude to my friend and my source of inspiration Sh. Suresh Kumar *I.A.S.* He stimulated my mind for the present project while I was serving at the NCERT, New Delhi.

My heart felt thanks are due to my better half Tejinder Kour, and my lovely sons, Gagandeep and Navdeep for the sacrifice they made during the writing of this book.

Sincere thanks are also due to members of various Libraries: JNU, New Delhi, NCERT, New Delhi and Delhi University for providing me reference material.

Last but not the least, I am thankful to Prof. Onkar Singh for helping me in proof reading and Sh. Ashok Kumar Mittal for publishing this book.

Nater Singh Raina

CONTENTS

1

CONTEMPORARY PHYSICAL GEOGRAPHY

The Geography which is taught in schools and colleges today is a comparatively new subject. Fifty years back students in the institutions probably spent most of their time learning by heart long list of the highest mountains, the longest rivers and the largest towns. Very little was taught about the various people of the world, their daily lives, their work, their food and their problems. Now question arises why should we study geography? Because we live on the surface of the earth and surrounded by the environment. Environment comprises Lithosphere, Hydrosphere and Atmosphere. We are standing on the surface of the earth (Lithosphere) and drinking water from the Hydrosphere and breathing oxygen from the Atmosphere. The study of these spheres and their relationship with human beings comes under the umbrella of Geography. The term geography was first coined by Eratosthenese, a Greek scholar (276-194 BC). The word "GEOGRAPHY" is derived from the Greek words 'Geo', meaning the earth and 'Graphos' meaning a description. Hence, Geography may be defined in its simplest form as the description of the earth.

The definition of Geography has a history of its own. In very early periods of its history geography was purely descriptive and its main concern was noting down the facts about the places and the people. It was sometimes referred to as gazetteer Geography. Although the knowledge of facts is very necessary but purely factual knowledge does not bring any fruit. The old fashioned geography with its long catalogues of names of capes and bays and towns and commodities was not much interesting or inspiring subject. Most of the maps prepared were also concerning the places, locating the mountains, rivers and towns etc.

The geographical discoveries of the fifteenth and sixteenth centuries and the explorations of the continents which followed had important effects upon the growth of geography. The improvement in scientific equipment and interest in mathematical geography led to the development of improved cartography. The opening up of new lands such as America and Australia brought a new urge in people to know the distant lands and a variety of new commodities.

Early Geographers concerned themselves with the collection, systematization and correlation of the vast wealth of new facts which discoveries and explorations had placed at the Geography's disposal. As the accumulation of facts increased, the geography narrowed its field of enquiry as it started dissociating from those aspects of terrestrial phenomena which more properly belonged to other sciences, such as physics, geology and biology.

During the nineteenth century quite a number of scientific explorers who were also great naturalists, added much to the knowledge and understanding of the distribution of natural, physical and human phenomena. Among them were Alexander von Humboldt, a German who travelled to South America and Siberia, Sir Joseph Hooker visited Antarctica and travelled to India, A.R. Wallace carried out important work in biology in South-East Asia and Charles Darwin voyaged around the world in Beagle.

Beginning of Scientific Approach

The *Origin of Species,* a well-known book by Charles Darwin, was published in 1859 which revolutionized all previous ideas and caused a great unrest in scientific circles by saying that forms of life had evolved from earlier more primitive form. What to speak of biology and anthropology it affected almost all the sciences. It started an unending era of rethinking, questioning and experimenting. Geographers now were no longer satisfied to learn that kiwis are found in New Zealand or that Dead Sea was extremely salty. They now wanted to know that why kiwis are found only in New Zealand and not anywhere else and why dead sea is more salty than other oceans. This inquisitive thinking leads to the development of a causal relationship in geography. Geographers not only started looking for the causes of the things but went further to note the effects. Even today the Geographers are interested in cause and effect approaches to various geographical phenomena.

Today Geography not only explains the surface features as they are but also investigates, correlates, organizes and rationalizes the phenomena of the earth.

Today it is a descriptive and analytical study. Moreover, it has its own particular point of view. Hence, Geography has been called simultaneously an art, a science and a philosophy. It is pertinent to look at some important definitions of Geography given by various scholars.

Many famous geographers and non-geographers have attempted to define the discipline in a few short words. The concept of geography has also changed through the ages, making a definition for such a dynamic and all-encompassing subject difficult. With the help of Gregg Wassmansdorf, here are some ideas about geography throughout the ages:

"The purpose of geography is to provide a view of the whole earth by mapping the location of places." —Ptolemy, 150 CE

"Synoptic discipline synthesizing findings of other sciences through the concept of Raum (area or space)."
—Immanuel Kant, c. 1780

"Synthesizing discipline to connect the general with the special through measurement, mapping, and a regional emphasis."
—Alexander von Humboldt, 1845

"Man in society and local variations in environment."
—Halford Mackinder, 1887

"How environment apparently controls human behaviour."
—Ellen Semple, c. 1911

"Study of human ecology; adjustment of man to natural surroundings." —Harland Barrows, 1923

"The science concerned with the formulation of the laws governing the spatial distribution of certain features on the surface of the earth." —Fred Schaefer, 1953

"To provide accurate, orderly, and rational description and interpretation of the variable character of the earth surface."
—Richard Hartshorne, 1959

"Geography is both science and art." —H.C. Darby, 1962

"To understand the earth as the world of man"
—J.O.M. Broek, 1965

"Geography is fundamentally the regional or chorological science of the surface of the earth." —Robert E. Dickinson, 1969

"Study of variations in phenomena from place to place."
—Holt-Jensen, 1980

"...concerned with the locational or spatial variation in both physical and human phenomena at the earth's surface."
—Martin Kenzer, 1989

"Geography is the study of earth as the home of people."
—Yi-Fu Tuan, 1991

"Geography is the study of the patterns and processes of human (built) and environmental (natural) landscapes, where landscapes comprise real (objective) and perceived (subjective) space."
—Gregg Wassmansdorf, 1995

Carl Ritter says that geography is concerned with the objects on the earth's surface around man.

Hettner considered geography as a chorological science in terms of spatial differences and their relationship.

Chorlley writes that Geography is concerned to know the earth, its total character in combination.

American college dictionary puts Geography as the study of the area differentiation of the earth's surface as shown in the character, arrangement and interrelation over the earth of various elements such as climate, relief, population etc.

"Geography is a science which studies the distribution of Phenomena"—Marthe. He considered Geography as a study which describes how the things are distributed over space. This was however an achievement in systematic approach.

Spykman writes, *"Geography is the strategy of men, space and resources."* He considers the locational factor having prime importance in the study of Geography.

James considered that geography seeks to interpret the importance of similarities and differences among places in terms of causes and consequences. He says that a geographer needs to explain what sort of similarities exist between the areas and why. He has to explore whether there exists a casual relationship and what will be the result of these relationships.

Richard Hortshorne explaining his viewpoint says that geography is a discipline that seeks to describe and interpret the spatial variations on the surface of the earth. In other words, he is seeking a description of variable character of the earth's surface. The variations in the physical characteristics, variation in cultural, political, social characteristics as envisaged in relation to man.

Geography, writes Hagget, is uniquely relevant to the current concern both with environment and ecology and with regional contrasts and imbalance in welfare. He says that geographers are concerned with the structure and interaction of two major systems, the ecological system that links man with his environment, and the spatial system that links one region with another in a complex interchange of flows. Modern geography lays a strong emphasis on the ways of analysis in its research problems, and many of these ways are quantitative.

Scope of Geography

The scope of Geography extends over where, why and how of things found on the earth. The subject has a wide range of its domain which has been sometimes considered too wide by some people. However, according to Robinson scope of the subject covers the following areas:

1. The size, shape and movements of the earth.
2. The position and distribution of land masses and water bodies.
3. The structure of rocks and reliefs of the earth's surface.
4. The waters of the oceans, seas and lakes and their movements.
5. The atmospheric conditions and the resultant climates.
6. Distribution of animals and the vegetation cover on the earth.
7. Human races and the distribution of population.
8. Activities of man to earn a livelihood.
9. Human shelter in the form of human settlements and their types.
10. Political organizations and their relationship with human groups.

To sum up, in simple words the scope of Geography is the world itself or earth surface with water and atmosphere. We can say it extends over three realms: Atmosphere, lithosphere and hydrosphere.

Divisions of Geography

It is an uphill task to determine clear cut dichotomies in Geography because all the fields in the discipline are interdependent and are difficult to dissociate from each other. There are two basic approaches in Geography, i.e. from particular to general and from general to particular. On the basis of the approach we can have Systematic and Regional Geography (Figs. 1.2 and 1.3). In systematic geography we study a particular phenomenon systematically over the world such as in climatology, geomorphology, oceanography, historical geography etc. In regional geography we take up a region and study all the phenomena in totality of a region such as in geography of India, geography of North America etc.

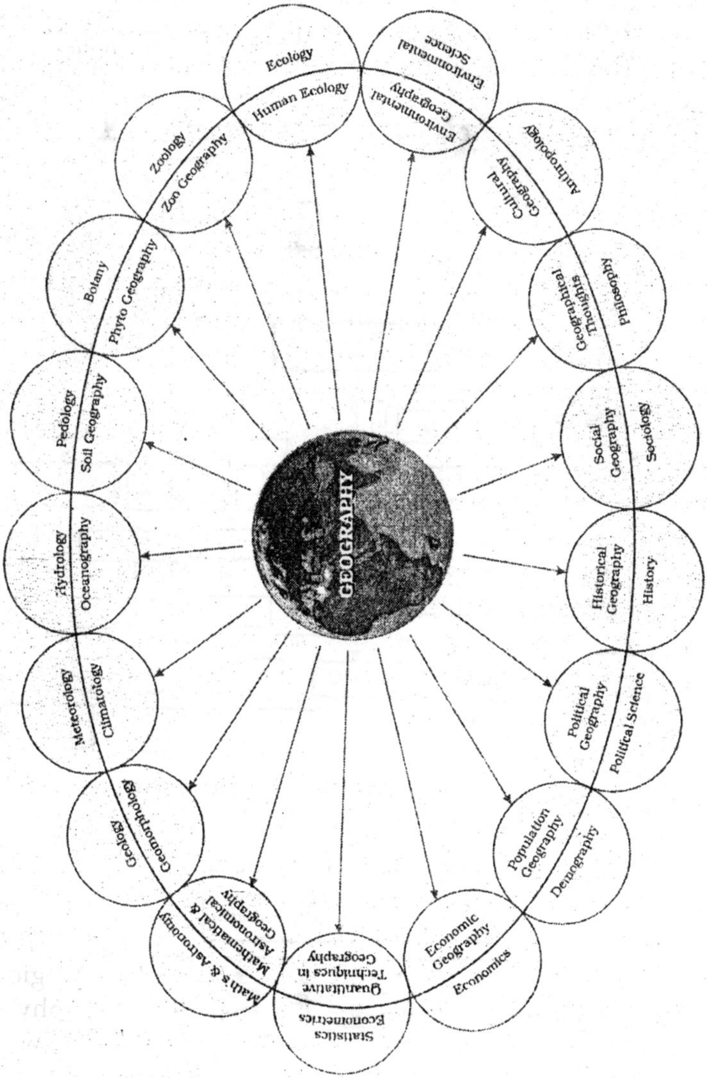

Fig. 1.1: Geography and its relation with other subjects

Fig. 1.1 : Linkage of geography with different natural and social sciences.

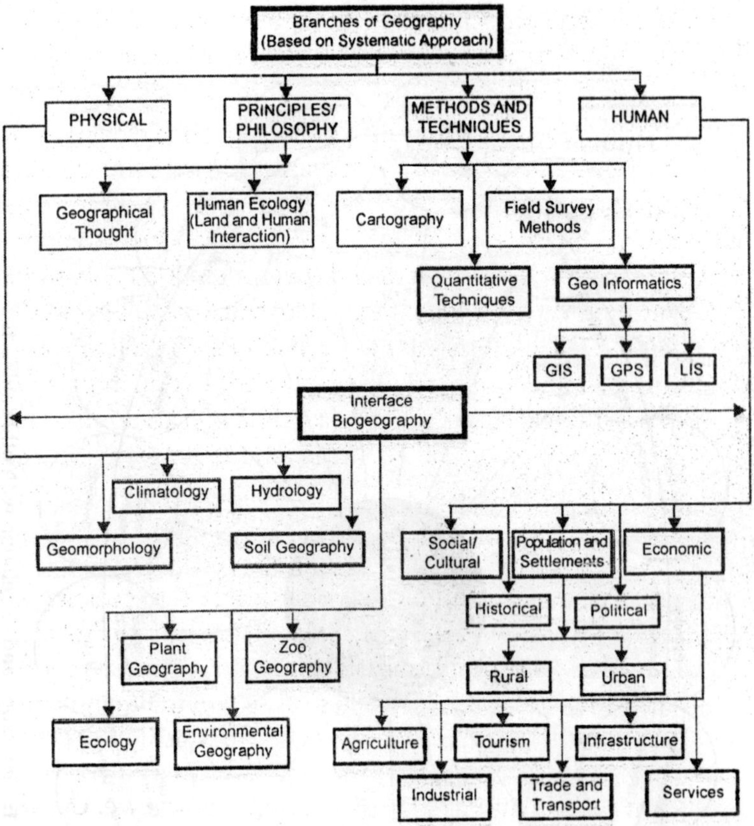

Fig. 1.2: Branches of Geography based on systematic approach.

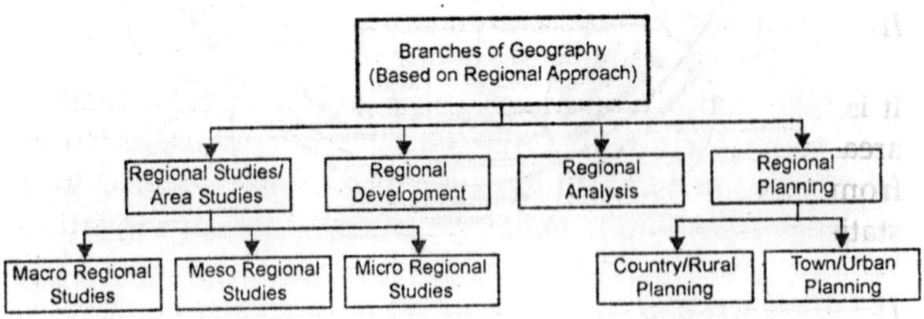

Fig. 1.3: Branches of Geography based on regional approach.

The orthodox division of geography still carries weight as it is more detailed and covers all the aspects:

1. Physical Geography
2. Human Geography

Physical Geography

Physical Geography is systematic in nature studying various physical aspects of the earth. However, it can be further subdivided into the following sub-divisions:

I. Geomorphology

Geomorphology is the description and interpretation of landforms. It is the study of physical features on the surface of the earth as envisaged in the light of endogenic and exogenic forces.

II. Oceanography

A systematic study of ocean waters in relation to life in water, the temperature, depth, movements and composition of ocean waters is known as oceanography. Oceanography is as important to mankind as any other branch of science. Oceans are the sources of food, cheapest highways for transportation, sources of energy etc.

III. Climatology

It is the scientific study of climates of a particular region or area. In other words, it explains the nature of climate as it differs from place to place. It is dependent upon meteorology for statistics regarding temperature pressure and precipitation.

IV. Biogeography

About three decades back Professor Wooldridge had written when a geographer is in 'blueprint mood' devising deal schemes for his subject, he is naturally inclined to recognize a sub-division called biogeography. The term includes the study of plant and animals. Here we study the habitat of various species of plants and animals and their relationship to man and geographical environment.

Human Geography

Human Geography may also be divided into a number of sub-disciplines:

I. Economic Geography

Economic geography deals with the economic activities, natural resources, industries, agriculture, etc. The subject of economic geography has become so vast that it has been subdivided into different fields:

1. Agricultural Geography
2. Industrial Geography (Geography of Manufacturing)
3. Geography of Resources
4. Geography of Transport
5. Commercial Geography

II. Historical Geography

As history has a geography on its back similarly geography too has a history of its own development. The concepts and ideas have changed with the passage of history. Every subject has a past without which it is difficult to project the future. In a way it is the study of historical present.

III. Social Geography

Social Geography studies the social aspects as caste, religion or stratification of a society. The study of blacks and whites in relation to the social implications. It includes the study of tribes, customs and traditions, etc.

IV. Political Geography

It is the study of political units. It focuses attention on both the external and internal relationships of states. Political geographer studies the political units from two different angles—firstly, the political action inside the unit and secondly, the role of geography in international politics.

V. Geography of Population

It describes and explains the numbers of people living in an area and their relationship with land area. It also explores scientifically the causes of their presence in a region. Population Geography seeks to explain different characteristics of human population such as sex ratio, age and sex composition, literacy, migration, manpower, density ratios, etc.

VI. Geography of Settlements

It deals with the nature of human shelter, its origin and various forms. The development of early settlements *vis-à-vis* geographical location. Horizontal expansion of settlements and their developments. The subject also deals with the relationship of nature of settlements and natural environment.

VII. Urban Geography

It is a systematic study of urban settlements. The scope of urban geography extends over the functions, morphology and concepts and planning of urban units.

Place of Geomorphology in the Discipline

Geomorphology is the core of Geography. It studies the landforms and is a scientific study of the geometric features of the earth's surface. It involves itself in explaining the nature of landform features, their evolution and the changes taking place due to geomorphic agents. Geomorphology has been taught by the geographers as well as the geologists and has remained a bone of contention between the two.

Geomorphology is the core concern of geographers whereas it is not the case with geologists. However in some countries the grip of geographers on the discipline is becoming loose. Geographers are shifting their interest in the peripheral areas such as economic geography, social geography, political geography, ecology, etc.

The element of objectivity is considered as a strong point of a discipline. In no other branch of geography, the objectivity is as strong as it is in geomorphology.

Some geographers believe that the shift of interest away from geomorphology is because our teachers do not have a geological base. Geographers started studying landforms without delving deep into history. A geographer is not supposed to trespass the domain of geology which deals with the history of the landforms.

1. Geographers are trying to analyse the problems of spatial variations in the landforms.
2. Quantitative analysis in geomorphology has become the basis of landform study today. A number of quantitative techniques have been developed to analyze the landforms, i.e. bifurcation ratio, number order ratio, stream length ratio, morphometric analysis, basin shape indices etc.
3. Analysis of landforms in relation to the economic development. Before planning for any region, a study of geomorphic stage is gaining importance.

REFERENCES

Allaby, A. and M. Allaby. (Editors). 1999. *Dictionary of Earth Sciences*. 2nd Edition. Oxford University Press, London. (Excellent dictionary of more than 6,000 earth science terms many of which are used in physical geography—browse/buy from amazon.com or amazon.ca)

Christopherson, R.W. 2005. *Geosystems: An Introduction to Physical Geography*. 5th Edition. Prentice Hall, Upper Saddle River, New

Jersey. (Excellent university-level textbook that focuses broadly on physical geography—browse/buy from amazon.com or amazon.ca)

Gregory, K.J. 2001. *The Changing Nature of Physical Geography.* Edward Arnold, London. (Very interesting university-level textbook that explores the historical evolution of thought and methods in physical geography—browse/buy from amazon.com or amazon.ca)

Harvey, M.E. and B.P. Holly (Editors). 1981. *Themes in Geographic Thought.* Croom Helm, London. (Very interesting university-level textbook that explores the historical evolution of geographical thought and methods—browse/buy from amazon.com or amazon.ca)

Holt-Jensen, A. 2000. *Geography—History and Concepts: A Student's Guide.* 3rd Edition. Sage Publications. (Very interesting university-level textbook that explores the historical evolution of geographical thought and methods—browse/buy from amazon.com or amazon.ca)

Johnston, R.J. 1997. *Geography and Geographers: Anglo-American Geography Since 1945.* 5th Edition. Arnold Publishers, London. (Very interesting university-level textbook that explores the historical evolution of geographical thought and methods—browse/buy from amazon.com or amazon.ca)

Lanegran, D.A. and R. Palm (Editors). 1978. *An Invitation to Geography.* Second Edition. McGraw-Hill Publishing Co., New York. (Very interesting university-level textbook that explores the historical evolution of geographical thought and methods—browse/buy from amazon.com or amazon.ca)

Livingston, D.N. 1992. *The Geographical Tradition: Episodes in the History of a Contested Enterprise.* Basil Blackwell, Cambridge, Mass. (Very interesting university-level textbook that explores the historical evolution of geographical thought and methods—browse/buy from amazon.com or amazon.ca)

Lukermann, F. 1964. Geography as a formal intellectual discipline and the way in which it contributes to human knowledge. *The Canadian Geographer* 8(4): 167-172.

Martin, G.J. and P.E. James. 1993. *All Possible Worlds: A History of Geographical Ideas.* 3rd Edition. John Wiley and Sons, New York. (Very interesting university-level textbook that explores the historical evolution of geographical thought and ideas—browse/buy from amazon.com or amazon.ca)

McKnight, T.L. and D. Hess. 2002. *Physical Geography: A Landscape Appreciation.* 7th Edition. Prentice Hall, Upper Saddle River, New Jersey. (Good university-level textbook that focuses broadly on physical geography—browse/buy from amazon.com or amazon.ca)

Morrill, R.L. 1983. The nature, unity and value of Geography. *Professional Geographer* 35(1): 1-9.

Minshull, Roger M. 1970. *The Changing Nature of Geography*. Hutchinson, London.

Northey, M. and D.B. Knight. 2000. *Making Sense: A Student's Guide to Research and Writing: Geography and Environmental Studies*. 2nd edition. Oxford University Press, Toronto. (University-level textbook that helps students doing research in Geography and Environmental Science—browse/buy from amazon.com or amazon.ca)

Parker, S.P. (Editor). 2003. *McGraw-Hill Dictionary of Earth Science*. Second Edition. McGraw-Hill, New York. (Excellent dictionary of more than 10,000 earth science terms many of which are used in physical geography—browse/buy from amazon.com or amazon.ca)

Peet, R. 1998. *Modern Geographical Thought*. Blackwell Publishers, Oxford. (Great university-level textbook that explores the historical evolution of geographical thought and ideas—browse/buy from amazon.com or amazon.ca)

Robinson, J.L. 1976. A new look at the four traditions of Geography. *Journal of Geography* 75: 520-530.

Rogers, A., H. Viles and A. Goudie (Editors). 2003. *The Student's Companion to Geography*. 2nd Edition. Blackwell Publishers, Oxford. (University-level textbook that helps students to broadly understand Geography as an academic discipline—browse/buy from amazon.com or amazon.ca)

Strahler, Alan H. and Arthur Strahler. 2003. *Physical Geography: Science and Systems of the Human Environment*. 2nd Edition. John Wiley and Sons, New York. (Excellent university-level textbook that focuses broadly on physical geography—browse/buy from amazon.com or amazon.ca)

Whitlow, J.B. 2001. *The Penguin Dictionary of Physical Geography*. Second Edition. Penguin, Harmondsworth. (Must have reference for those wanting to learn physical geography terms—browse/buy from amazon.com or amazon.ca)

2

GEOLOGICAL TIME SCALE

The words of James Hutton, 'the present is the key to the past' have a value when we study the geological past of the earth's surface. The study of the present surface features cannot be explained effectively and meaningfully with a study of the geological past of the surface features. A.N. Strahler writes that development of relief features of the continents requires at least occasional brief reference to events of the geological past. Hence, the knowledge of the length and sequences of the geologic time units is necessary.

The knowledge of the occurrence of mineral deposits in different areas of the world can be better understood with the help of historical geology. Hence, it becomes all the more important for a geographer.

It is a sequential arrangement of geologic time units, as currently understood, and it is subdivided into four major eons, principally on the basis of the observable presence or absence and type of life-forms that characterize each.

Standard names evolved for the subdivisions of geologic time are Eons, Eras, Periods and Epochs. Figure 2.1 represents geological time scale.

Eons: An eon is the largest time interval into which geologic time is divided. The term **Hadean** (Greek for beneath part of the earth) is given to the oldest eon. This is the earliest part of the earth's history, an interval for which no rock record is known.

Archean (Greek for ancient) rocks the oldest we know on the earth, contain microscopic life-form of the bacterial character. **Proterozoic** (Greek for earlier life) rocks include

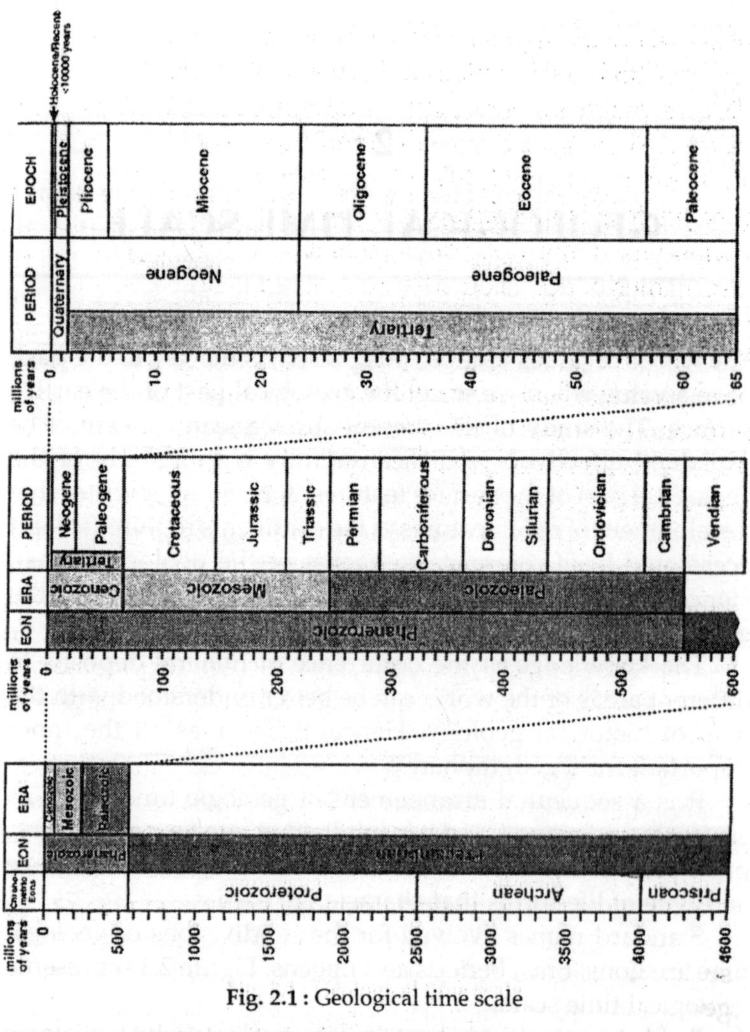

Fig. 2.1 : Geological time scale

evidence of multi-celled organisms that lack preservable hard parts. **Phanerozoic** (Greek for visible life) rocks often contain plentiful evidence of past life in the form of well preserved hard parts. Most example of museum fossils.

Eras: Geologic eras cover major spans of time that also are defined on the basis of the life forms found in the corresponding rocks. No formal eras are yet widely recognized for Archean and Proterozoic rocks, but the Phanerozoic. Eon is divided into the **Paleozoic** (Greek for old life). **Mesozoic** (Greek for middle life) and **Cenozoic** (Greek for recent life) eras each name reflecting the relative stage of development of the life of these intervals.

Paleozoic forms of life progress from marine invertebrates to fishes, amphibians and reptiles. Early land plants also appeared expanded and evolved. The Mesozoic saw the rise of the dinosaurs, which became the dominant vertebrates on land. Towards the end of that era, the primitive mammals appeared and later dominated the Cenozoic. The Mesozoic witnessed the evolution of flowering plants, while during the Cenozoic Era grasses appeared and became an important food for grazing mammals.

Periods: The geologic periods have a more haphazard nomenclature. They were defined over an interval of nearly 100 years on the basis of strata that crop out in U.K., Germany, Switzerland, Russia and the United States. The names of the periods and corresponding systems within the Paleozoic and Mesozoic eras are partly geographic in origin, but in some cases they are based on the characteristic of the strata in the place of original study.

Epochs: The epochs of the tertiary period were defined in a piecemeal fashion. Studies of marine strata in sedimentary basin of France and Italy led English geologist Charles Lyell to subdivide the rocks into grouping based on percentage of their fossils that are represented by still living species. Each of various periods of the Paleozoic and Mesozoic eras are also subdivided into epochs, the names of which are primarily geographic in origin.

Measuring Geologic time: At the end of the nineteenth century (1896), the discovery of radioactivity provided needed method. By 1904, the American scientist Benjamin Boltwood had demonstrated that the radioactivity of uranium could be used to determine the time of formation of uranium minerals. That discovery opened the door to radiometric dating, new and reliable means of measuring geologic time.

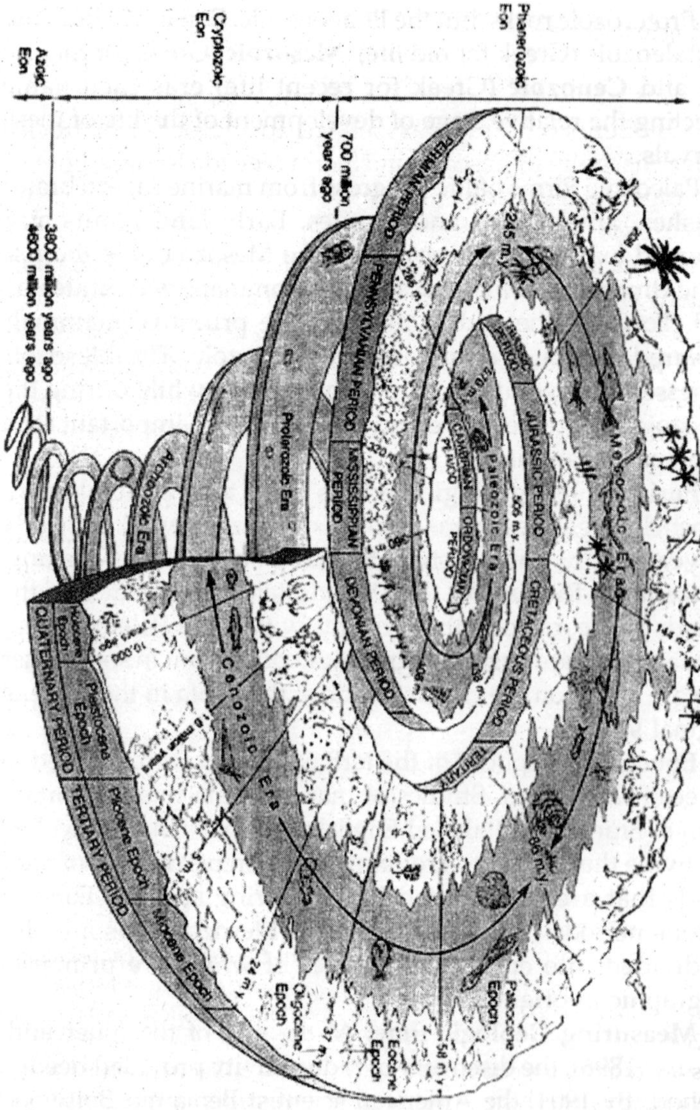
Phanerozoic Eon
Cryptozoic Eon
Azoic Eon
700 million years ago
3800 million years ago
4600 million years ago
PERMIAN PERIOD
PENNSYLVANIAN PERIOD
MISSISSIPPIAN PERIOD
DEVONIAN PERIOD
TRIASSIC PERIOD
JURASSIC PERIOD
CRETACEOUS PERIOD
CAMBRIAN PERIOD
ORDOVICIAN PERIOD
Paleozoic Era
Mesozoic Era
Cenozoic Era
Proterozoic Era
Archeozoic Era
QUATERNARY PERIOD
TERTIARY PERIOD
Holocene Epoch
Pleistocene Epoch
Pliocene Epoch
Miocene Epoch
Oligocene Epoch
Eocene Epoch
Paleocene Epoch
10,000 years ago
245 m.y.
505 m.y.
66 m.y.
58 m.y.
37 m.y.
24 m.y.

Fig. 2.2

Geological Events

Different geological events have taken place in different periods. Cenozoic Era is associated to the uplift of Rocky mountains, uplift of Cascade and Sierra Nevada Ranges, spread of continental ice sheets.

Mesozoic Era witnessed the deposition of coal bearing strata in eastern United States, Acadian Orogeny, Taconian Orogeny, folding in eastern United States, Nevada and Utah.

Cambrian and Ordovician periods are considered to be the age of marine invertebrates. First land plants and air breathing animals and fish appeared in Silurian period. First reptiles and some insects appeared in Permian period. Mammals and dinosaurs appeared in Triassic period. Jurassic period is associated to the culmination of dinosaurs and appearance of first birds. Dinosaurs became extinct in Cretaceous period and the flowering plants appeared. Earliest horse is associated with Paleocene epoch of tertiary period. Fruits, cereals, primitive mammals and grasses appeared in Eocene epoch. Whales, bats and monkeys developed in Miocene.

Early evolution of man along with elephants, horses and large carnivores took place in Pliocene epoch. Rise of civilization started only 10,000 years back. (Fig. 2.2).

REFERENCES

Correlating Earth's History, Paul R. Janke.

Cox & Richard, *A formal model for the geologic time scale and global stratotype section and point, compatible with geospatial informatin transfer standards,* Geosphere, volume 1, pp. 119-137, Geological Society of America, 2005.

Dates are slightly uncertain with differences of a few per cent between various sources being common. This is largely due to uncertainties in *radiometric dating* and the problem that deposits suitable for radiometric dating seldom occur exactly at the places in the geologic column where they would be most useful. The dates and errors quoted above are according to the International Commission on Stratigraphy 2004 time scale. Dates labelled with a * indicate boundaries where a Global Boundary Stratotype Section and Point has been internationally agreed upon: see List of Global Boundary Stratotype Sections and Points for a complete list.

Felix M. Gradstein, James G. Ogg, Alan G. Smith (Editors); *A Geologic Time Scale 2004*, Cambridge University Press, 2005.

Geologic Time Scale.

Great Soviet Encyclopedia (3rd ed.). Moscow: Sovetskaya Enciklopediya. 1974. Vol. 16, p. 50.

Historically, the *Cenozoic* has been divided up into the *Quaternary* and *Tertiary* sub-eras, as well as the *Neogene* and *Paleogene* periods. However, the International Commission on Stratigraphy has recently decided to stop endorsing the terms Quaternary and Tertiary as part of the formal nomenclature.

How the discovery of geologic time changed our view of the world, Bristol University.

International Commission on Stratigraphy. "International Stratigraphic Chart". Retrieved on 2008-06-17.

Intute Timeline—Fossils (Subject: Science Engineering and Technology).

John McPhee, *Basin and Range*, New York: Farrar, Straus and Giroux, 1981, pp. 95-100.

John McPhee, *Basin and Range*, pp. 113-114.

Munim M. Al-Rawi and Salim Al-Hassani (November 2002). "The Contribution of Ibn Sina (Avicenna) to the development of Earth Sciences". FSTC. Retrieved on 2008-07-01.

Paleontologists often refer to faunal stages rather than geologic (geological) periods. The stage nomenclature is quite complex. See "The Paleobiology Database". Retrieved on 2006-03-19 for an excellent time ordered list of faunal stages.

Scheppler, Bill (2006), *Al-Biruni: Master Astronomer and Muslim Scholar of the Eleventh Century*, The Rosen Publishing Group, 86, ISBN 1404205128.

Sivin, Nathan (1995). *Science in Ancient China: Researches and Reflections*. Brookfield, Vermont: Ashgate Publishing Variorum series. III, 23-24.

Stephen Toulmin and June Goodfield (1965), *The Ancestry of Science: The Discovery of Time*, p. 64, University of Chicago Press (cf. The Contribution of Ibn Sina to the development of Earth Sciences).

3

THE SOLAR SYSTEM

For I dipped into the Future, as far as human eye could see.
Saw the vision of the world, and all the wonder that would be.

—Alfred Lord Tennyson, 1842

Our solar system consists of an average star we call the Sun, the planets Mercury, Venus, Earth, Mars, Jupiter, Saturn, Uranus, and Neptune. It includes: the satellites of the planets; numerous comets, asteroids, and meteoroids; and the interplanetary medium. The Sun is the richest source of electromagnetic energy (mostly in the form of heat and light) in the solar system. The Sun's nearest known stellar neighbour is a red dwarf star called Proxima Centauri, at a distance of 4.3 light years away. The whole solar system, together with the local stars visible on a clear night, orbits the centre of our home galaxy, a spiral disk of 200 billion stars we call the Milky Way. The Milky Way has two small galaxies orbiting it nearby, which are visible from the southern hemisphere. They are called the Large Magellanic Cloud and the Small Magellanic Cloud. The nearest large galaxy is the Andromeda Galaxy. It is a spiral galaxy like the Milky Way but is 4 times as massive and is two million light years away. Our galaxy, one of billions of galaxies known, is travelling through intergalactic space.

The planets, most of the satellites of the planets and the asteroids revolve around the Sun in the same direction, in nearly circular orbits. When looking down from above the Sun's North Pole, the planets orbit in a counter-clockwise direction. The planets orbit the Sun in or near the same plane, called the ecliptic. The axis of rotation for most of the planets is nearly perpendicular to the ecliptic. The exception is Uranus which is tipped on his sides.

Composition of the Solar System

The Sun contains 99.85 per cent of all the matter in the Solar System. The planets, which condensed out of the same disk of material that formed the Sun, contain only 0.135 per cent of the mass of the solar system. Jupiter contains more than twice the matter of all the other planets combined. Satellites of the planets, comets, asteroids, meteoroids, and the interplanetary medium constitute the remaining 0.015 per cent. The following table is a list of the mass distribution within our Solar System :

- Sun: 99.85%
- Planets: 0.135%
- Comets: 0.01% ?
- Satellites: 0.00005%
- *Minor Planets*: 0.0000002% ?
- Meteoroids: 0.0000001% ?
- Interplanetary Medium: 0.0000001% ?

Interplanetary Space

Nearly all the solar system by volume appears to be an empty void. Far from being nothingness, this vacuum of "space" comprises the interplanetary medium. It includes various forms of energy and at least two material components: interplanetary dust and interplanetary gas. Interplanetary dust consists of microscopic solid particles. Interplanetary gas is a tenuous flow of gas and charged particles, mostly protons and electrons—*plasma*—which stream from the *Sun*, called the *solar wind*.

The solar wind can be measured by spacecraft, and it has a large effect on comet tails. It also has a measurable effect on the motion of spacecraft. The speed of the solar wind is about 400 kilometres (250 miles) per second in the vicinity of Earth's orbit. The point at which the solar wind meets the interstellar medium, which is the "solar" wind from other stars, is called

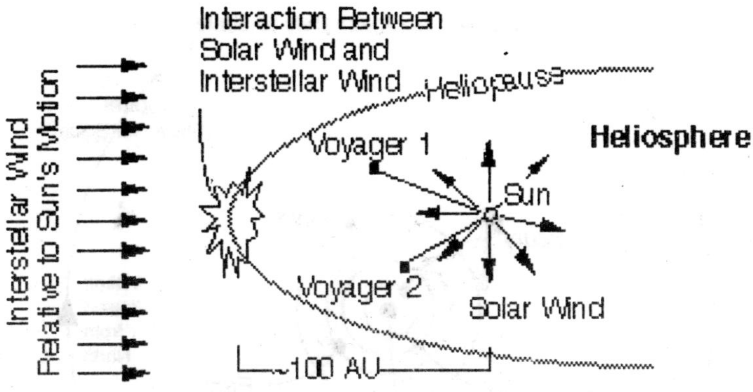

Fig. 3.1 : Heliosphere.

the heliopause. It is a boundary theorized to be roughly circular or teardrop-shaped, marking the edge of the Sun's influence perhaps 100 AU from the Sun. The space within the boundary of the heliopause, containing the Sun and solar system, is referred to as the heliosphere. (Fig. 3.1).

The solar magnetic field extends outward into interplanetary space; it can be measured on Earth and by spacecraft. The solar magnetic field is the dominating magnetic field throughout the interplanetary regions of the solar system, except in the immediate environment of planets which have their own magnetic fields.

The Terrestrial Planets

The terrestrial planets are the four innermost planets in the solar system—Mercury, Venus, Earth and Mars. They are called terrestrial because they have a compact, rocky surface like the Earth's. The planets, Venus, Earth, and Mars have significant atmospheres while Mercury has almost none. Figure 3.2 shows the approximate distance of the terrestrial planets to the Sun.

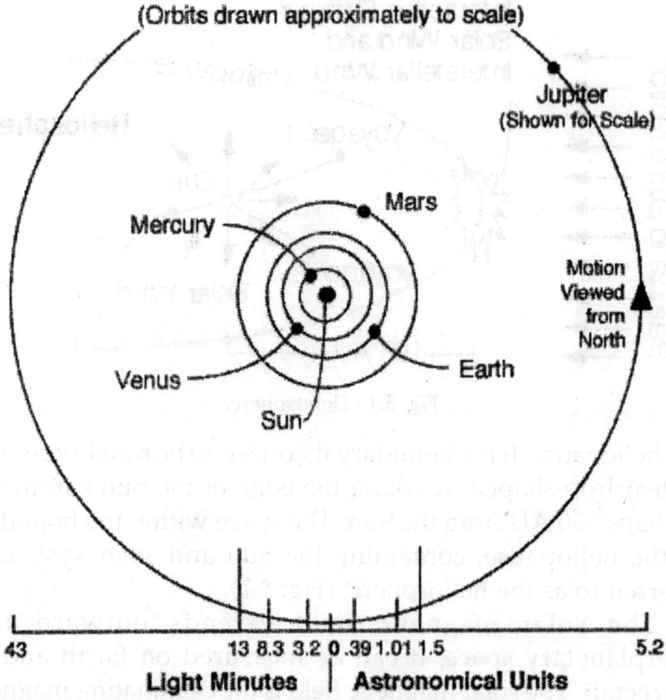

(By Courtesy of NASA)

Fig. 3.2 : Mean Distances of the Terrestrial Planets from the Sun.

The Jovian Planets

Jupiter, Saturn, Uranus, and Neptune are known as the Jovian (Jupiter-like) planets, because they are all gigantic compared with Earth, and they have a gaseous nature like Jupiter's. The Jovian planets are also referred to as the *gas giants*, although some or all of them might have small solid cores. Figure 3.3 shows the approximate distance of the Jovian planets to the Sun.

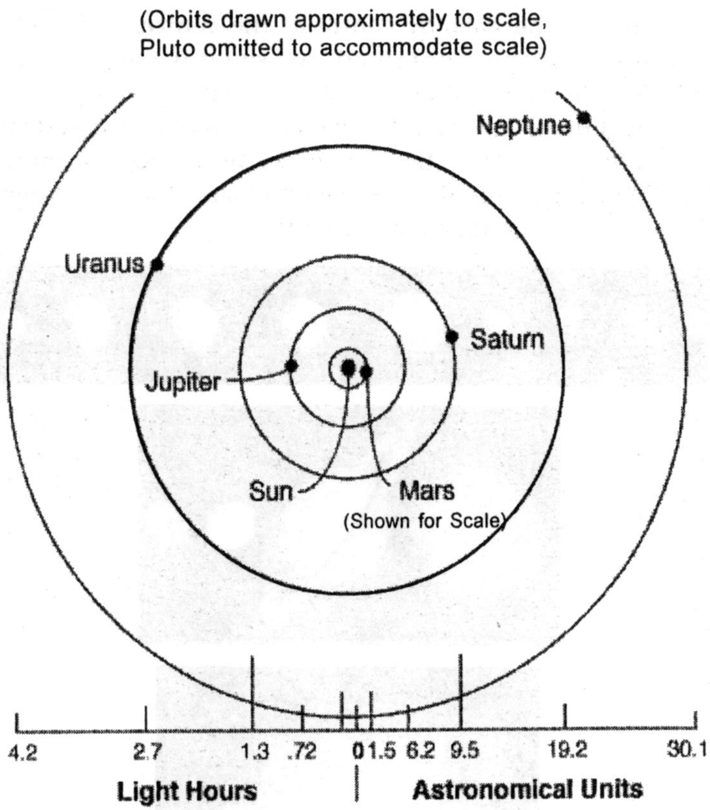

(By Courtesy of NASA)

Fig. 3.3 : Mean Distances of the Jovian Planets from the Sun.

- Formation of the Solar System.

Views of the Solar System

Our Milky Way Galaxy

This image of our galaxy, the Milky Way, was taken with NASA's Cosmic Background Explorer's (COBE) Diffuse Infrared Background Experiment (DIRBE). This never-before-seen view shows the Milky Way from an edge-on perspective with the galactic North Pole at the top, the South Pole at the bottom and the galactic centre at the centre. The picture combines images obtained at several near-infrared wavelengths. Stars within our galaxy are the dominant source of light at these wavelengths. Even though our solar system is part of the Milky Way, the view looks distant because most of the light comes from the population of stars that are closer to the galactic centre than our own Sun.

Fig. 3.4

(By Courtesy of NASA)

Obliquity of the eight Planets

This illustration shows the obliquity of the nine planets. Obliquity is the angle between a planet's equatorial plane and its orbital plane. By International Astronomical Union (IAU) convention, a planet's North Pole lies above the ecliptic plane. By this convention, Venus, and Uranus have a retrograde rotation, or a rotation that is in the opposite direction from other planets.

The Solar System

During the past three decades a myriad of space explorers have escaped the confines of planet Earth and have set out to discover our planetary neighbours. Fig. 3.6 shows the Sun and all eight planets of the solar system as seen by the space explorers. Starting at the top-left corner is the *Sun* followed by the planets *Mercury, Venus, Earth, Mars, Jupiter, Saturn, Uranus,* and *Neptune*.

Fig. 3.5

Sun and Planets

Fig. 3.6

Fig. 3.5 shows planets more distance from Sun are much larger are called **Jovian Planets.**

Jovian Planets

This image shows the Jovian planets *Jupiter, Saturn, Uranus* and *Neptune* approximately to scale. The Jovian planets are named because of their gigantic Jupiter-like appearance (*Courtesy NASA*).

The Largest Moons and Smallest Planets

There are more than fifty largest moons and the smallest planets in the solar system. The largest satellites are: *Ganymede* (5262 km), *Titan* (5150 km), *Callisto* (4806 km), *Io* (3642 km), the *Moon* (3476 km), *Europa* (3138 km), *Triton* (2706 km), and *Titania* (1580 km). Both Ganymede and Titan are larger than planet *Mercury* followed by Io, the Moon, Europa, and Triton.

Portrait of the Solar System

These six narrow-angle colour images were made from the first ever "portrait" of the solar system taken by *Voyager 1*, which was more than 6.4 billion kilometres (4 billion miles) from *Earth* and about 32° above the *ecliptic*. *Mercury* is too close to the *Sun* to be seen. *Mars* was not detectable by the Voyager cameras due to scattered sunlight in the optics, and *Pluto* was not included in the mosaic because of its small size and distance from the Sun. These blown-up images, left to right and top to bottom are *Venus, Earth, Jupiter, Saturn, Uranus,* and *Neptune.* *(Courtesy NASA/JPL)*

Sun and Planet Summary

Table 3.1 lists statistical information for the Sun and planets.

Pluto Controversy

Pluto, the little runt among the nine planets though to make up our solar system is no longer to be counted among the Sun's immediate family. The International Astronomical Union, at its meeting in Prague, decided that a celestial body must satisfy three criteria to be counted as a planet in solar system :

(1) It must orbit the sun.
(2) It must be sufficiently big so that its gravity pulls it into a nearly round shape, and

Table 3.1

	Distance (AU)	Radius (Earth's)	Mass (Earth's)	Rotation (Earth's)	# Moons	Orbital Inclination	Orbital Eccentricity	Obliquity	Density (g/cm^3)
Sun	0	109	332,800	25-36*	9	—	—	—	1.410
Mercury	0.39	0.38	0.05	58.8	0	7	0.2056	0.1°	5.43
Venus	0.72	0.95	0.89	244	0	3.394	0.0068	177.4°	5.25
Earth	1.0	1.00	1.00	1.00	1	0.000	0.0167	23.45°	5.52
Mars	1.5	0.53	0.11	1.029	2	1.850	0.0934	25.19°	3.95
Jupiter	5.2	11	318	0.411	16	1.308	0.0483	3.12°	1.33
Saturn	9.5	9	95	0.428	18	2.488	0.0560	26.73°	0.69
Uranus	19.2	4	17	0.748	15	0.774	0.0461	97.86°	1.29
Neptune	30.1	4	17	0.802	8	1.774	0.0097	29.56°	1.64

* The Sun's period of rotation at the surface varies from approximately 25 days at the equator to 36 days at the poles. Deep down, below the *convective* zone, everything appears to rotate with a period of 27 days.

(3) It should have cleared other objects from the vicinity of its orbit.

Pluto gets demoted because it does not meet the last criterion. Even the Astronomical Union was not so hard-hearted as to deny Pluto a special place in its new scheme. Pluto is to be recognized as the prototype of a new category of objects that lie beyond the orbit of the planet Neptune. In 2005, California Institute of Technology astronomer Michel Brown claimed that the rocky object 2003 UB313, which he discovered and named Xena, should be considered tenth planet. It was after all bigger than Pluto. The controversy of the Pluto is still going on among the astronomers of the world.

ORIGIN OF EARTH

This beautiful planet, the home of man, the earth is about 10 billion years old (10,000,000,000 years). There are a number of theories about its origin but all the theories explain that it has come out from the body of the sun. The theories about its origin can be divided into three groups, e.g. (i) Monolistic Theories, (ii) Dualistic Theories, and (iii) Multi Reactional Theories. The monolistic theories explain that the earth was thrown out from the body of the sun due to rotation of the sun. Kant and Laplace are the champions of monolistic theories. The dualistic theories explain this origin by the competitive pull of two stars one of them was decidedly the sun, e.g. Jean and Jeffrey and Chamberlain and Moulton propounded the major theories which explain the origin of the earth by the involvement of two stars.

Nebular Hypothesis

Nebular hypothesis is one of the early theories. It was suggested by a number of men at different times. Pierre S. de Laplace, a French Mathematician detailed upon this hypothesis in 1784. Laplace said, "Long before the birth of the solar system there was an anormous pan cake-shaped mass of hot gases turning in space."

The diameter of this disc was bigger than the diameter of today's solar system. Due to the rapid pace of rotation a process of cooling started on the outer surface of the disc. Cooling resulted in contraction and shrinkage of outer surface. The rotating disc became smaller. As it became smaller its speed further increased. As the speed increased some of the gaseous material near the edge of the disc was flung out, like drops of water from a spinning wheel. The thrown gas formed a ring around the disc. Still turning the ring began to shrink. After a long time its particles of gas were drawn together by gravitation and formed a globe—which continued to move on around the central body and became the outermost planet. Similarly, the other planets were formed with the passage of time. But the great central mass of gas was held together by its own gravitation. Still turning it contracted more and more, finally forming the sun and planets we see today.

Criticism

This hypothesis has been criticized on the following grounds :

1. The ring cannot condense into planets.
2. There is an anomaly in the distribution of angular velocity between the sun and the planets. Laplace is in favour of greater velocity in sun which is not possible under the rules of rotation and revolution.

Tidal Hypothesis of Jeans and Jeffrey

According to Jeans and Jeffrey sun was a huge gaseous mass wandering in the universe.

In its journey through the universe another star came near the sun. Both these stars exerted a gravitational pull on the surfaces of each other trying to snatch out some material. Since the approaching star was bigger than the sun it has higher gravitational force. Hence a tide occurred on the surface of the sun and went on becoming bigger and bigger as the approaching star came nearer and nearer. When they came very near, the material protruding from the sun came out of the body of the sun. The approaching star ran away and the material jutted out from the body of the sun continued to revolve around the sun in the form of spirals. Later these spiral arms condensed into planets (Fig. 3.7a).

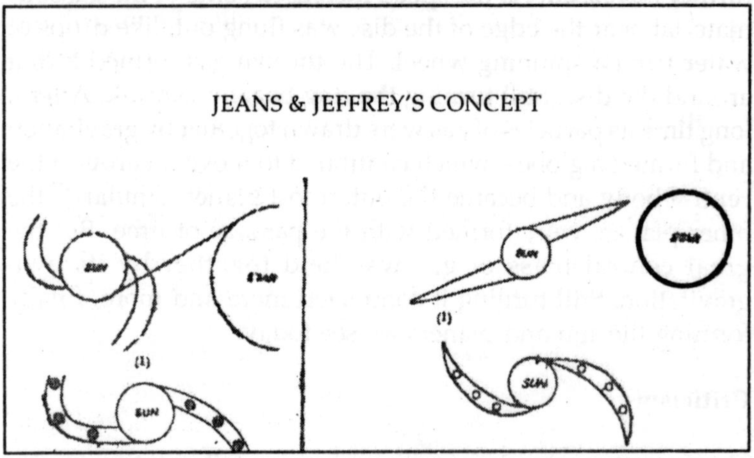

Fig. 3.7a

Supporting Points

1. The tidal protuberances should have been cigar shaped, i.e. thick in the middle and thinner in the sides.
2. The arrangement of the satellites is also cigar shaped as observed in case of the satellites of Jupiter and Saturn.
3. According to this theory the bigger planets remained in gaseous state for a longer period hence they have larger number of satellites. The smaller planets cooled early hence they have fewer satellites.

Chamberlain and Moulton's Hypothesis

It is also a dualistic hypothesis but differs from that propounded by Jeans and Jeffrey. Moulton and Chamberlain write that due to certain unknown reasons periodic eruptions has already been taking place on the surface of the sun. Another star came near and produced tidal action over and above, the eruption arm became longer. When the approaching star retreated the eruption arm was bent and became spiral more or less like a spiral nebula. The bigger eruption condensed into planets, whereas the smaller ones gave satellites because these were solid in the very beginning at the period of eruption.

The Big Bang Theory

The Big Bang Theory is the dominant scientific theory about the origin of the universe. According to the big bang, the universe was created sometimes between 10 billion and 20 billion years ago from a cosmic explosion that hurled matter in all directions.

In 1927, the Belgian priest Georges Lemaitre was the first to propose that the universe began with the explosion of a primeval atom. His proposal came after observing the red shift in distant nebulas by astronomers to a model of the universe based on relativity. Years later, Edwin Hubble found experimental evidence to help justify Lemaitre's theory. He found that distant galaxies in every direction are going away from us with speeds proportional to their distance (Fig. 3.7b).

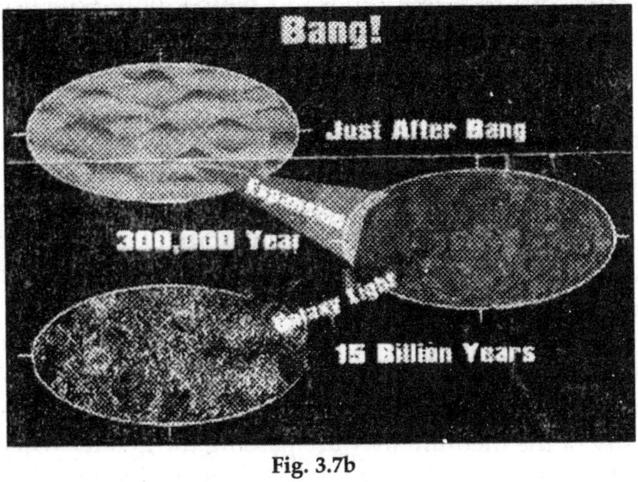

Fig. 3.7b

The big bang was initially suggested because it explains why distant galaxies are travelling away from us at great speeds. The theory also predicts the existence of cosmic background radiation (the glow left over from the explosion itself). The Big Bang Theory received its strongest confirmation when this radiation was discovered in 1964 by Arno Penzias and Robert Wilson, who later won the Nobel Prize for this discovery.

Although the Big Bang Theory is widely accepted, it probably will never be proved; consequentially, leaving a number of tough, unanswered questions (Fig. 3.7b).

INTERIOR OF THE EARTH

The knowledge of the interior of the earth is important to understand the origin of the earth, causes of mountain building, development of deep trenches in ocean, etc. The interior of the earth can be defined as that part of our planet which lies below a depth of 10 miles. The interior can be visualized as being composed of concentric shells, e.g. crust at the top, mantle intermediate and core innermost shell.

Since the interior of the earth is very hot our knowledge is based on indirect sources of information such as earthquake waves and volcanoes. Direct drilling by petroleum companies is limited up to a few thousand metres only which barely scratch the outermost layer. The core is situated at a distance of 3959 miles from the centre of the earth to the surface.

Although modern technology has to some extent enabled man to unfold the mystery of the interior being liquid or solid or gaseous, yet it is still based on indirect sources of information. Man is sometimes confused when he observes big boulders, gas and liquid coming out of the volcanoes. Nevertheless, the seismological evidences show that it is in liquid form.

The earthquakes set in motion three types of waves, e.g. Longitudinal Waves, Transverse Waves, and Surface Waves. A study of their movement and speed reveals that higher the density of the medium they are passing the greater will be the speed of the waves. The study also reveals that up to a depth of a 2880 kilometres (1900 miles), their speed increases, thereafter it changes peculiarly. The transverse waves are unable to penetrate below this depth. The character of these waves is such that they cannot enter a liquid. Hence, it is inferred that below a depth of 2880 kilometres these waves have been interrupted by a liquid. Secondly, the longitudinal waves do enter the central mass but their speed is greatly reduced. Longitudinal waves under rule cannot directly travel through the liquid central of mass of 6400 km. across. Therefore,

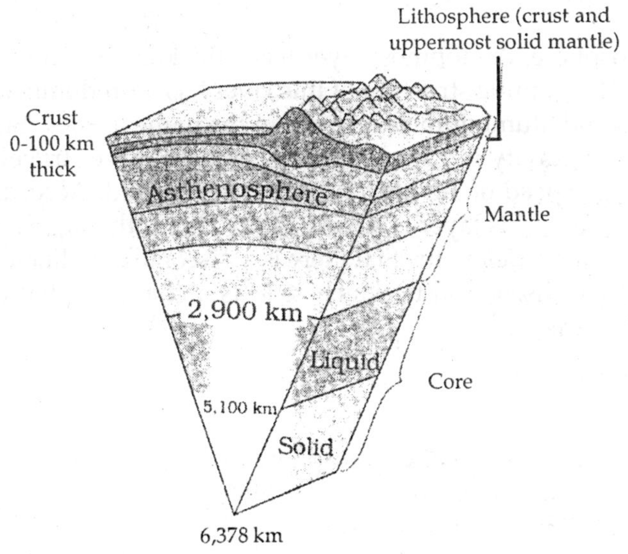

Fig. 3.8: The interior of the earth.

the longitudinal waves turn back after entering into the central liquid part. Hence, we can conclude that the interior below 2800 km. is in liquid form.

Layers of the Earth's Body

On the basis of seismological evidences the body of the earth can be divided into different layers and zones. Ordinarily, it can be divided into 3 layers known as:

1. Lithosphere 2. Pyrosphere and 3. Barysphere.

German Scientist Gracht named these layers as Sial, Sima and Nife. Jeffrey recognizes Sial and Lithosphere as upper layer, Barysphere and Nife as central layer of the earth but subdivides Sima into intermediate layer and lower layer. Similarly, Prof. Homes recognizes Barysphere or Nife as metallic core and lower layer or lower Sima as substratum but includes intermediate layer and upper layer (Sial and Upper Sima) in the crust.

Sial

Lithosphere, the topmost layer has a thickness of about 20-25 km. The granitic structure of this layer has a predominance of silica and aluminium as referred in the name itself— 'Sial.' Its specific gravity is 2.7 and the longitudinal waves travel at an average speed of 5-6 km. (3½ miles) per second. According to Gutenberg, this layer is 40 km. thick. This is the highest layer of the earth. Below the Sial there is a discontinuity line known as Moho discontinuity which reflects a change in basic rock characteristics.

Sima

Below the Sial lies the pyrosphere or Sima which is heavier than the crust or Sial having a specific gravity of 3.5. The upper portion of this layer is a mixture of Sial and Sima and the lower portion is more specifically substratum which is elastic in nature. This layer is having wax like rigidity. Silica and magnesium are the dominant elements in this layer.

The longitudinal waves travel at a speed of 7.2 km. per second whereas, transverse waves move at a speed of 3 to 4 km. per second. This layer extends up to a depth of 2880 km.

Nife

Below 2880 km. lies the heaviest layer of the body of the earth comprising nickel and iron in liquid form. This layer is known as the core of the earth or Barysphere or Nife (Nickel + Ferrous).

This layer has the highest specific gravity, i.e. 8. The core is about 4/10th of the radius of the earth.

The material of the core is ten times denser than water. The longitudinal waves are unable to penetrate it.

However, a lot of efforts are being made in the direction of getting correct information about the interior but our knowledge is still wanting (Fig. 3.9).

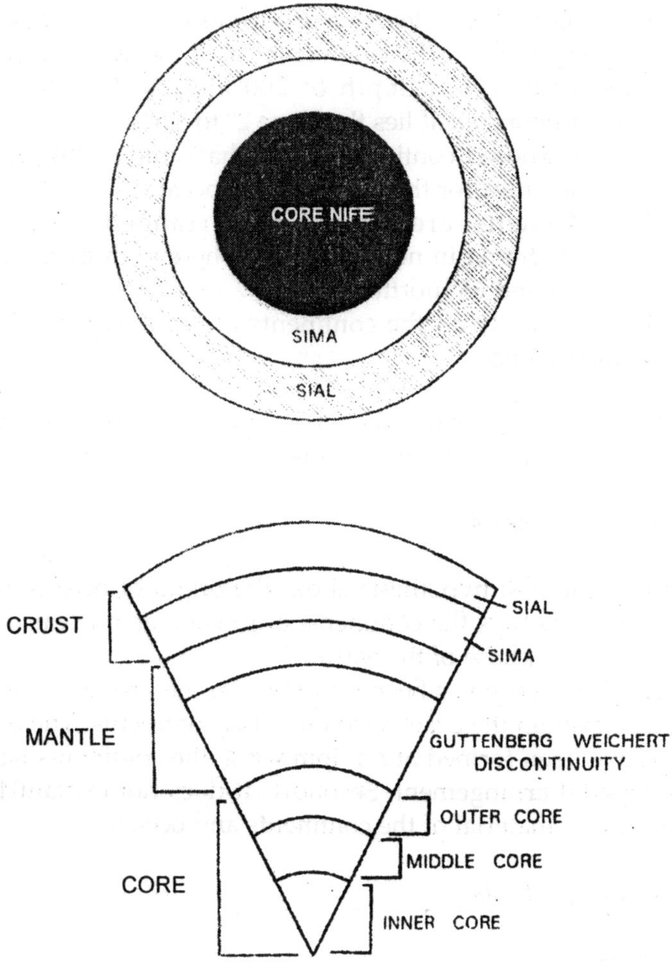

Fig. 3.9: Layers of the Earth.

Origin of Continents and Oceans

The continents and oceans are the major features of the surface of the earth. The ocean and continents can be differentiated on the basis of topographical elevation, chemical and mineralogical composition, peculiar arrangement patterns and geological structure :

1. Topographical elevation: Concentration of continents along edges of the continental shelf whereas the edge is located at a depth of 200 metres. Beyond the continental shelf lies the slope 2° to 3.5°.
2. The exposed continental rocks have more silica than the average for the continents or oceans.
3. Continents are irregularly arranged and are concentrated in northern hemisphere with maximum lot around 66° north latitude.
4. Structurally all the continents are basins fringed by mountains.

A number of theories have been put forward to explain the origin of continents and oceans.

Kelvis's Hypothesis

This was the first hypothesis about the origin of oceans and continents. He says the continents originated as nuclear dots in the gaseous phase of the earth.

But this theory has been criticized on the grounds that it does not explain the exact location of the continents. The dots must have been formed at random while the continents have an antipodal arrangement. Secondly, it does not explain the difference in material of the continents and oceans.

Sollas's Hypothesis

Basic facts

(i) Earth originally was molten.

(ii) It could be pressed at several places to produce bulges at other.

(iii) These hollows and bulges are produced by unequal pressure acting against the then existing moulten conditions.

(iv) The depression became ocean and the bulge became continent.

Criticism

The theory faces following criticism:

1. The anticline areas are randomly distributed hence the hollows should also be randomly distributed. But Green has proved that oceans and continents have an antipodal arrangement.
2. The original differences in the atmospheric pressure could not have produced so large and deep oceanic basins separated by so large continental areas.

Green's Theory

Basic Facts

(i) A sphere has the smallest surface in relation to its volume. The tetrahedron has the largest surface area in proportion. The surface of a tetrahedron, having the same volume as that of a sphere, is longer than the surface of the sphere.

(ii) The earth was a cooling and sinking globe. Therefore, there was an excess of cooled crust. This cooled crust had to accumulate itself to a grinding nucleus.

(iii) Thus the crust could not fit value. There was a misfit condition between the crust and the nucleus.

(iv) The crust developed irregularities and deformities. In such deformities the higher part became continent and the lower part became the ocean.

Supporting Points

(i) There is an excess of land in northern hemisphere.
(ii) The major continental units do have a triangular shape and taper towards the south.
(iii) The ocean units taper northward because their basis is in the continuous spreading water bodies, the southern hemisphere.
(iv) The continents have a triradiate deposition in relation to the poles.
(v) The land and sea have antipodal arrangement.

Chamberlain and Moulton's Hypothesis

This is a planetesimal hypothesis based on the assumption that the solar system has evolved from a spiral nebula whose matter was either very finely-divided solid or a liquid before it became compact. Earth's initial view was due to an unequal infall of planetesimal material. The excess of the material fall resulted in continents and other became oceans.

Criticism

1. The random infall of planetesimal material cannot display an antipodal arrangement.
2. If the infall is continuous the irregularities should also be continuous. But oceans and continents once formed contemporaneously have by and large remained unchanged.

Love's Theory

Love worked out a hypothetical process. Basic facts are:

(i) The centre of gravity does not coincide with the geometric centre of the earth. The centre of gravity depends upon the mass of the material. Hence, it shifts towards the area containing heavier material.

(ii) There is a lack of coincidence between the centre of gravity and the geometric centre. This lack of co-incidence generates forces. These forces deform the earth's body so that it can gain stability and equilibrium.

(iii) The deformation resulted in the creation of oceans and continents. The sagged parts became the oceans and the protrusions became continents.

Criticism

There is no agreement in this theory between the geophysical conception and hypothetical process.

Continental Drift

According to this theory, the continental portions of the earth's crust consist chiefly of the lighter and more acid rocks, beneath this there was a layer of denser and more basic rocks. Wegener proposed that the lighter continents are floating on the denser underlying material. About 200,000,000 years ago this single land mass which was known as PANGAEA broke apart under the influence of the tidal force and force generated by the movements of the earth's axis of rotation and revolution.

There were two drifts. One caused the drift towards the equator and the other towards the west. According to the theory, there was only one continent and one ocean. Wegener named this continent Pangaea (meaning all lands) and the ocean Panthalassa (meaning Universal Ocean). Pangaea covered an area of about 150 million square km.

The northern drift cut Pangaea from east to west, along a line slightly north of the equator, creating Laurasia in the north and Gondwana in the south (Fig. 3.10). Laurasia consisted of north America and Greenland and Eurasia (without India and Arabia). While Gondwana contained Africa with Arabia, South America, Australia, Antarctica and India. This rift opened up the Atlantic Ocean. Atlantic Ocean was created between north and South America in the west and Europe and Africa on the

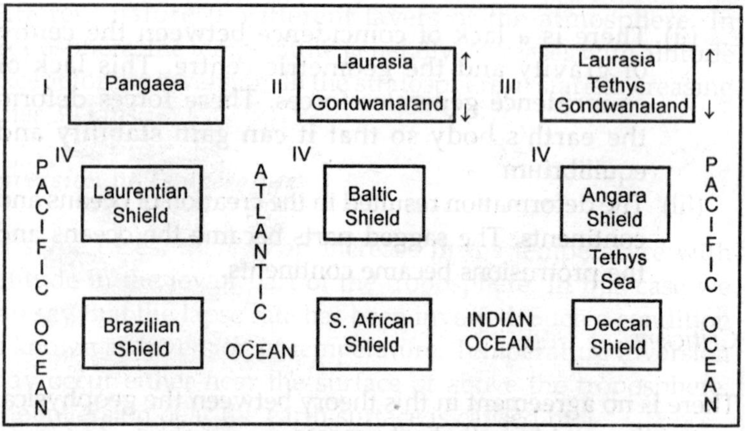

Fig. 3.10: Sequences in Continental Drift.

east. While between Eurasia and Africa lay a long large bay, the Tethys Sea, the ancestor of the Mediterranean.

By the compression of Indian part against the Angara (present Siberian part) Shield, the mountain chains of Himalayas were created. The western end of the Tethys Sea remained unaffected and subsequently emerged as Mediterranean Sea. The land mass which replaced the eastern end of the Tethys Sea formed a depression between Deccan plateau and the high rising Himalaya.

Evidences or Arguments in Support of this Theory

1. The Jig saw fit, which means eastern north and south America would fit well into the outline of western Africa and Europe, Australia fits in Indian Ocean etc., e.g. they have juxtaposition.
2. The sediments deposited during the permo-carboniferous glaciations are found in parts of South Africa, Australia, Antarctica and India the Talchir glacial deposit was discovered in Orissa in 1856.
3. The continuity of tectonic trends of the blocks of the continents across their present boundaries.

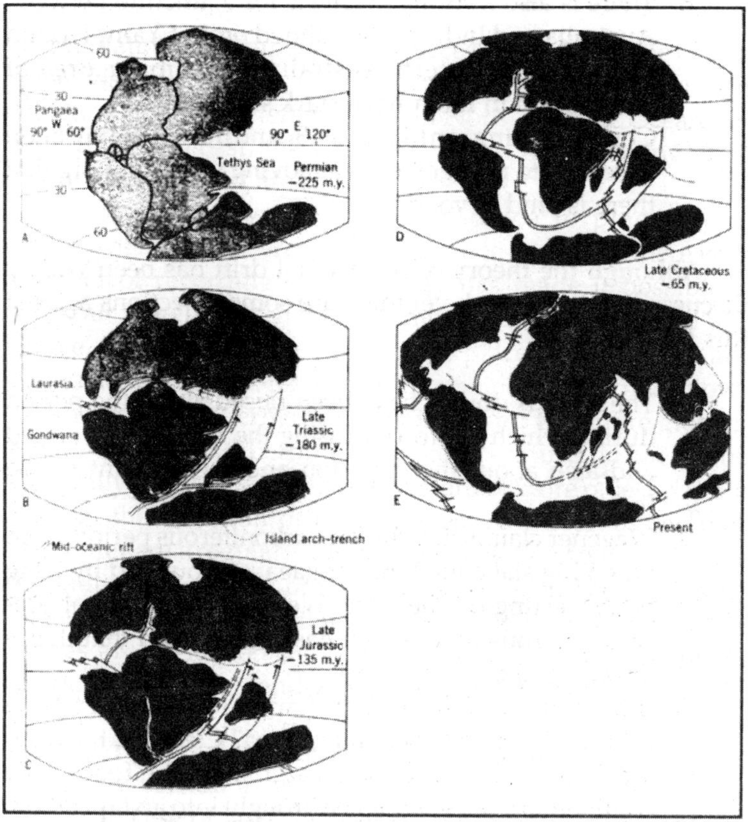

Fig. 3.11: The breakup Pangaea is shown in five stages. Inferred motion of lithospheric plates is indicated by arrows. (Re-drawn and simplified from maps by R.S. Dietz and J.C. Holden, *Journal of Geophysical Research*, vol. 75, pp. 4943-51, Figure 2 to 6, copyrighted by the American Geophysical Union.)

4. The distribution of the plant fossils in Argentina, South Africa, Western Australia and Antarctica, has a marked resemblance.
5. The increasing angle of the triangular gap between Arabia and the west coast of India indicates that these parts are still drifting. Indian subcontinent is drifting at the rate of 7 cm. per year.

6. There is also a strong evidence for a juncture between Australia and India as the banded ores of Yampi found in the northwestern Australia and similar ores of Singhbhum in India have stark similarity.
7. Certain living animals of Scandinavia have been observed to plunge into sea moving west thinking that there is land beyond.

Although the theory of continental drift has been widely discussed and accepted, yet there are some objections against this hypothesis:

1. The serious doubts have been raised about the period during which the force causing the drift had operated and also about the direction and the amount of the force.
2. Wegener claims that during carboniferous period there was a big sial continent Pangaea, surrounded by a big ocean resting on the sima. Wegener says little of pre-carboniferous times and it has been argued that this is a weak point in his theory.
3. If the continents have drifted apart how are they to be refitted together to obtain Pangaea? Clearly if the deep sea floor is the upper surface of sima, the foot of the continental slopes should be brought into juxtaposition. This point has not been explained by Wegener.
4. Willis pointed out that it is hard to see how compression could be possible (to form the Andes and Rockies due to the westward drift of the Americas). Bowie on the other hand deprive the sima of all strength and so cannot see how the sial could be crumpled at all.
5. Du Toit had made a geological study of South America and South Africa. According to geologic history the two continents cannot be brought together because the gap of some 400-800 km. separating them is too large.
6. Betweeen the Sahara and opposite shores of America no close fit is possible nor it is attempted. Wegener's Fig. 3.11 shows that this intervening area was occupied

by a shallow sea. According to Wegener, if Europe and North America were joined together the lines of the outer moraines of the Quaternary glaciation would coincide. It is explained with orthodox views. Because carboniferous glaciation was containing the Glassopteris flora. This flora is found in India, South Africa, South America, the Falkland Islands, Antarctica and Australia. In addition to the places mentioned above the flora is also found in Kashmir, northwestern Afghanistan, northeastern Persia, Tonquin and Siberia. These places are not mentioned by Wegener.

7. Wegener shows Tillites of upper Palaeozoic period in the salt range of India within 30° of the then equator. According to Schuchert similarities occur in northwest. Africa at Boston, U.S.A. (which was then on the Equator and in Alaska). Tillites existence on two different places is not explained by Wegener.
8. Hess proposed that the continents do not move as self-contained entities. Wegener held that as a continent moves, new sea floor forms along their retreating shores, no such activity has actually been observed by recent research.

Plate Tectonics

This theory was propounded by Mckenzzie and Parker in 1967. The word plate was first coined by Juzo Wilson. In general plates are the broken part of the lithosphere with relative movement over partly liquid asthenosphere. Word Tectonics means the study of rock structure underneath the surface involved in the earth movements. Plate Tectonics stands for the study of certain blocks of land, found in the shape of big discs which are constantly in motion. These plates have an average thickness of about 10 kms. floating on the upper mantle and carry the continents on their backs.

Depending upon the nature of the movement, there are three types of plates, e.g. Diverging Plates, Converging Plates and Parallel Plates.

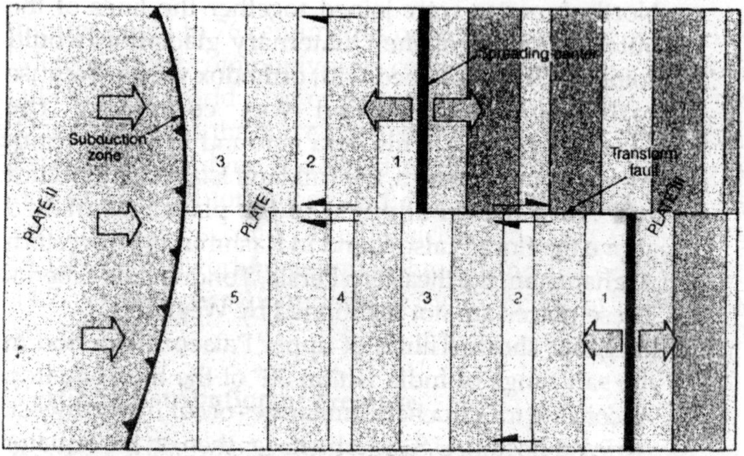

Fig. 3.12: Simplified structure of a plate. The spreading centre is offset by the transform fault that divides the plate. Bars are magnetic time lines of oceanic crust. Broad arrows show the direction of plate motion.

(i) When two adjacent plates move away from each other they are called diverging plates. As the two plates move away hot magma finds a way out and solidifies forming symmetrical ridges. The result is the formation of a new crust and may be sub-marine ridges and mountains. Mid Atlantic Ridge, East Pacific Ridge and Chagos-Lacadives Ridge are the best examples. If the divergence occurs under the continents a split may take place in the crust forming new oceans. (Fig. 3.12)

(ii) When the two plates, moving from different directions, collide with each other they are called converging plates. These collision boundaries are called subduction zones. Three important activities are associated to the subduction zone, e.g. melting of older plates, uplift earthquakes and volcanicity.

(iii) The paralleled plates slide past each other along a common boundary line. They usually create fractures in the adjoining plate areas. The San Andreas fault in California is the common front between the plates carrying North America and the one carrying the

Pacific Ocean. It is about 700 miles in length running in northwest direction. Both these plates are moving towards the northwest. (Fig. 3.13)

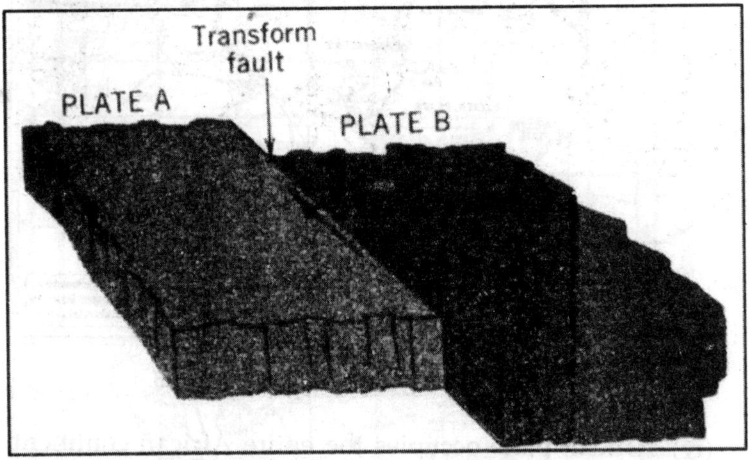

Fig. 3.13: A transform fault involves the horizontal motion of two adjacent lithospheric plates, one sliding past the other. (From A.N. Strahler, *Physical Geology*, Harper & Row, Publishers. Copyright © by Arthur N. Strahler.)

Major Plates

Dietz and Holden have divided the lithosphere into 10 plates whereas Morgan has given 20 plates. Le Pichon has given this number as six major and several minor plates which is a widely accepted view. The major plates are: (Fig. 3.14)

(i) Pacific occupying most of the Pacific Ocean basin. Its relative motion is recorded in northwestern direction.

(ii) American plate covers mostly the continental lithosphere of north and south Americas as well as entire Atlantic Ocean lithosphere laying west of Mid-Atlantic Ridge. Its movement is from east to west.

(iii) Eurasian plate is largely known as continental plate of Eurasian lithosphere with oceanic lithosphere on its north and east. Its relative movement is from northeast to southwest.

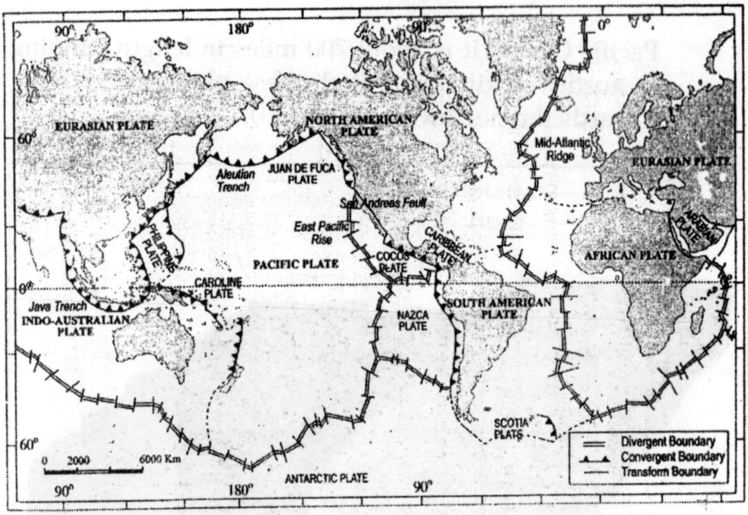

Fig. 3.14: Major and minor plates of the world.

(iv) African plate occupies the entire African continent, oceanic lithosphere east of Mid Atlantic Ridge, northwestern part of the southwest Indian Ocean Ridge and western part of Charlesberg Ridge. Its relative movement is from southwest to northeast direction.

(v) Australian plate is rectangular in shape mostly of oceanic lithosphere. It also covers Indian and Australian continental lithosphere. Its movement is from southwest to northeast.

(vi) Antarctica plate is elliptical in shape and occupies the entire Antarctic continental lithosphere and the surrounding oceanic lithosphere.

There are other minor plates also, e.g. Nazco plate (west of S. America), Cocos plate, Philippine plate, Caribbean plate, Arabian plate, Scotia plate, etc.

Movement of the Indian Plate

The Indian plate includes Peninsular India and the Australian continental portions. The subduction zone along the Himalayas forms the northern plate boundary in the form of continent—

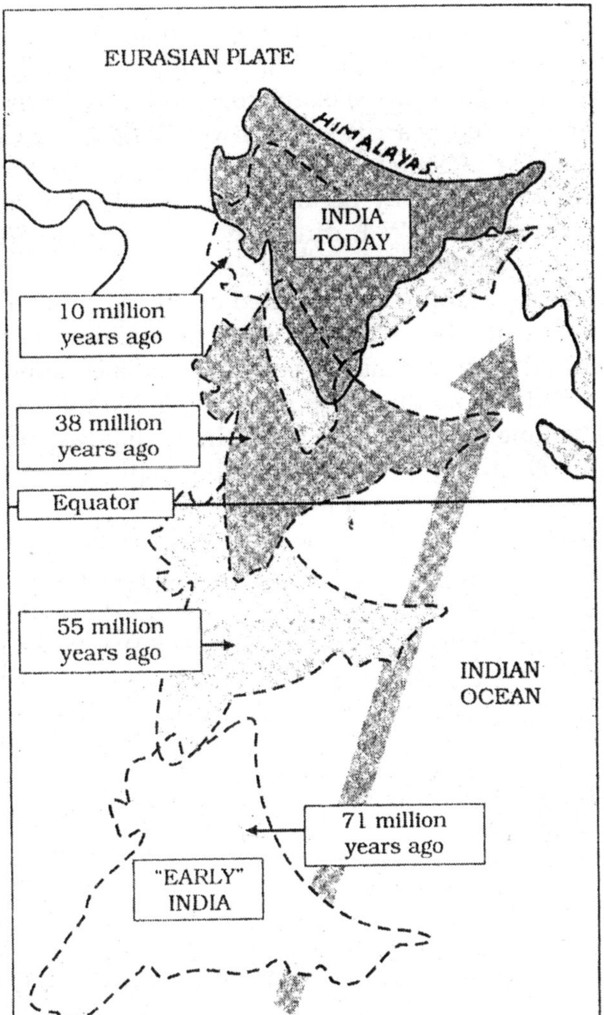

Fig. 3.15: Movement of the Indian plate.

continent convergence. In the east, it extends through Rakinyoma Mountains of Myanmar towards the island arc along the Java Trench. The eastern margin is a *spreading site* lying to the east of Australia in the form of an oceanic ridge in SW Pacific. The Western margin follows Kirthar Mountain of Pakistan. It further extends along the Makrana coast and joins the *spreading site* from the Red Sea rift sutheastward along the Chagos Archipelago. The boundary between India and the Antarctic plate is also marked by oceanic ridge (*divergent boundary*) running in roughly W-E direction and merging into the *spreading site,* a little south of New Zealand.

India was a large island situated off the Australian coast, in a vast ocean. The Tethys Sea separated it from the Asian continent till about 225 million years ago. India is supposed to have started her northward journey about 200 million years ago at the time when Pangaea broke. India collided with Asia about 40-50 million years ago causing rapid uplift of the Himalayas. The positions of India since about 71 million years till the present are shown in the Figure 3.15. It also shows the position of the Indian subcontinent and the Eurasian plate. About 140 million years before the present, the subcontinent was located as south as 50° S. latitude. The two major plates were separated by the Tethys Sea and the Tibetan block was closer to the Asiatic landmass. During the movement of the Indian plate towards the Asiatic plate, a major event that occurred was the outpouring of lava and formation of the Deccan Traps. This started somewhere around 60 million years ago and continued for a long period of time. Note that the subcontinent was still close to the equator. From 40 million years ago and thereafter, the event of formation of the Himalayas took place. Scientists believe that the process is still continuing and the height of the Himalayas is rising even to this date.

REFERENCES

The Solar System by Thérèse Encrenaz, Jean-Pierre Bibring, Michel Blanc, Maria-Antonietta Barucci, Francoise Roques, Storm Dunlop, Philippe Zarka-*Science*-2004-512 pages. This text treats our knowledge of the solar system from an astrophysical point of view.

View of the Solar System by Calvin J. Hamilton-*Education*-1996. View of the Solar System contains a special teacher's section with lesson plans, activities, and NSTA journal articles covering key processes of space.

Solar System by Golden Books Staff, Julia Hillyard-*Juvenile Nonfiction*-1999-24 pages. Introduces the solar system, describes each of its parts, and discusses space flight.

The NASA Atlas of the Solar System, by Ronald Greeley, Raymond M. Batson-*Science*-1997-369 pages. This is truly a 'road map' of our solar system.

The New Solar System, by J. Kelly Beatty, Carolyn Collins Petersen, Andrew Chaikin-*Science*-1999-421 pages. New to this edition are descriptions of collisions in the solar system, full scientific results from Galileo's mission to Jupiter and its moons, and the...

Solar System by Kendrick Frazier-*Solar System*-1985-176 pages.

Our Solar System, by Seymour Simon-*Juvenile Nonfiction*-1992-64 pages. Our Solar System was born almost 5 billion years ago at the edge of the Milky Way galaxy.

Solar System by Colorprint Staff-*Science*-1996.

The Solar System by Roger Smith-*Science*-1998-1151 pages.

The Solar System by Michael A. Seeds-*Science*-2006-721 pages. *Astronomy: The Solar System and Beyond* with its accompany online learning tools, Ace, Astronomy and Virtual Astronomy Labs shows you your place in the universe.

Penobscot Bay Watch *Earth First! Treesitting Blockages Begin*. Spring/ Summer 1985.

fojb.org *Friends of Judi Bari-FOJB*.

Anderson Valley Advertiser *Who Bombed Judi Bari?* Interviewed by Bruce Anderson.

law.com *Cal Law-FBI, Oakland Handed Defeats in Bari Lawsuit* 020612b.

Sierra Nevada Earth First! *History of Earth First*.

Federal Bureau of Investigation, *Congressional Testimony—The Threat of Eco-Terrorism*.

Mail Tribune, *Forest speaker says he's no terrorist*—By Paul Fattig.

Further Reading: Books about the Early Earth First!

Davis, John, ed. *The Earth First! Reader: Ten Years of Radical Environmentalism* (1991) (ISBN 0-87905-387-9).

Foreman, Dave. *Confessions of an Eco-Warrior* (1991) (ISBN 0-517-88058-X).

Foreman, Dave. *Ecodefense: A Field Guide to Monkeywrenching* (1985) (ISBN 0-9637751-0-3).

Manes, Christopher. *Green Rage: Radical Environmentalism and the Unmaking of Civilization* (1990) (ISBN 0-316-54532-5).
Scarce, Rik. *Eco-Warriors: Understanding the Radical Environmental Movement* (2006) (ISBN 1-59874-028-8).
Zakin, Susan. *Coyotes and Town Dogs: Earth First! and the Environmental Movement* (1993) (ISBN 0-8165-2185-9).

Books about the Post-1990 Earth First

Bari, Judi. *Timber Wars* (1994) (ISBN 1-56751-026-4).
Lee, Martha. *Earth First!: Environmental Apocalypse* (1995) (ISBN 0-8156-0365-7).
Scarce, Rik. *Eco-Warriors: Understanding the Radical Environmental Movement* (2006) (ISBN 1-59874-028-8).
Wall, Derek *Earth First and the Anti-Roads Movement* (1999) (ISBN 0-415-19064-9).
Chadwick, Paul "Concrete: Think Like A Mountain".
King, Elli (Editors) *LISTEN: The Story of the People at Taku Waken Tipi and the Reroute of Highway 55 or The Minnehaha Free State* (2006).

Books Critical of Earth First

Arnold, Ron. *Ecoterror: The Violent Agenda to Save Nature* (1997) (ISBN 0-939571-18-8).
Bradford, George. *How Deep is Deep Ecology?* (1989) (ISBN 0-87810-035-0).
Clausen, Barry. *Walking on the Edge: How I Infiltrated Earth First!* (1994) (ISBN 0-936783-12-5).
Coleman, Kate. *The Secret Wars of Judi Bari* (2005) (ISBN 1-893554-74-0).

4

EARTH MOVEMENTS

The landforms on the surface of the earth are the result of a continuous struggle between endogenic and exogenic movements. The endogenic forces act from within the body of the earth, whereas the exogenic forces react from above. Endogenic movements are mostly associated with diastrophism which act suddenly. They are also, sometimes, called tectonic forces. The endogenic movements can be divided into two categories, e.g. (i) Epeirogenic and (ii) Orogenic.

1. Epeirogenic Movements: These are also known as continent building earth movements. These are vertical movements which result in subsidence or elevation of an area. These movements do not disturb the horizontal arrangements of the crust. The best example of subsidence is that of land towards the west of the Western Ghats. The elevation of certain coastal areas of Florida, Sweden and West Indies are the best examples of uplift. The scientists have found that the entire Scandinavian Peninsula is gradually inclining from north to south while its northern part is rising.

2. Orogenic Movements: As the name suggests these movements are related to the orogeny or the mountain building process. These forces work horizontally and cause lot of crumpling, faulting, folding etc. on the surface of the earth.

These mountain building movements can further be subdivided into two types:

1. Movements of Contraction, and
2. Movements of Expansion.

Movements of Contraction are forced by compression. Here the forces are converging from different directions at a common centre. The intervening landmass is compressed, folded and raised upwards forming fold mountains or related structures such as nappes, over folds, synclinorium, anticlinorium, etc.

Movements of Expansion: Such movements take place at the points of tension. In this case the forces are moving in opposite direction away from a common centre, resulting in downthrowing of the strata giving rise to faults and joints.

Structures Produced by Earth Movements

However, these forces are related to each other—compression in one region is bound to produce tension in another. These movements produce the following structures: Folds, Faults and Joints.

Folds: When the forces of compression press a crustal block from both sides its rock strata are pinched in a number of bends which are known as folds. In the formation of these folds the surface layout results in troughs and arches. The tops of such folds are called anticlines and the ebbs or crests are called synclines. The dropping legs of the anticlines are called limbs.

The line between syncline and anticline is known as Axial Plane. (Fig. 4.1)

Types of Folds: Depending upon the intensity of the forces of compression and resistance from the rocks the folds may take different shapes. Depending upon the shape they may be classified as under:

(i) Symmetrical Fold: If axial plane of the fold bisects the fold equally it is known as a Symmetrical Fold or upright Fold.

(ii) Asymmetrical Fold: If the axial plane is tilted the fold is known as inclined or Asymmetrical Fold.

(iii) Nappe or Recumbent Fold: If the pressure from one side is higher and the other side is resistant the strata is overthrown making the axial plane parallel to the horizontal strata. Such a fold is known as Recumbent fold or Nappe.

(iv) Overturned Fold: If in an inclined fold the two limbs tend to tilt in the same direction at different amounts it forms an overturned fold.

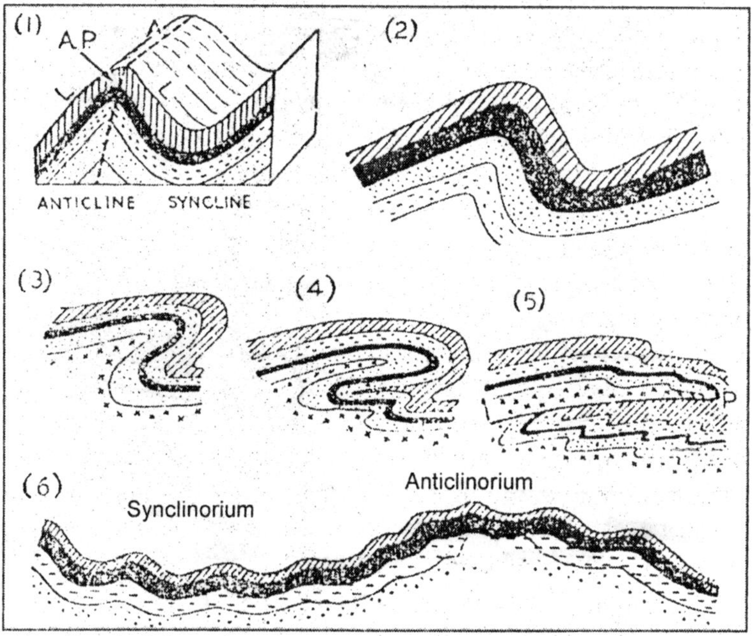

Fig. 4.1: Types of Folds
(1) Simple folding (A = Axial; AP = Axial plane; L = Limb); (2) asymmetrical folding; (3) overfold; (4) recumbent anticline; (5) overthrust fold or nappe (TP = thrust-plane); (6) an anticlinorium (an anticlinal complex of folds) and a synclinorium (a synclinal complex of folds).

(v) Open Folds: If the limbs of a fold make an angle exceeding 70° it is called an *Open Fold,* whereas when the angle between the limbs ranges from 30° to 70° it is called a Closed Fold. In case the angle between the two limbs of a fold is less than 30° it is a Tight Fold.

(vi) Isoclinal Fold: If the axial plane or limbs are parallel or nearly parallel an Isoclinal Fold is formed.

(vii) When a fold has got only one limb a step like fold is formed which is known as Monoclinal Fold.

(viii) Synclinorium: When a large Syncline is marked by a series of smaller folds it is known as an Synclinorium.

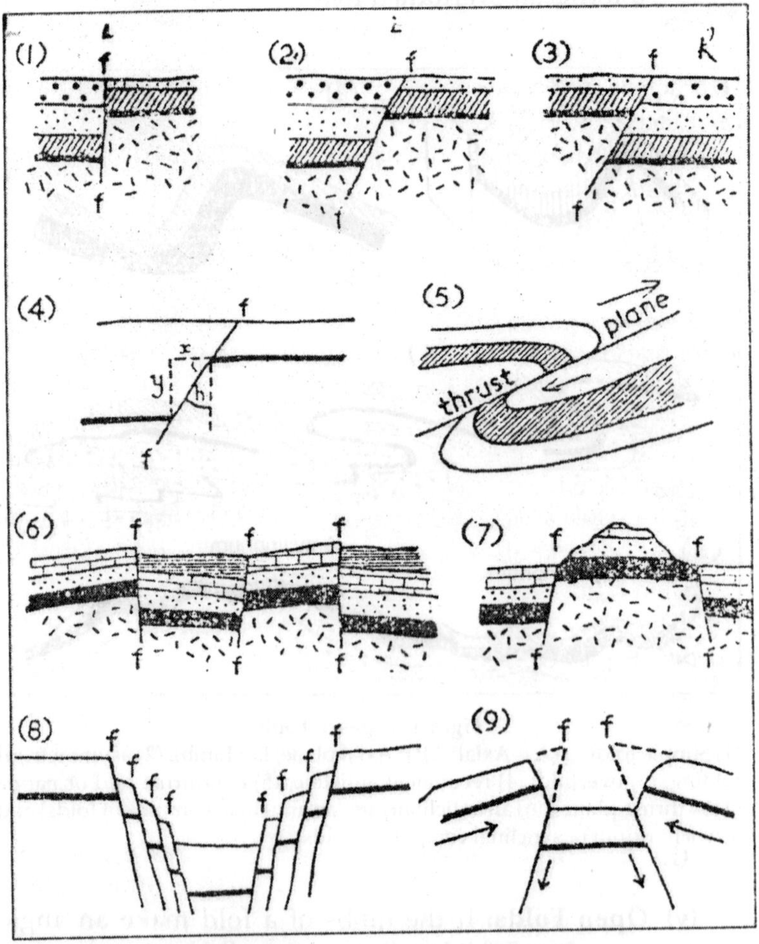

Fig. 4.2: Types of Faults and Rift-valleys

f indicates the line of the fault.

(1) Normal fault, verticle; (2) normal fault; inclined; (3) reversed fault; (4) terminology of a fault h = hade (in degrees); y = throw (in feet); x = heave (in feet); (5) thrust-fault in recumbent fold; (6) 'basin-range' faulting; (7) a horst; (8) step faulted rift-valley; (9) compressional rift-valley.

(ix) **Anticlinorium:** When tangential compression is increased further the rocks break and movement takes place on the gently inclined planes of rupture forming Over Thrust Folds.

Faults

A fracture takes place due to tension. If the rock strata are displaced on either side of this fracture in relation to one another the resultant feature is called a fault. This displacement of the strata can be vertical as well as horizontal.

Components of a Fault

The line along which the displacement takes place is known as fault plane. The surface expression of the fault plane is a fault scarp. The lower part of this fault scarp is called foot wall and the upper part is called hanging wall. The vertical change in the level of strata is known as throw which ranges from a fraction of an inch to thousands of feet. When the fault is inclined there must be a horizontal displacement too known as heave. The angle of inclination of fault plane from vertical is called hade. (Fig. 4.2)

Types of Faults

(i) **Normal Fault:** When the inclination of the fault plane and the direction of the downthrow are both to the left or both to the right the resultant feature is a Normal Fault (Fig. 4.2).

(ii) **Reverse Fault:** Where the beds on one side of the fault plane are thrust over those on the other side by compressional forces a reverse fault is formed.

(iii) **Tear Fault or Strike-Slip Fault:** Basically it is a vertical fracture but here the displacement of strata is horizontal. The blocks move horizontally along the line of the fault. Sometimes the roads and railway lines are cut across by the Tear Faults. An example of tear fault in Britain is that of Great Glen of Scotland with a gap of 65 miles.

(iv) **Oblique Fault or Oblique-Slip Fault:** Where none of the major axis is vertical, faults of any possible orientation and direction of relative motion can be formed known as Oblique Faults.

(v) **Hinge Fault:** In this case the displacement increases from zero to a maximum along the strike.

(vi) **Point Fault:** Here one block appears to have rotated about a point on the fault plane. It is also called Scissor Fault.

(vii) **Step Faulting:** When a number of parallel faults are thrown down in the same direction the whole pattern is known as Step Faulting.

(viii) **Rift Valley or Graben:** These are formed in an area where the parallel faults happen to be arranged in pairs. If the rock strata between the two parallel faults are thrown down it results in the formation of a rift valley.

(ix) **Horst:** This too is formed where the faults are arranged in pairs. In this case the block of material is thrown upwards.

Joints

In case a brittle rock is subjected to strain it may break or fracture. Where the actual movement or displacement is absent or very little a local tensional stress may produce cracks on the surface known as Joints. The joints may be formed due to other reasons also such as the tensile stress caused due to cooling in igneous rocks. Joints may also be formed in sedimentary rocks during their period of consolidation.

Isostasy

In 1889 C.E. Dutton, a geologist coined the term Isostasy from Greek Isostasies (In "equipoise" compare isos "Equal" and Statikos "Stable") means equal balance or equilibrium. This doctrine states that wherever equilibrium exists on the earth's surface, equal mass must underlie equal surface area. Lighter continental material must extend downwards to some distance under the continent and maintain a isostatic balance in order that unit area beneath the ocean and continents may remain in stable equilibrium. The extra amount of lighter material piled up into a great chain must be compensated by a proportionate amount penetrating down into the denser material. Large features on the continent are locally compensated, e.g. the Himalayas, the Rockies, the Appalachians great plains.

During a survey of the Indo-Gangetic plain about a hundred years ago by the Survey of India under the guidance of Sir George Airy, the difference in latitude between Kalianpur and Kaliana (600 kms due north) was determined astronomically as well as by direct triangulation on the ground. The two values differed by 5°23' second due to attraction exerted on the plumb line by the enormous rock mass composing the Himalayas.

A few years later Archdeacon Pratt made an estimate, assuming that average density of mountains was 2.75 approximately that of continental rocks as a whole. The calculated difference was 15°8'85" which was more than three times the observed deflection 5°23". The explanation for this anomaly is that the density of the rocks must be comparatively low down to considerable depths.

In fact, the principle is really nothing else than that of ordinary floating, and the example so often taken of an iceberg floating in the water, may help to stress the fundamental concept of the doctrine. Ice floats in such a way that approximately for every one part above water level there are nine parts below. That is to say, the ratio of free board to draught is 1 to 9, that means a mountain one mile in height on the surface of the earth can remain in equilibrium only if it extends nine miles below the surface.

Isostasy is neither a force nor a process. It is a condition of gravitational balance between crustal segments of different thickness. Isostasy also refers to a tendency towards restoration of balance once it has been disturbed by some other force or process. When the crustal segments are in an Isostatic balance the high and the low lying relief features of the earth's surface are also in the state of balance. This is made possible by the fact that the high relief features have lighter materials, while the large low relief features such as ocean basin have heavier material under them.

The great features of the crust such as continents, mountains, plateaus and ocean basins are in isostatic equilibrium. Volumes of eroded material are washed down from mountain ranges and deposited in the sea. The mountains become lighter and column under the sea becomes heavier. In order to maintain equilibrium vertical movement takes place adjusting the balance. (Fig. 4.3)

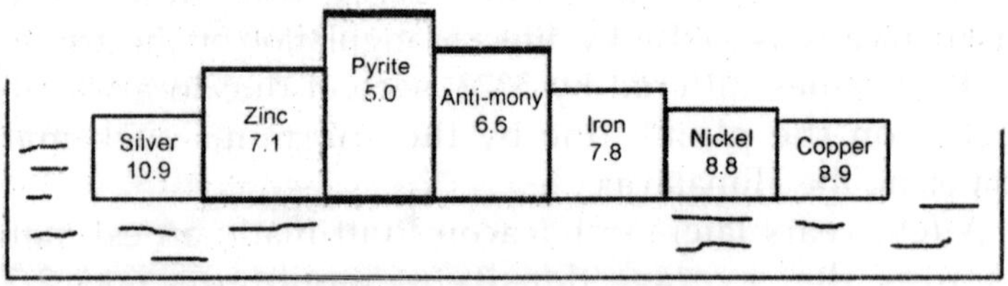

Blocks of different materials with equal cross-section and weight sink to equal depths in mercury but their tops rise to different heights.

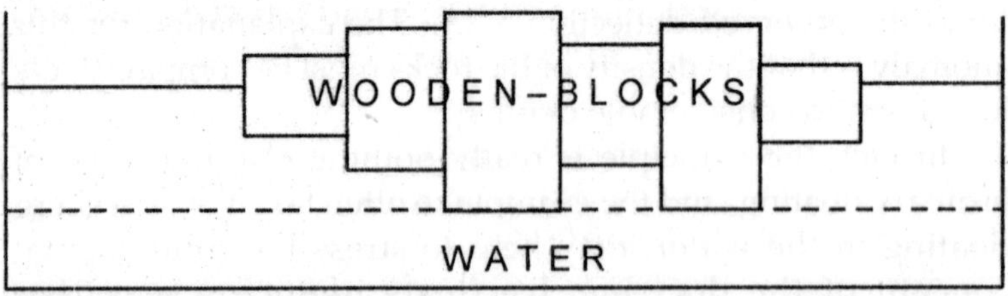

Same kind of wooden blocks of equal cross-section but of different heights float in water with longer blocks rising higher and extending deeper than the smaller blocks. High mountains behave in the same manner.

Fig. 4.3: Illustrations of isostasy.

Below the earth's surface there is a certain level beneath which the densities were the same throughout. This level is called compensation level. There was a controversy regarding the compensation level and the densities. Sir George Airy says that there is a uniform density, with varying thickness. On the other hand, Pratt explains that there is uniform depth with varying density.

Later on another investigation of Hayford assumed a layer about ten miles thick underlying the continent. All the density variations giving rise to isostatic phenomena were supposed to occur in this layer.

Hayford and Bowie first assumed that crustal densities varied with elevation above this plane. Under the mountain, the crust is lighter than under the ocean. Below the plane of compensation, densities were taken as uniform in lateral direction.

Earthquakes

An earthquake is a sudden, temporary motion or series of motions which originate in a particular region and spread out from the place of origin in all directions. The point of origin of earthquake is known as a focus and a point immediately above the focus on the surface is called epicenter. The scientific study of earthquakes is called seismology. The word seismology is derived from the Greek word 'Seismas' means earthquake and 'logos' means reason or speech. The earthquake motion has definite beginning. It continues for a while and then gradually dies out. It does not die out suddenly. The duration as well as the direction of motion can be estimated precisely with an instrument known as Seismograph. (Fig. 4.4)

Causes of Earthquakes

Normally, there are two main causes of earthquakes:

(i) Tectonic Forces: Tectonic forces cause earthquakes by the shift in the rock structure of the earth. With the widespread acceptance of the theory of plate tectonics it has become evident that most earthquakes occur along the margins of plates, where one plate comes into contact with another and stresses develop. The release of energy results in earthquakes along the subduction zones.

(ii) Volcanic Eruption: A sudden eruption of volcano may also result in tremors in the surrounding areas. There emerges

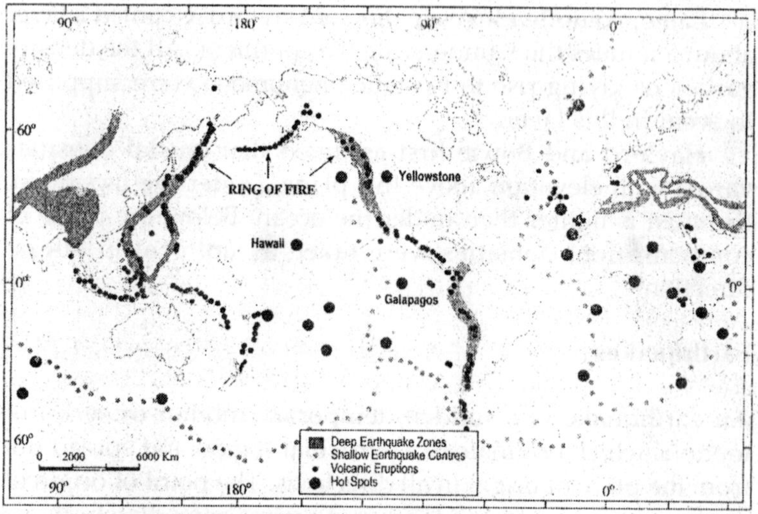

Fig. 4.5: Distribution of earthquakes and volcanoes.

a realization that volcanoes and earthquakes may have a common ultimate cause in deep movements of mantle material and the coincidence of belts of major earthquakes activity with belts that include active volcanoes supports the idea. Another cause of earthquake is associated with the Reids' theory.

(iii) Reids' (1906) Elastic Rebound Theory: The deposition of debris at a particular place disturbs the balance and if the mantle layer underlying the crust is weak the large blocks of Sial are broken and sent down into the liquid. Since it is lighter material it rebounds upwards and strikes the Lithosphere to send shock waves above.

Earthquake Waves

When an earthquake occurs tremors spread outward in different types of waves. The point at which the shock occurs is known as the origin or the Focus. A point on the surface of the earth vertically above the origin is known as Epicenter. [Word Epicenter is derived from Greek word (Epi) means above and centre.] It is passable to plate places suffering from equal degrees of damage with the help of Isoseismic Line.

Isoseismic Line is that line which joins points of equal intensity of earthquake shock. The energy released from an earthquake travels away by means of earth waves. There are three main waves which travel from the focus. First is primary wave or P waves. The primary waves can travel through any material, solid, liquid or gas. They move the particles forward and backward, consequently the materials in the path of these waves are alternately compressed and rarefied. For the proper understanding we can give an example of 'tuning fork.' When we strike a tuning fork sharply the prongs vibrate back and forth, first pushing and then pulling the molecules of air with which they come in contact. Each molecule jumps the next one and a wave of pressure is set in motion through the air. Like this P waves move with the pushing and pulling movements. Secondary or, S waves can travel only through solid media. These waves move the particles in their path at right angle to the direction of their advance. For example, you are holding one end of a rope fastened to a wall. If you move your hand up and down regularly a series of waves will travel along the rope to the wall. Each wave moves along the particles as the rope moves up and down. In other words, the particles move at right angle to the direction of the waves advance (Fig. 4.5B).

The surface waves usually termed as Ryleigh Waves move along the boundaries between solid, liquid and gaseous media. Their intensity decreases rapidly with depth but with low velocity. They may travel deep into earth's mantle over a long period of time.

The movement of surface wave is again discussed with an example of surface of water. If you stand on the shore and throw a stone in quiet lake setting up surface waves the water seems to be moving in all directions in the form of waves. It is energy which is moving in the form of waves. The particles of water move in a definite pattern as each wave advances up, forward, down and back in small circle.

When surface waves are generated in rock one common type of particle motion is just the type reverse of the water particle motion that is forward, up, back and down. These waves reach the epicenter after making a full round of the

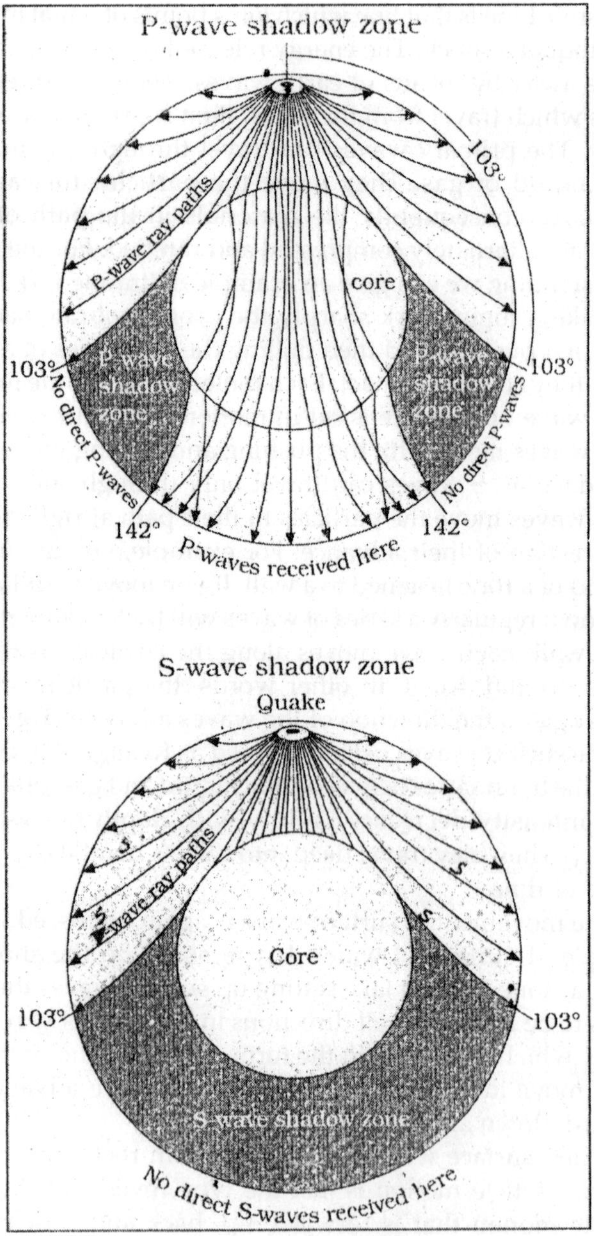

Fig. 4.5B: Earthquake Shadow Zones (*Courtesy: Monkrous*).

earth's surface. They cause maximum damage on the surface of the earth.

The Richter Scale

In 1935 a leading seismologist, Charles F. Richter devised a rating scale of earthquake magnitude. The quantity of energy released during an earthquake at the place where the fault slippage originates is known as the earthquake magnitude (M). The Richter scale consists of numbers ranging from less than 0 (negative numbers) to more than 8.5. Values are given to the nearest one-tenth, thus 2.5, 4.9, 6.2, 7.8, 8.5. There is neither a fixed maximum nor a minimum, but the highest magnitude earthquake thus for measure has been rated as 8.9 on the Richter scale. Earthquake of 2 magnitudes are the smaller detected by human sense, but Seismograph (Instrument) can detect quake as small as 3.0. The scale is logarithmic, which is to say that the amplitude of the recorded wave increases ten fold for each integer in Richter magnitude. An earthquake of magnitude: 5.0 is ten times larger than one of magnitude 4.0.

For every increase of one integer of Richter magnitude, the quantity of energy increases by a factor of about 32 times. Thus an earthquake of the magnitude 7.0 releases about 32 times as much energy as one of magnitude 6.0. An earthquake of magnitude 8.0 releases about 1024 times as much energy as one of magnitude 6.0 (32×32=1024).

The effects of earthquakes are also measured by the Modified Mercalli Intensity scale. The intensity of the earthquake is evaluated according to the observed severity of quake at specific location. The Mercalli scale rates the intensity on a Roman number scale of range from I to XII. But at present Richter scale is most recognized in the world.

Types of Earthquakes

There are three main categories of earthquake:

(i) Volcanic earthquakes,
(ii) Tectonic earthquakes, and

(iii) Impact earthquakes.

I. Volcanic Earthquakes are caused by eruption on volcanoes. On the land they are localized around the volcanoes. But in the sea these earthquakes can give rise to tidal waves, which travel across the ocean and destroy shoreline far away.

II. Tectonic Earthquakes: Tectonic earthquakes are those earthquakes which are caused by shifts in the rock structure of the earth. They affect vast areas and cause catastrophic changes over the surface of the earth. When the parts of the strata or basin structures are subjected to stress, they bend as far their elasticity would permit beyond which they break. After being broken the rocks are thrown downwards under pressure. Being elastic the down throws rocks rebound. They try to come back to original position with the result of vibratory movement starts beneath the earth's crust.

The modern Theory of plate movement has given the idea that earthquakes are also resulting from plate movements. As different plates slide past each other, enormous amount of stress builds up between the plates. These stresses are released by sudden slipping action, producing earthquakes along the subduction zones. These earthquakes are also included in tectonic earthquakes. Fig. 4.6 shows seismic zones of India.

III. Impact Earthqauakes: These are associated to surface explosions, falling of meteors etc., cavities, rockslides and similar phenomena. Surface explosion cause the earthquakes covering a large area on the surface but they are shallow. For example open explosion in Germany during 1921 was felt throughout western Europe. Large meteorites strike the earth, sometimes which set up impact waves. The distance in which the vibrations can be recorded is usually small. During 1908 a giant meteorite (Podkamannaya Tunguska) landed in Siberia 150 miles north of lake Baikal sent shock waves in large area. Sometimes roofs of the cavern collapse and produce shock waves. Large landslides also produce enough energy to cause an earthquake. People of Ramban and Khilani in Doda district stand witness to the idea.

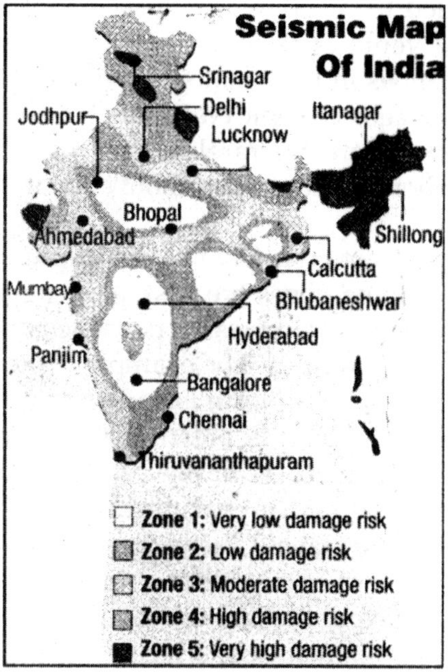

Fig. 4.6: Seismic Map of India.

What is a Volcano?

One of the most difficult problems of standardization has been the varying usage of the word "volcano". Definitions of "volcano" range from individual vents, measured in metres, through volcanic edifices measured in kilometres or tens of kilometres, to volcanic fields measured in hundreds of kilometres. In a database compilation, the disadvantage of the narrowest definition is not so much the multiplicity of names introduced, as the dismembering of a single volcanic plumbing system's history into apparently unrelated separate records. The interiors of ancient volcanoes, now eroded and exposed for geologic study, show us that most subsurface magma chambers—the suppliers of lavas to overlying volcanoes—are at least several kilometres in diameter. We also know that many contemporary volcanoes grow by additions from countless flank vents as well as activity at a central crater. Consequently, we have tended to group closely spaced "volcanoes" such as the historical vents of the Canary Islands (many listed as separate volcanoes in the Catalog of Active Volcanoes of the World) by the major volcanic edifice on which they are found. Volcanoes listed here are rarely closer than 10 km to their nearest neighbour, and are commonly separated by at least 20 km.

Another problem is simply the identification of volcanoes. Prominent, steaming cones are easy to recognize, but water, ice, erosion, collapse processes, or dense vegetation can mask very dangerous volcanoes. For example, Lake Taupo, in the centre of New Zealand's North Island, is beautifully tranquil, with no obvious features alerting non-geologists to its particularly violent history. In the Alaskan summer of 1975, two volcanologists traced an ever-thickening ash layer to a vent now covered by the Hayes Glacier, and a "new" volcano was added to the NE end of the Aleutian arc. Also in Alaska, five decades passed before the true source of this century's largest eruption was recognized: subsurface magma connections led to prominent collapse of Mount Katmai in 1912, and this was assumed to be the eruption's source until careful fieldwork showed it to be Katmai's inconspicuous neighbour, Novarupta. These examples illustrate why the listings generated from this database must be recognized as incomplete. Inclusion in this compilation may depend on thoroughness of mapping—quite variable through the world's volcanic regions—and the most dangerous volcanoes may be those not yet recognized. (Fig. 4.7)

What is an Eruption?

The arrival of volcanic products at the Earth's surface is termed an eruption. Some definitions of the word include purely gaseous expulsions, but we confine the term to events that involve explosive ejections of fragmental material, the effusion of liquid lava, or both. This fragmental material may be old as well as new; the explosive interaction of volcanically generated heat and near-surface water can cause dramatic eruptions without any fresh volcanic material reaching the surface. (Fig. 4.8)

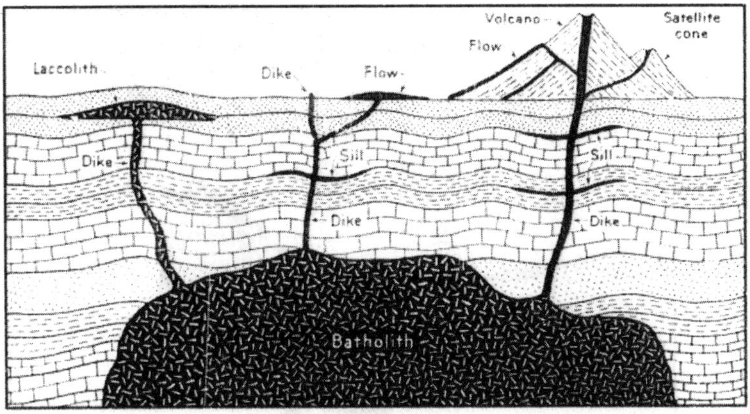

Fig. 4.7: Plutons and Landforms associated with igneous activity.

Types and Processes: Pyroclastic Flows

Pyroclastic flow is a term that deserves a place in the popular lexicon. One of the most destructive volcanic processes, pyroclastic flows are mixtures of hot gas and rock that travel gravitationally down the flanks of a volcano at highway velocities. Their elevated temperatures and great mobility make them lethal to anything in their path. These flows include a basal component containing blocks of pumice or dense rocks and a dilute cloud of hot gas and ash that rises convectively above the moving flow. They can be formed by collapse of dense materials within a rising eruption column or directly from the volcanic vent. Pyroclastic flows include vesicle-rich pumice flows and dense block-and-ash flows formed by gravitational collapse of growing lava domes and active lava-flow fronts. Their travel paths typically follow topography. Pyroclastic surges, in contrast, are more dilute gas-rich flows that can have broader lateral extent less sensitive to topography. The dilute part of pyroclastic flows can become detached from the bedload porticn and travel long distances over water. Extremely large volume pyroclastic flows known as ignimbrites are produced during catastrophic eruptions and can travel long distances from a volcano, producing deposits tens to a few hundred metres thick. (Fig. 4.8)

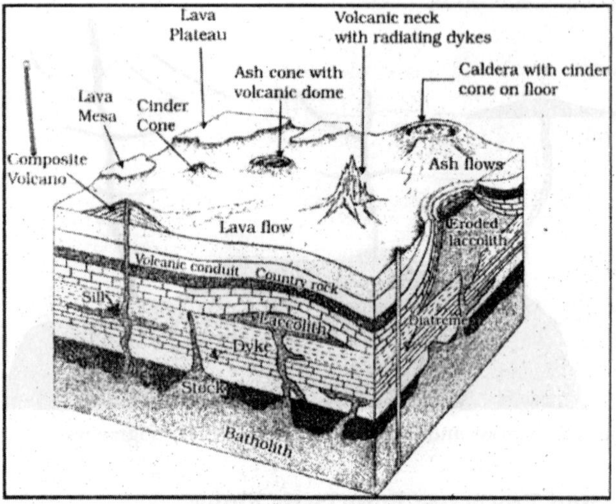

Fig. 4.8: Volcanic Landforms.

Santorini

This spectacular outcrop shows light-coloured deposits from the 3500-year-old Minoan eruption of Santorini volcano filling a valley cut in darker, bedded ashfall layers of Pleistocene age. The lower, beige-coloured unit filling the ancient valley is a pumice-fall deposit from vertical explosions early in the

Fig. 4.9: Santorini.

eruption. It is overlain by laminated pyroclastic-surge deposits produced when water gained access to the magma reservoir as the volcano collapsed into the sea. The upper whitish layer truncating both these deposits is a pyrocloastic surge (Fig. 4.9).

Pyroclastic Fall

Pyroclastic-fall deposits are produced by the gravitational fall and accumulation of fragmental material ejected by explosive eruptions. Tephra is a generic term referring to fragmental material of all size classes. Explosive eruptions are one of Earth's most dramatic geological processes and have been the subject of the brush and pen for thousands of years. Eruptions can involve new magma (referred to as juvenile material) or the ejection of pre-existing rock and debris from around the vent. They can range from minor, relatively passive gas-and-ash emissions reaching only a hundred metres or so above the vent to powerful magmatic explosions that can eject large volumes of material and produce eruption columns that reach tens of kilometers above the volcano. Smaller eruptions eject material that accumulates near the vent, but large explosions eject material that accumulates near the vent, but large explosions inject ash and gases into Earth's stratosphere, where it can circle the planet and have major impact on global climate. Pyroclastic-fall deposits vary widely in particle size, but typically have a narrow range of particle sizes at a given location and become finer-grained away from the vent. They provide valuable time layers useful in unravelling the eruptive history of a volcano and a volcanic region.

White Island

This 4-m-wide, water-filled impact crater was formed when the block in background, with volcanologist Ian Nairn providing scale, was ejected during a strombolian eruption from White Island in New Zealand in late March 1977. The block, composed of pre-existing wall rock of the crater, bounced, forming the impact crater, and then slid to its present

Fig. 4.10: White Island.

location 250 m from the source vent. (Fig. 4.10). [*Photo by Bruce Houghton, 1977 (Wairakei Research Center)*].

Pyroclastic Cones

Pyroclastic cones are the most abundant of volcanic structures. These small cones, named from a combination of the Greek words for fire (pyro) and broken (klastos), are built of fragmental material. They are found both as subsidiary features on larger stratovolcanoes and shield volcanoes or as independent volcanic structures. Pyroclastic cones typically range from a few tens of metres to a few hundred metres in height and are most often formed during single eruptions, when explosively ejected material accumulates around the vent. The slopes of pyroclastic cones consequently approach the angle of repose of its constituent materials. Pyroclastic cones, also known as cinder cones or scoria cones, can form rapidly, but remain active for long periods of time. Paricutin, the famous volcano born in Mexican cornfield in 1943, reached 80 per cent of its 424 m height during its first year, but remained active for 9 years. Tuff cones are a variety of pyroclastic cone formed by more energetic explosions involving groundwater that extensively fragment ejecta, producing finer-grained ash deposits; they typically have steeper slopes and wider craters than cinder cones.

Barren Island

The 1991 eruption dramatically modified the morphology of the central cinder cone on the small Barren Island, one of the

Fig. 4.11: Barren Island.

Andaman Islands, north of Sumatra. During the eruption the height of the cinder cone was lowered from 305 m to 225 m, and the diameter of its crater increased from 60 m to about 200 m. Two small cinder cones also formed in the lava field west of the main cone. (Fig. 4.11) [*Photo courtesy of D. Haldar, 1992 (Geological Survey of India)*]

Aso

Kometsuka scoria cone, on the NW side of the central cone complex of Aso volcano on the island of Kyushu, was constructed about 1800 years ago. Explosive eruptions building the cinder cone were accompanied by lava flows that travelled down the flank of this complex of stratovolcanoes in the centre of Aso caldera. A broad moat separates the central cone complex from the caldera walls, which form the horizon

Fig. 4.12: Aso.

(Fig. 4.12). *Copyrighted photo by Dick Stoiber, 1981 (Dartmouth College).*

Lava Flows

Lava flows are one of the most visually dramatic manifestations of volcanoes. Their incandescent trace across the landscape has inspired countless paintings and photographs. Two common lava flow types are often produced during the same eruption. Aa and pahoehoe are Hawaiian terms referring respectively to clinkery, rough-textured, higher-velocity flows, and less viscous, smooth-textured, lower-velocity flows. More silica-rich magmas produce block-lava flows, whose surfaces consist of dense solidified blocks the size of televisions to small houses, and obsidian flows, which contain large masses of glassy, crystal-poor rocks. Lava flows vary widely in viscosity and volume, and are distinguished from lava domes by their elongated extent downslope. Most lava flows travel less than 10 km, but fluid pahoehoe flows can reach a length of 50 km or more. Overlapping lava flow lobes in low-relief terrain can form broad lava fields. Flow velocities are typically low enough that lava flows most often do not pose a direct hazard to human observers, but long-term, long-duration flows such of those currently occurring at Kilauea volcano can cause extensive damage to houses and other structures. In 2002, fluid lavas from Nyiragongo volcano in the Democratic Republic of Congo flowed through the business district of the city of Goma.

Lengai, Ol Doinyo

The chemically exotic lava flows of Tanzania's Ol Doinyo Lengai volcano have unusually low temperatures and viscosities. Its lava flows are not incandescent during daylight hours. The flows are usually only a few cm thick and travel

Fig. 4.13: Lengai, Ol Doinyo.

only a few tens of metres from their source. This 1966 photo shows a small lava flow, only a metre wide, with miniature channels and levees. Ol Doinyo Lengai is Earth's only volcano known to erupt carbonatitic (sodium carbonate) lava flows in historical time (Fig. 4.13). *Photo by Gordon Davies, 1966 (courtesy of Celia Nyamweru Kenyatta University).*

Lava Domes

Lava domes are formed when viscous magma slowly extrudes from a vent and piles up around it. Silicon promotes rigidity in magmas because it has a +4 charge and forms multiple bonds with other elements. Domes are consequently often produced by magmas of more silica-rich dacitic-to-rhyolitic compositions. Domes are steep-sided structures typically a few tens of metres to a few hundred metres high and can form at the summit of a volcano, on its flanks, or as independent volcanic centres. Domes can form during single eruptive episodes or by periodic lava extrusion, such as occurred during the 1980-1986 eruption of Mount St. Helens. Compound lava domes, such as currently forming at Santiaguito at the base of Santa Maria volcano in Guatemala, are common. Some volcanoes, such as Augustine in Alaska, entirely consist of a complex of overlapping summit lava domes surrounded by fragmental material produced during growth and collapse of the domes. Dome formation is often preceded or accompanied by explosive eruptions, and dome collapse can produce pyroclastic flows and debris avalanches.

Merapi

A blocky lava dome produced during eruptions from 1972 to 1985, lies within the summit crater of Merapi volcano in central Java. A small spine appears on the right in this July 4, 1986 photo, which was taken during a period of inactivity prior to

Fig. 4.14: Merapi.

destruction of the dome on October 10, 1986. Merapi's historical eruptions have been characterized by repeated growth and collapse of the summit lava dome, periodically producing pyroclastic flows that have affected populated areas on the volcano's western and southern flanks (Fig. 4.14). [*Photo by Tom Casadevall, 1986 (U.S. Geological Survey)*].

Stratovolcanoes

Stratovolcanoes (also known as composite volcanoes) are what most people associate with the word volcano. These towering peaks rise hundreds to several thousand metres above their surroundings, often visually dominating the landscape around them. As their name implies, they are formed of stratified layers of both viscous lava flows and fragmental material. Classic symmetrical stratovolcanoes such as Shishaldin in the Aleutian Islands and Mayon in the Philippines are the exception rather than the rule. Most stratovolcanoes are complex structures formed by multiple eruptions from summit and flank vents. Compound stratovolcanoes may form when the focus of eruptions shifts, forming multiple overlapping edifices. Some stratovolcanoes may form in a few thousand years, but may remain active for tens to hundreds of thousands of years. During their lifespans, dormant intervals may also last tens of thousands of years. Perhaps the most common rock type of stratovolcanoes is andesite (named after the Andes Mountains), but stratovolcanoes form in a wide variety of tectonic settings and erupt a wide range of rocks. Their steep-sided profiles are often exaggerated even further by artists for dramatic effect. Slopes of stratovolcanoes rarely exceed the angle of repose for fragmental material at about 30-35 degrees, and eruptions from multiple vents often produce complex volcanoes with greatly variable morphology.

Mayon

Mayon volcano in the Philippines is one of Earth's best examples of a classic, conical stratovolcano. Its symmetrical

Fig. 4.15: Mayon.

morphology is the exception rather than the rule, and is the result of eruptions that are restricted to a single central conduit at the summit of the volcano. Eruptions are frequent enough at Mayon, the most active volcano in the Philippines, to overcome erosive forces that quickly modify the slopes of most volcanoes. (Fig. 4.15)

Shield Volcanoes

Shield volcanoes contrast with stratovolcanoes both in origin and morphology. Their name derives from their low-angle profile, which resembles the broad shields used by Hawaiian warriors. They are formed primarily by the succesive accumulation of fluid lava flows, which descend from summit or flank fissure systems. Although shield volcanoes are not as visually dramatic as stratovolcanoes, they are of often much larger features. Oceanic shield volcanoes such as those in the Hawaiian Islands can rise as much as 8000 m above the surrounding sea floor and 12,000 m above their actual bases, which have sagged due to the immense mass of the volcano. Their volumes can exceed that of stratovolcanoes by several orders of magnitude. Most shield volcanoes are formed of fluid basaltic lava flows. Shield volcanoes can vary widely in size, and much smaller Icelandic-type shield volcanoes are common in many volcanic regions. These volcanoes can form during single long-term effusive eruptions. A less common type of shield volcano is a pyroclastic shield, whose broad low-angle slopes are formed by accumulation of fragmental material from powerful explosive eruptions.

Auckland Field

Rangitoto shield volcano, the youngest volcanic centre of New Zealand's Auckland volcanic field, forms a 5.5-km-wide island.

Fig. 4.16: Auckland Field.

The volcano, seen here from the NW erupted about 600 years ago and is capped by a scoria cone containing a deep crater. The 140 sq km Auckland volcanic field contains more than 50 maars, tuff rings, and scoria cones. Of the 19 eruptions known to have occurred during the past 20,000 years, only Rangitoto has erupted during the Holocene. (Fig. 4.16)

Calderas

Calderas are large volcanic depressions formed by collapse of the summit or flanks of a volcano into underlying chambers evacuated by very large explosive eruptions or the effusion of large volumes of lava flows. Earth's calderas range from a kilometre to as much as about 100 kilometres in width; many contain scenic caldera lakes. Calderas may be simple structures formed during an eruption that truncates either the summit of a single stratovolcano or a complex of multiple overlapping volcanoes, such as at Crater Lake in Oregon. Other calderas are compound structures formed incrementally as a result of multiple eruption-related collapses, such as the massive 30 × 100 km wide Toba caldera in Sumatra, which was formed during four powerful explosive eruptions during the Pleistocene. Calderas are most often defined as depressions produced as a result of large-scale eruptions, but the word has also been used as a morphological term that encompasses volcanic depressions formed by erosion or large volcanic landslides. Calderas may be ephemeral features that are partially or totally obscured by post-caldera eruptions.

Toba

The 35 × 100 km wide Toba caldera, partially filled by waters of Sumatra's Lake Toba, is Earth's largest Quaternary caldera. This view looks west towards the northern end of Samosir Island, which is part of a massive inclined block uplifted after eruption of the Young Toba Tuff (YTT) about 74,000 years ago.

Fig. 4.17: Toba.

The island, once entirely covered by Lake Toba, is formed of caldera-fill deposits of YTT capped by lake sediments (Fig. 4.17). [*Photo by Tom Casadevall, 1987* (*U.S. Geological Survey*)].

Banda Api

The arcuate islands of Neira and Lonthor, seen here looking east from the summit of Indonesia's Banda Api volcano, are

Fig. 4.18: Banda Api.

remnants of two largely submarine calderas that preceded the construction of the Banda Api stratovolcano. The outer caldera has a diameter of 7 km, the nested inner caldera is 3 km wide. Neira, the largest town in the Banda Islands, occupies the southern tip of Neira Island (Fig. 4.18). [*Photo courtesy Tom Casadevall, 1988* (*U.S. Geological Survey*)].

Taal

A small, 3-km-wide caldera is located at the centre of Volcano Island, in the Philippines' southern Luzon Island. The 5-km-wide Volcano Island lies within a much larger, 15 × 20 km Taal caldera, whose low, western wall is seen across Lake Taal in the distance. The small island in the centre of the photo, is a

Fig. 4.19: Toba.

remnant of historical eruptions on Volcano Island, is a geographical oddity—an island in a lake on an island in a lake on an island (Fig. 4.19). [*Photo by Chris Newhall, 1989* (*U.S. Geological Survey*)].

Craters

Craters differ from calderas both in size and origin. Craters are much smaller features than calderas and are typically defined as being less than 1 km in diameter. Although both craters and calderas are most often associated with explosive eruptions, craters are typically formed by the explosive ejection of material in and surrounding the upper part of the conduit, rather than by collapse. Steep-walled pit craters, in contrast, often found on shield volcanoes, are more passive features formed when magma drains from a fissure, leaving overlying lava flows unsupported. Multiple explosive eruptions can form overlapping or nested craters, and adjacent craters may reflect localized areas of eruption along fissures, as seen in crater's rows in Iceland that extend for tens of kilometres. Maars and tuff rings are broad, low-rimmed craters formed during powerful explosive eruptions involving magma-water interaction. The accumulation of material ejected from craters contributes to the formation of their surrounding rims.

West Eifel Volc Field

The Mehrener maar is one of about 80 maars of the West Eifel volcanic field. The village of Mehrener is located on the shore of a lake partially filling the crater, whose rim lies behind the

Fig. 4.20: West Eifel Volc Field.

village. Maars, scoria cones, and small stratovolcanoes cover an area of 600 sq. km west of the Rhine River. Most originated during eruptions between about 730,000 and 10,000 years ago (Fig. 4.20). [*Photo by Richard Waitt, 1990 (U.S. Geological Survey)*].

Lengai, Ol Doinyo

The northern crater of Ol Doinyo Lengai is seen here in February 1980 from the summit of the volcano. The inner crater is a steep-walled pit that remained after the powerful explosive eruptions of 1966 and 1967. Three years after this photo was taken, another eruption began. Slow lava effusion completely filled in the inner crater. By December 1988, lava had overflowed the near southern crater rim, at the lower centre of the photo (Fig. 4.21). [*Photo by Peterson, 1980 (courtesy of Celia Nyamweru, Kenyatta University)*].

Fig. 4.21: Lengai, Ol Doinyo.

Kelimutu

A double crater lake of Kelimutu volcano on Indonesia's Flores Island is seen in this aerial view from the SW. Tiwu Nua Muri Kooh Tai (Lake of Young Men and Maiden) on the left and Tiwu Ata Polo (Bewitched, or Enchanted Lake) are separated by a narrow septum about 35 m high. Phreatic eruptions have occurred from Tiwu Nua Muri Kooh Tai in the 19th and 20th centuries, and continuous upwelling occurs at both lakes (Fig. 4.22). [*Photo by Tom Casadevall, 1986 (U.S. Geological Survey)*].

Fig. 4.22: Kelimutu.

Magma Intrusion

The intrusion of magma into an edifice can produce large fractures, or fissures that locally become the focus of explosive activity or lava effusion. Fissures often extend radially away from the summit and can reach from the upper part of the

Fig. 4.23: Magma Intrusion.

volcano to its base. Fissures can also reflect regional tectonic trends independent of the volcanoes themselves, such as in Iceland, where the mid-Atlantic Ridge rises above sea level, or along continental rift zones such as those in eastern Africa. Volcanoes in the Galápagos Islands are notable for circumferential fissures that parallel the rims of large sumit calderas. Although explosive eruptions and lava effusion can occur simultaneously over long portions of a fissure, activity often migrates along a fissure and is subsequently localized at a few specific points, where cones or craters can form. Much larger scale fissures of Tertiary or older age fed voluminous eruptions of lavas known as flood basalts. These massive eruptions have produced extensive lava fields covering tens of thousands of square kilometres in both continental areas, such as the Columbia River Plateau in the Pacific Northwest of the U.S. and the Deccan Traps of India, and oceanic areas such as the Ontong-Java Plateau in the SW Pacific. The Deccan Traps and Siberian Traps flood basalt provinces have been linked temporarily to major Cretaceous/Tertiary and end-Permian faunal extinctions, respectively.

Volcanic Landslides

Volcanoes appear to be permanent fixtures on the landscape, but in fact are inherently unstable structures composed of both strong (thick lava flows) and weak (fragmental and hydrothermally altered) materials. Large-scale collapse of volcanic edifices, first witnessed and documented at the start of the 1980 Mount St. Helens eruption, is now known to be a common volcanic process. Large volcanic landslides can occur with volumes exceeding a cubic kilometre at continental volcanoes and several orders of magnitude larger at oceanic shield volcanoes. These collapses can produce extremely mobile debris avalanches that can travel at high velocities in some cases for tens of kilometres beyond the base of a volcano. This process, once thought to be extremely rare, has been documented at hundreds of volcanoes worldwide. Repeated episodes of growth and collapse have occurred at many volcanoes, and large-volume volcanic landslides have been found to be the most common catastrophic destructive process at volcanoes.

Taranaki (Egmont)

Grass-covered conical hills dot the plains around Mount Taranaki (Egmont), on New Zealand's North Island. Small hills such as these, often located in lowland areas well beyond the flanks of a volcano, were once thought to be cinder cones or small secondary vents produced by explosions when a lava flow passed over a body of water. They now are known to be hummocks of massive debris avalanches produced by volcanic landslides. Debris-avalanche deposits originating from repetitive collapse surround the volcano to distances of about

Fig. 4.24: Taranaki (Egmont).

40 km (Fig. 4.24). [*Photo by Don Swanson, 1984 (U.S. Geological Survey)*].

How Many Active Volcanoes are there in the World?

The answer to this common question depends upon use of the word "active". At least 20 volcanoes will probably be erupting as you read these words (Italy's Stromboli, for example, has been erupting for more than a thousand years); roughly 60 erupted each year through the 1990s; 154 in the full decade 1990-1999; about 550 have had historically documented eruptions; about 1300 (and perhaps more than 1500) have erupted in the Holocene (past 10,000 years); and some estimates of young seafloor volcanoes exceed a million. Because dormant intervals between major eruptions at a single volcano may last hundreds to thousands of years, dwarfing the relatively short historical record in many regions, it is misleading to restrict usage of "active volcano" to recorded human memories: we prefer to add another identifying word (e.g. "historically active" or "Holocene volcano").

The definition of "volcano" is as important in answering the number of questions as the definition of "active". Usage has varied widely, with "volcano" applied to individual vents, measured in metres, through volcanic edifices measured in tens of kilometres, to volcanic fields measured in hundreds of kilometres. We have tended towards the broader definition in our compilations, allowing the record of a single large plumbing system to be viewed as a whole, but this approach often requires careful work in field and laboratory to establish the integrity of a group's common magmatic link. The problem is particularly difficult in Iceland, where eruptions separated by many tens of kilometres along a single rift may share the same magmatic system. A "volcanic field", such as Mexico's Michoacán-Guanajuato field (comprising nearly 1,400 cinder cones, maars, and shield volcanoes derived from a single magmatic system, dotting a 200 × 250 km area) may be counted the same as a single volcanic edifice. Perhaps the most honest answer to the number of questions is that we do not really have an accurate count of the world's volcanoes, but that there are at least a thousand identified magma systems—on land alone—likely to erupt in the future.

How many active volcanoes known?

Erupting now:	perhaps 20
Each year:	50-70
Each decade:	about 160
Historical eruptions:	about 550
Known Holocene eruptions (last 10,000 years):	about 1300
Known (and possible) Holocene eruptions:	about 1500

Note that these figures do *not* include the large number of eruptions (and undescribed volcanoes) on the deep sea floor. Estimates of global magma budgets suggest that roughly ¾ of the lava reaching Earth's surface does so unnoticed at submarine midocean ridges. (see Fig. 4.25)

Volcanism distributed by tectonic setting. Pie diagram on left shows proportion of documented historical eruptions from subduction zones (black), midocean ridges (Fig. 4.25) (stipple), and hot spot setting (white). Pie diagram on right shows proportion of annual magma budget in the same settings (with same symbols) to emphasize the dramatic contrast between the volcanism that we see and that we don't. Data for right diagram from Crisp (1984) and left from Smithsonian.

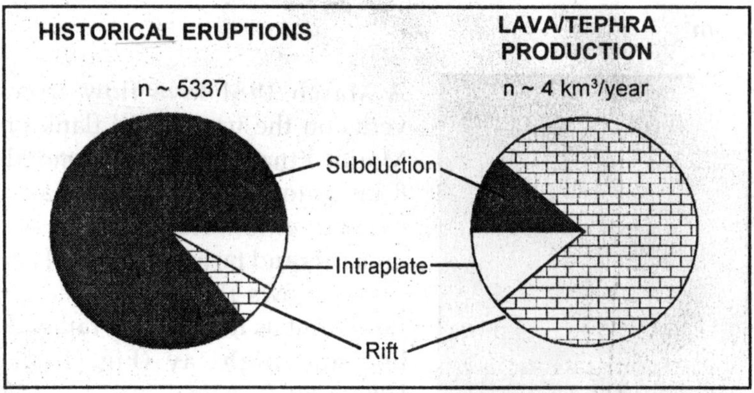

Fig. 4.25

Volcanoes and Humans

Volcanoes and humans have an uneasy coexistence. Volcanic eruptions can cause severe destruction. More than three dozen historical eruptions have caused more than 1000 fatalities. The heavy weight of ashfall can demolish the roofs of houses and buildings. Lahars, debris avalanches, and tsunamis can extend the devastating effects of an eruption well beyond the volcano. The beneficial aspects of volcanoes, in contrast, are less well known. Volcanoes provide essential soil nutrients for agriculture; volcanic soils cover more than 1.5 million sq. km and form a significant component of the surface area in countries along the Pacific Rim of Fire. It is no accident that some of the most densely populated areas on Earth are in proximity to volcanoes. Volcanic materials are a source of aggregate for construction and additives used in a wide variety of commercial products. Geothermal heat is a major source of electrical power. Volcanoes provide spectacular scenic destinations that are featured in many national parks and whose aesthetic attractions are of great economic benefit to surrounding communities.

Etna

Fig. 4.26: Etna.

A March 1981 lava flow from vents on the upper NW flank of Mount Etna in Sicily descended 8 km into populated areas and caused extensive damage to orchards and farm buildings. This photo shows the steaming lava-flow front as it overran a railroad bed and highway (Fig. 4.26). [*Photo by Romolo Romano, 1981 (IIV-CNR, Catania, Italy)*].

A major south-flank eruption of Mount Etna beginning March 28, 1983, produced lava flows that damaged tourist facilities. Extensive efforts at lava diversion, such as the gray embankment at the left of this photo, were made to divert the flows from populated areas. The flows eventually descended to 1080 m altitude, but stopped short of several south-flank villages (Fig. 4.27). [*Photo by Romolo Romano, 1983 (IIV-CNR, Catania, Italy)*].

Fig. 4.27: Etna.

Fresh black lava flows overrun a highway and tourist facility buildings the day after the start of an eruption at Sicily's Mount Etna on March 28, 1983. The lava flows originated from a fissure 4 km south of the summit of Mount Etna and over a 5-month period formed a complex lava field as much as 1 km wide (Fig. 4.28). [*Photo by Romolo Romano, 1983 (IIV-CNR, Catania, Italy)*].

Fig. 4.28: Etna.

Dieng Volc Complex

A stone monument is inscribed with a list of eruptions from the Dieng volcanic complex in central Java. The second column from the right lists fatalities (most recently 149 in 1979), which have occurred many times as a result of phreatic explosions and toxic gas emissions. The designers of the monument have somewhat ominously left abundant space to record future events.

Fig. 4.29: Dieng Volc Complex.

Has Volcanic Activity been Increasing?

We don't think so.

A look at the number of volcanoes active per year, over the last few centuries, shows a dramatic increase, but one that is closely related to increases in the world's human population and communication. We believe that this represents an increased reporting of eruptions, rather than increased frequency of global volcanism: more observers, in wider geographic distribution, with better communication, and broader publication. The past 200 years (see Fig. 4.30) show this generally increasing trend along with some major "peaks and valleys" which suggest global pulsations. A closer look at the two largest valleys, however, shows that they coincide with the two World Wars, when people (including editors) were preoccupied with other things. Many more eruptions were probably witnessed during those times, but reports do not survive in the scientific literature.

If these apparent drops in global volcanism are caused by decreased human attention to volcanoes, then it is reasonable to expect that increased attention after major, newsworthy eruptions should result in higher-than-average number of volcanoes being reported in the historical literature. The 1902 disasters at Mont Pelee, St. Vincent, and Santa Maria (see 1902 arrow: Fig. 4.31) were highly newsworthy events. They represent a genuine pulse in Caribbean volcanism, but we believe that the higher numbers in following years (and following Krakatau in 1883) result from increased human interest in volcanism. People reported events that they might not otherwise have reported and editors were more likely to print those reports.

Additional strong evidence that the historical increase in global volcanism is more apparent than real comes from the lower plot below. Here only the larger eruptions (generating at least 0.1 km^3 of tephra, the fragmental products of explosive eruptions) are plotted. The effects of these larger events are often regional, and therefore, less likely to escape documentation even in remote areas. The frequency of these

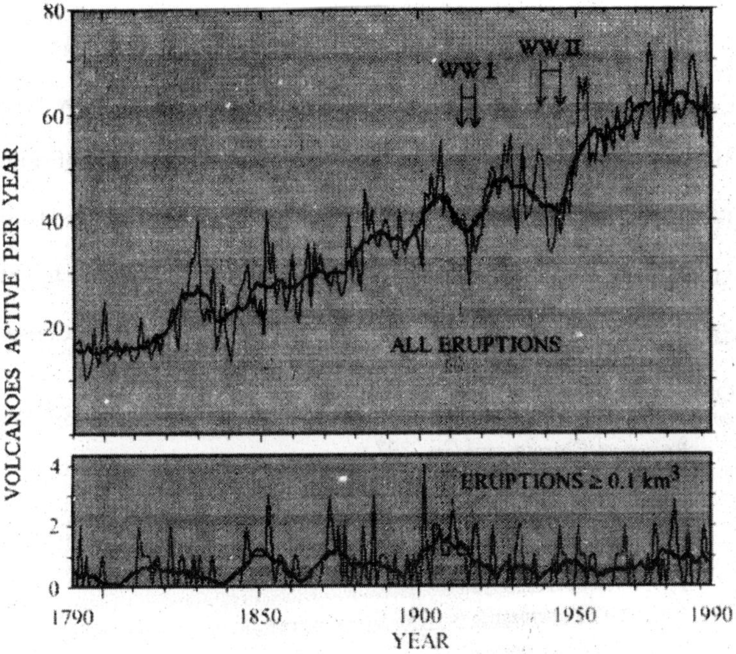

Fig. 4.30

events has remained impressively constant for more than a century, and contrasts strongly with the apparent increase of smaller eruptions with time. (Fig. 4.30)

Finally, we plot below the record since reasonably comprehensive reporting of global volcanism began in the 1960s. (Fig. 4.31) Note that the number of confirmed erupting volcanoes has levelled off between 50 and 70 per year through the past four decades, and a linear regression line through the data indicates that volcanism has been virtually constant.

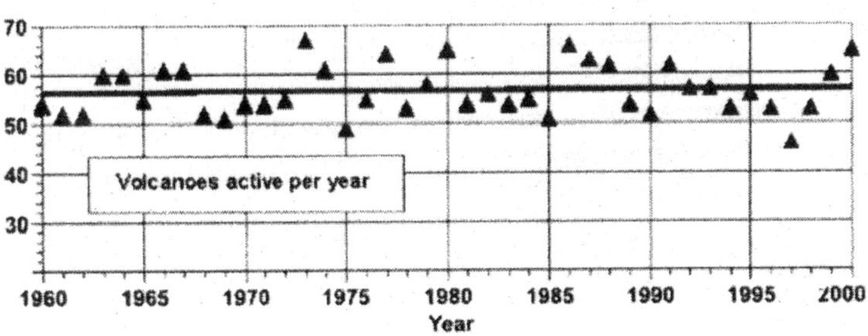

Fig. 4.31

REFERENCES

Cotton, C.A., 1969. *Volcanoes as Landscape Forms.* 2nd ed. New York: Hafner, 416 p.

Decker, R.W. and Decker, B.B., 1991. *Mountains of Fire.* Cambridge: Cambridge University Press, 198 p.

Francis, P., 1993. *Volcanoes: A Planetary Perspective.* Oxford: Oxford Univ. Press, 443 p.

Macdonald, G.A., 1972. *Volcanoes.* Englewood Cliffs, New Jersey, Prentice-Hall, 510 p.

Marti, J. and Aranya, V., 1993. *La Volcanologia Actual.* Madrid: Consejo Superior de Investigations Cientificas, 578 p.

Ollier, C.D., 1969. *Volcanoes.* Cambridge: M.I.T. Press, 177 p.

Williams, H. and McBirney, A.R., 1979. *Volcanology.* San Francisco: Freeman Cooper and Co., 397 p.

5

MODELS OF LANDSCAPE EVOLUTION

Summary: Intellectual Structures are Critical in Geomorphology because

1. Landforms develop over longer timespans (usually much longer) than human lifespans,
   - therefore, we resort to analogues either in the present or historical records to explain the evolution of landscapes,
   - the extrapolation of short records over long timespans requires a theoretical basis.
2. There are different methodologies and objectives depending on the relevant theory,
   - this is most apparent in the contrast between research which links characteristic form to present day processes (and thus limits retrodiction and prediction) *versus* research that focuses on relaxation forms but does not address relationships between contemporary forms and processes,
   - appreciation of theory leads to recognition that there is an artificial entry point to this circular argument and that the starting point will often depend on the nature of a landscape and objectives of the study.
3. By its vary nature, erosion eliminates evidence of landscape evolution,
   - thus we depend very much on depositional records (stratigraphy),

- however rarely are depositional records continuous and gaps present significant challenges and opportunities for interpretation,
- just as we need a strong theoretical basis for extrapolating from short process records, the same applies to the use of fragmented depositional records.

4. The most profitable approach to landform studies is to derive hypotheses from process research and corroborate or falsify them with morphometric research,
   - however for logistical and methodological reasons, process studies are local,
   - once again a conceptual framework is required to link scales and permit the testing of hypotheses about landscapes.

Thus far we have considered a number of specific concepts within the context of time and space in geomorphology. Now we visit concepts in the context of models of landscape evolution. Thorn (1988: 121) uses the analogy of fabric; the concepts are individual threads in the fabric of geomorphology. When the concepts are woven together, a pattern merges, but they can be woven in a variety of designs at the weaver's (modellers) discretion. Thus models are like fabric created by weaving together concepts in a pattern that embraces the various temporal and spatial scales of landform development.

W.M. Davis: The Geographical Cycle

- The first model of landscape evolution to gain widespread acceptance within the discipline,
- was remarkably influential and persistent but no longer dominates research thinking like it did, but still used as a teaching tool and residual influence reflected in the way geomorphologists cling to cyclical models.

W.M. Davis was professor of geography at Harvard University. He wrote about his model from 1880-1938, travelled and spoke widely. Like his contemporaries in natural science he was strongly influenced by Charles Darwin (On the Origin of the Species) and Charles Lyell (Principles of Geology), although used evolution as a notion of history (inevitable progress or change over time) rather than a process and took a deterministic rather than probabilistic view of evolution like Darwin.

Thus Davis aspired to a deductive, theoretical, genetic model of landscape evolution. The concepts of structure, process and time were his theoretical framework:

- Structure was regional and considered as an initial condition (beyond the scope of his model),
- Process was the sum of weathering and transport rather than specific processes or mechanisms, although since his cycle was based on the assumption of a normal climate, *i.e.* humid temperate, fluvial processes predominate,
- Time was the central theme, but time in the sense of landscape development relative to the completion of the entire geographical cycle, *i.e.*, extent of landscape development or stage.

Walter Penck: Relating Landforms to Crustal Movements

- Similar to Davis in terms of morphological and deductive but failure to relate hypothesized forms and field observations,
- But whereas Davis related form to stage, Penck related it to rate of uplift and he rejected Davis' assumption that uplift is followed by erosion of a stable crust,
- He didn't see a sequence of landform development but rather various possible sequences according to differing rates of uplift and erosion,
- He was careful to define the domain of his model as sub-aerial, but excluding eolian and glacial processes and climatic variability.

The best known manifestation of Penck's model is the retreating slope profile, where evolution of the profile is controlled by rate of output (river erosion) at the base and rate of uplift of the land. He was able to deduce various slope profiles for different combinations of river erosion, uplift and rock resistance, by assuming that stronger rock requires steeper slopes for the same rate of denudation. He also modelled stream longitudinal profile as controlled by uplift, rock type and stream discharge.

Another expression of his model was three categories of landform assemblage according to tectonic history (*versus* normal climate):

- great folding from lateral forces (orogenic),
- dome formation without folding (epeirogenic),
- stable regions.

Penck also envisaged three landscapes resulting from slow, intermediate and rapid rates of uplift. Morphologically, they were similar to Davis' old, mature and young stages, but whereas Davis ascribed morphology to age (time-dependent), Penck's model was largely time-independent based on tectonic history.

Evaluation of Penck's model is hindered by its hurried writing, posthumous publication and confused representation in English, including misrepresentation by Davis who was defending his own ideas. Although there were important flows and contradictions in Penck's work, and it was poorly translated or misrepresented especially by adherents to the Davisian school, it was the one comprehensive alternative to the cycle of erosion and thus was a focus for contrary ideas, such as emphasis on process rate (both endogenic and exogenic) and greater attention to slope retreat.

Lester King: Global Geomorphology during the Era of Process Studies

- King was trained in the Davisian school but eventually rejected many of Davis' notions as he attempted to apply the cycle of erosion to an interpretation of the landscape of South Africa.

- His model is based on the notion of parallel retreat of scarps and the resulting constraints on the downwearing of the surface of the scarp.

King's model was based on a slope profile consisting of four segments, anyone of which may be entirely absent:

1. *Waxing slope*: convex segment at the crest, dominated by soil creep of a weathered mantle; increasing slope angle is required to transport the greater quantity of slope debris with distance downslope.
2. *Free face*: bedrock outcrop; retreats parallel with weathering and uniform removal; may be absent in areas of low relief.
3. *Debris slope*: debris from the free face resting at the angle of repose; does not bury the free face but retreats with it.
4. *Waning*: gentle concave profile controlled by sheetwash and transport of sediment over an eroded bedrock surface (pediment):

    - The corresponding sequence of landscape evolution would be expansion and coalescence of pediments (pediplanation to form pediplains) as the free face retreats eliminating the higher surfaces (older pediments) and creating isolated erosional remnants.
    - King suggested that this evolution of landscapes would occur everywhere, but regarded the arid cycle as dominant given its extent geographically and over geological time.
    - His model implies that slopes lacking the free face (*i.e.*, having the classic sigmoidal slope) are inactive and a degenerative form of his classic profile.

Systems Modelling

Systems modelling swept though geomorphology in the 1960s with the shift to process and quantitative geomorphology. Although not directly related to process geomorphology, the two methodologies are so compatible that they are closely associated in the same manner as Davis' cycle of erosion and denudation chronology.

The formal structure of systms modelling (inputs, outputs, throughputs), and its application either conceptually or quantitatively make it a popular framework for research and teaching (hydrological cycle, continuity equations).

System: a set of objects together with relationships between the objects and their attributes, or a set of objects and the common processes by which they interact

In geomorphology, objects are usually landscape elements at a particular scale, relationships are geomorphic processes and attributes are physical properties like slope gradient, soil texture or drainage density (also depending on scale).

The components of the system will depend on the relationships that the researcher or teacher considers important or relevant. A model is a simplification of a system that is assumed to exist in the real world. Since, it will be based on selected parts of a system, the model will always be incorrect to some degree.

Scientists can agree on the definition of a particular system, but it can never be fully known. Conversely, a model can be simple and completely known, but always will be incomplete and useful only for specific purposes.

Types of Systems

1. *Isolated*: assumed to have boundaries that prevent the import or export of energy and mass; besides the universe, the only isolated systems are in labs or assumed for convenience.
2. *Closed*: import or export of energy but not mass; the earth is closed to closed and some geomorphic system can be modelled as if they are (*e.g.*, an internal drainage basin).
3. *Open*: both energy and mass move freely across system boundaries, includes most of the natural world.

The isolated system is the basis for the concept of maximum entropy where free energy decreases, as it becomes more evenly distributed and moves along inceasing lower gradients. However, in open systems, energy is input and output. Therefore, natural systems tend to adjust by self-regulation thereby approaching or remaining near an equilibrium (*e.g.*, circulation of the ocean and atmosphere in response to the global solar energy balance).

Types of System Models according to Complexity

1. Qualitative data, verbal description, graphs to describe a system; does not permit true modelling but is a first stage;
2. Identification and analysis of morphological variables (*e.g.* free face and talus); simplest model, only functional information is relative position of the components, but no inference or understanding of mass or energy exchange is possible;
3. Identification and analysis of energy and mass transfer; cascading system (direction specified; energy and mass move from the free face to the talus and not *vice versa*);
4. Integration of form and process; unlike 3 these involve feedback, output from one part is input to another, opposing or slowing the general trend of the system (negative feedback) or magnifying the main trend (positive feedback); natural systems (except glacial systems) are characterized by negative feedback so that changes are resisted and tendency towards an equilibrium is maintained; thus change in relationships over time can be modelled, for example, thus talus grows to bury the free face resulting in negative feedback and decreasing rates of mass wasting and talus slope growth over time; and
5. *Control systems*: conscious or deliberate human intervention in the system; useful for addressing engineering or planning issues (impact assessment).

Types of System Models according to Degree of Understanding of the System

1. *Black-box*: nothing known except the relationship between input and output.
2. *Grey box*: structure is known (subsystems considered), but no detailed knowledge or investigation.
3. *White box*: subsystems, storages and flows known and investigated in detail.

Construction of a System Model

- System Model requires identification of the important morphological variables, and linkages among these in terms of direction and strength.
- One of the simplest and most common quantitative techniques is correlation: that is establishing the covariation among pairs of variables using either nominal/ordinal data (non-parametric) or interval/ratio data (parametric).
- Correlation does not distinguish between dependent and independent variables because variance in both variables is examined; therefore, cause and effect can only be determined by interpreting other evidences; linkages established in this manner represent only statistical relationships and not the flow of energy or matter.
- The principal objective of science is to predict the behaviour of systems by determining the behaviour of individual components with absolute certainty using deterministic mathematics; this usually is not possible, so the behaviour of groups of objects is predicted with some known margin of error using the statistical methods of stochastic mathematics.
- A fundamental between rational and empirical equations is that the former balance dimensionally, that is, when dimensions of mass, length, time and temperature are summed the units are either the same

on both sides of an equation or result in dimensionless quantities, including angles; the advantage of deterministic modelling is prediction with certainty and dimensionless numbers apply to a range of scales, however, energy and mass transfer models cannot be mathematically integrated given the dimensional incompatibility.

- Besides equations, relationships within and among subsystems are depicted with canonical structures usually as a schematic diagram as a geomorphic interpretation of a system model; these diagrams are able to show the nature (direct or inverse), direction (dependent/independent), and strength of linkages within and among subsets of variables; by following pathways it is possible to identify feedback.
- The special significance of certain variables can be shown symbolically, in particular, regulators are variables that tend to stabilize the system internally (*e.g.* thresholds variables—slope angle, infiltration capacity; those that influence energy or mass transfer—vegetation; those that are important in terms of presence/absence—basal erosion), while stores are components that retain and then release energy, mass or information (*e.g.* vegetation, soil water, sediment sinks, energy sinks).

Evaluation of Systems Modelling

- The greatest strength is the consistent structure across a range of application, complexity and techniques.
- The use of symbols is powerful means of simplification and abstraction and thereby enables the use of mathematical symbols.
- Greatest weakness if the choice was system components which is arbitrary and unscientific (there is no standard approach) and influences are further analysis and interpretation.

- Similar to the problem of distinguishing between core and peripheral variables is identifying the limits of the system, that is, defining boundaries.
- Systems models often are used just to present what is already known, that is, to summarize relationships by presenting them diagrammatically; this is problem of application and not the methodology; systems models should be related to theory by displaying new relationships and serving to generate hypotheses.
- System models are abstractions not reality, they must be tied to theory and treated as simplifications with limitations.

John T. Hack: A Time-Independent Model

The interactions of time and space add much complexity to modelling. Thus usually one theme is dominant and the other produces variability or noise. Traditionally, time-dependent models were more readily accepted because they were compatible with the popular notion of evolution and slow change over time and qualitative descriptive investigation, and because the approach was advocated by W.M. Davis, the pre-eminent American geomorphologist. Davis, King and Penck (in that order) placed most, less and lesser emphasis on time, respectively. King also focused on process while Penck emphasized structure and process. These themes have a spatial component, although King and to a lesser extent Penck related landscapes to initiating (past) events (*e.g.* tectonic history). Davis' contemporary G.K. Gilbert had a methodology that was founded on principles of physics and engineering and was more suited to a time-independent approach. However, Gilbert made no attempt to develop or champion a comprehensive model. With the quantitative revolution and shift to process geomorphology, there was renewed interest in Gilbert and attention to a time-independent perspective.

John T. Hack is the champion of a time-independent model where landscape variability due to age is not modelled, but rather considered a source of variability in landscape from related to contemporary process. This approach assumes a dynamic equilibrium between contemporary surficial processes and the surface upon which they are acting. Hack chose dynamic equilibrium as his conceptual and methodological framework. He derived this perspective directly from G.K. Gilbert who worked in the western US, where the dramatic semi-arid landscape seems youthful and dynamic. Hack applied dynamic equilibrium to reinterpretation of the Appalachian Mountains, the landscape that lead Davis to think in terms of change over time.

The time-independent perspective of Gilbert is reflected in his "laws" of:

- Uniform slope: non-linear increase in rate of erosion with slope angle.
- Structure: differential erosion on resistant and non-resistant substrata.
- Divides of increasing acclivity: stream and slope gradients increase towards divides.
- Tendency of equality of action: same rates of erosion on hard and soft rocks through adjustment of slope angles; steep and high relief in strong rocks, low gradient and relief in weak rocks.

These laws all reflect a perspective of spatial variation and dynamic equilibrium between forces and resistance.

The basic features of dynamic equilibrium as applied to spatial relations within a drainage basin by Hack:

- All elements of the topography are mutually adjusted so that they are downwasting at the same rate.
- Forms and processes are in a steady state of balance.
- Differences and characteristics of form are explainable in terms of spatial relations in which geologic patterns are the primary consideration rather than a theoretical evolutionary development.
- Opposing forces (inputs and outputs) are in a state of balance where their effects cancel out to produce a steady state.

Hack also maintained that his model is not comprehensive, that time can also be invoked to explain landscape features, but it does apply to the entire range of spatial scales of interest to geomorphologists. Under dynamic equilibrium, landscapes evolve without obvious change, unless there is a change in energy inputs (climatic change, tectonism) or surface resistance. Examples of the latter include the denudation of surface materials to expose harder or softer materials, or the accumulation of coarse materials in valley bottoms. The consequent adjustment to these changes represents a disequilibrium but does not conflict with the time-independent perspective. Hack argued like Penck that rates of uplift and erosion are linked, although he related erosion plus relief to uplift and rock resistance and had a thin database to support this relationship.

Evaluation of Dynamic Equilibrium

1. The past usually is poorly or only partly known, thus a model based on current conditions has a definite advantage.
2. Mutual relationship with process geomorphology.
3. Time-independent is an end-member of the distribution of systems and system models; these are fairly easily identified (*e.g.* and underfit stream is time-dependent relative to valley form but time-independent with respect to channel form).
4. However, it is not usually this easy to resolve the complex forms that represents both time-independent and time-dependent behaviour; attention to spatial and temporal scale help, for example time-independent behaviour is more likely at more local scales, and the influence of past processes is proportinal to their intensity and inversely proportional to time elapsed since the event.
5. Dynamic equilibrium implies characteristic forms as opposed to relaxation forms.
6. Situations where form is not maintained include uplift exceeding rates of erosion or increasing relief controlled

by difference in rock resistance (*e.g.*, inversion of topography).

7. When small segments of landscape evolution are sampled it becomes difficult or impossible to distinguish between dynamic equilibrium (trending mean), steady state (constant mean) and dynamic metastable equilibrium (two scales of oscillations). Hack referred to both steady state and dynamic equilibrium, however, a trending mean is much more likely in geomorphic systems.
8. Dynamic equilibrium is more of a conceptual framework than a fully tested corroborated model, which will require much more extensive data bases. However, it is a very useful framework in that the reduced role of time is replaced with an expansion of spatial variability and the integration of parts of landscapes. In this respect it is tied to a systems approach and the notion that systems move towards equilibrium at a rate proportional to their distance from it. Thus those far from equilibrium change quickly (time-independent) and thus near equilibrium change slowly (time-dependent). This systems perspective unites both the time-dependent and time-independent viewpoints.

Mathematical Modelling

Mathematics allows ideas to be expressed with great precision but this also demands a great deal of knowledge about the phenomenon. With this precision and lack of ambiguity, mathematical models have internal logic (one step leads logically to another), and thus the primary source of error is external logic, *i.e.* the assumptions required so that the model can be expressed mathematically. In geomorphology, these assumptions are often very constraining or even unrealistic given our incomplete understanding of geomorphic systems.

The use of mathematics provides the option of modelling which is deterministic (where one physical or chemical condition leads uniquely from another) or probabilistic (a degree of randomness and therefore variability is introduced), with the recognition the natural world is deterministic but too complex to be modelled as such. Mathematical modelling has been out of the mainstream of geomorphological research with the emphasis on process observation and an inferential inductive methodology.

Following is a review of the mathematical models of two contemporary geomorphologists.

Frank Ahnert: A Mathematical Model of Slope Evolution

- a German born and educated geomorphologist who was a professor at the University of Maryland for many years and so his approach to geomorphology was typical of the North American/British school of process geomorphology.

Ahnert developed based on a central concept of equilibrium of debris supply and removal, rather than Hack's equilibrium between process and form. He modelled the spatially and temporally variable rates of waste production, transport and redeposition at points in the landscape.

Ahnert's model is discrete: comprised of a network of points, where computations are made individually and sequentially with small temporal and spatial increments. Waste thickness is related to form by modelling changes along slope profiles (2-D) or on a grid of points (3-D). He differentiated between direct removal (no redeposition), and point-to-point transfer (the norm) which results in accumulation of debris at the foot of the slope. The computations are of the change in waste cover thickness (mass balance) in response to mechanical and chemical weathering, slope wash, rain splash, soil creep and landsliding. Rational equations simulate each process.

The model user is able to simulate the influence of climate by varying the relative importance of mechanical and chemical weathering. The dynamic equilibrium between waste thickness and rates of includes decreasing weathering as waste thickens and increasing weathering as the wasted cover thins, although a thin waste cover is most effective for chemical weathering and so rate of weathering initially increases with increasing waste cover until a threshold thickness beyond which weathering rates decline.

Splash transport is a non-linear function of slope gradient while wash transport is a non-linear function of slope gradient, surface runoff and soil erodibility. Wash transport can be specified as point-to-point or complete removal. Surface runoff is derived from a slope hydrology submodel based on the mass balance between precipitation at a point, infiltration capacity and runoff from upslope.

Ahnert modelled slow mass movement as either viscous (non-threshold) or plastic (threshold), depending on slope gradient and fluidity of the material. In both cases, there is a critical slope angle but the value is zero for non-threshold viscous mass wasting. Beyond a threshold slope, waste does not accumulate, so Ahnert has the material move to the next lowest point where slope gradient is less than the threshold. However, this is not a realistic description of rapid mass wasting, because (1) momentum usually carries the debris further than this model suggests and (2) the critical threshold for rapid mass wasting will vary from point-to-point as kinetic energy is converted to kinetic energy (*i.e.*, in a moving mass the angle of dynamic friction is less than the angle of static friction).

Ahnert approach to modelling is numerical simulation, that is, the outcome depends not just on the mathematical functions but also the numerical values that the user enters to simulate, for example, rock structure (resistance) at a point, nature of base level change (reference data for iterations of the model), relative importance of weathering processes and dominant process of mechanical transport.

By varying all these boundary conditions, Ahnert concluded that the strongest control on slope profile form was the mode of removal (point-to-point or direct).

The greatest strengths of his model are the use of waste thickness as a control and response with respect to slope processes and as a parameter than can be verified empirically. Also his model is geographic; it can accommodate spatial variability between point locations. The two main weaknesses are that he does not model feedback among processes and some relationships are suggested by observation and not physically based.

Michael J. Kirkby: Deterministic Modelling of Process-Response

Kirkby slope model is based on a mass balance equation formulated using differential calculus so that, unlike Ahnert, the slope is modelled simultaneously as a continuous profile. It is structured to accept submodels and the analytical use is based on vectors, although a co-ordinate system must be specified in order to solve the equations. Kirkby models change in elevation as a function of mechanical and chemical transport processes.

Kirkby's initial and preferred approach was analytical, that is, expressing theory symbolically and deriving new equations by manipulating the symbols; very much a deductive methodology. He eventually resorted to numerical modelling for the modelling of complex problems where the theory is inadequate, however, once empirical data (numbers) are inserted into equations, only numbers can be derived and not new equations.

Kirkby based the simulation of weathering effects on the concept of a soil deficit:

- The amount of material required to reconvert soil into bedrock ("everywhere within a soil some material has been removed chemically in comparison to the bedrock from which it is derived").
- This soil profile model is primarily chemical in nature and strongly dependent on the effects of hillslope hydrology on chemical weathering and solute transport.
- Rate of change in chemical removal is constant with respect to changes in the volume of subsurface flow.

Other important concepts are:

- Erosion-limited removal from transport-limited where the actual transport rate is equal to the transporting capacity of the process *versus* weathering-limited where the capacity of the transporting process greatly exceeds the transport rate.
- A hydrological submodel based on annual precipitation, means rain per day, surface storage capacity and evapotranspiration combined in a mass balance.
- Treating landsliding as a threshold-based, erosion-limited process and including distance travelled by the landslide as a funciton of slope gradient.

Major Geomorphological Concepts

The term 'Geomorphology' is derived from three Greek words 'ge'—Earth, 'Morpho'—form and 'Logas'—a discourse. In simple form Geomorphology is the scientific study of geometric features of the earth's surface. Some geomorphologists think that 'geomorphology is the science of landforms'. Another definition preferred by the liberals is: "Geomorphology is the interpretation and description of the relief features of the earth". In other words, geomorphology is the science that describes the surface of the lithosphere, explains its origin and interprets its history. The study includes the form of features of earth surface and geological evolution of the landform.

In order to study these things successfully, it is necessary to know a good deal about the composition and structure of rocks of the earth and also to understand the geomorphic processes. To understand the forms and processes of geomorphology some concepts were developed. These concepts are based on various scientific studies which were conducted by different geomorphologists. Thornbury (1969) listed following 10 concepts relevant to geomorphology :

Concept 1

"The same physical processes and laws that operate today operated throughout geologic time, although not necessarily always at the same intensity as now".

The above said concept is given by James Hutton (1726-1797) of Scotland. This concept was given in a paper which was presented in Royal Society of Edinburgh in 1785. The same paper appeared in print of volume I of the transact in 1788. He has not consided the climatic changes in the past. For example, Kashmir valley was covered with ice during Pleistocene. There were the glaciers in pleistocene, but today we have stream dominance in the valley.

Thar desert had been a humid part of the earth but at present it is only hot desert. The same physical processes could not have operated there. The intensity of various geologic processes has varied throughout geologic time. For example, stream did not cut valleys in the past as they do now. But the physical processes and laws which are operating today were operating throughout geologic time. For example, with passage of time the intensity of glacier changed but physical processes remained the same. No doubt there are some loop-holes in this principle of uniformitarianism but without this principle there could hardly be a science of geology that was more than pure description. For the geological study we have to depend on this concept.

Concept 2

"Geologic structure is a dominant control factor in the evolution of landforms and is reflected in them".

All the modern geologists are working on this principle. They give more importance to process and structure. According to this principle, the structural features of rocks are much older than geomorphic development upon them. Structure means fold, fault, joints, bedding plane, mineral assemblage, permeability of rocks, susceptibility of rock, rock sequences and hardness of rock. There is a control of structure on geomorphic process. For example, the drainage pattern of river Tawi of Jammu is controlled by the Udhampur thrust and Panjal thrust, e.g. the flow of its tributaries like Ramnagar Khad and Dudar Nal are flowing with the dip of Udhampur thrust.

River Narmada and Tapti are controlled by geologic structure in the region. Some areas may not represent any structural control. But it does not mean that where the effect of geologic structure is neither obvious nor striking its influence is lacking. The effects are there very much only we may lack the ability to see them. The increasing application of geomorphic interpretation of Aerial photographs is likely to bring us to a fuller appreciation of the broad applicability of this principle.

Concept 3

"To a large degree the earth's surface possesses variations in relief because the geomorphic processes operate at differential rates."

The central idea of this concept is related with variation of geomorphic processes. On the crust of the earth the composition of the rocks and structure is different with the result there is variation in geomorphic processes. For example, the process of weathering is higher on the weak rocks than the hard rocks. Similarly, joints and fractures are more vulnerable to weathering. It means that physical and chemical weathering processes will be higher on joints, fractures and beds of sedimentary rocks than on the metamorphic rocks.

The intensity of geomorphic processes may be dependent on climate, altitude, exposure, location, vegetation cover, composition of rock or assemblage of minerals. The micro-climatic conditions may vary between a valley floor and a hill top, between northern and a southern exposure and between bare ground and that with a forest cover. Suppose weak and resistant rocks are existing in a area, the weathering and erosion processes will be different on both. But the weak rocks will form low lands and resistant rocks stand in elevation. It is because the geomorphic processes operated at a differential rates on two rocks.

Concept 4

"Geomorphic processes leave their distinctive imprint upon landform and each geomorphic process develops its own characteristic assemblage of landforms."

The interpretation of this concept is related with complexity of geomorphic processes. On the earth's crust two types of forces operate, namely, endogenetic or internal forces and exogenetic or external forces. Some areas would have complex landforms because in these areas endogenetic forces operated on that part of land which had been worn down by exogenetic forces. Naturally, complex type of landforms will develop. No doubt, that each landform has its individual distinguishing features dependent upon the geomorphic process responsible for its development. For example, moraines and drumlins are associated with the erosional and depositional processes of glacier. On the same landforms later on another geomorphic process operates which develops its own features. But there will be a relationship with first geomorphic process and second geomorphic process. In that area glacial as well as fluvial landforms exist together. The complex of geomorphic processes and agents which operate under a particular set of climatic conditions has been termed by morphogenetics Triccart and Cailleux in 1955. Many landscapes of the world are developed under changing morphogenetic systems associated with the changing climates of ice age.

Concept 5

"As the different erosional agents act upon the earth's surface, there is produced an orderly sequence of landforms."

The main follower of this concept was W.M. Davis (1850-1934). He will be remembered in future for his concept of geomorphic cycle. The central idea of this concept is that in the evolution of landscapes there is a systematic sequence of landforms which makes possible the recognition of stages of development in a sequence as youth, maturity and old age. The diferences in landforms are largely explainable in term of differences in geologic structure, geomorphic process and stage of development.

No doubt this idea of orderly sequences of landforms was criticised on many grounds. But it is difficult to find a substitute term or for that matter a substitute concept of landscape evolution that does not have its shortcoming. We shall, therefore, continue to apply this concept for geomorphic studies. The geomorphologists of the world still recognise that the different erosional agents act upon earth's surface and produce an orderly sequence of landforms.

Concept 6

"Complexity of geomorphic evolution is more common than simplicity."

According to this concept, many landscapes of the earth's crust are the product of complex or many dominant geomorphic processes. It is a rare thing to find landscape assemblages which can be attributed solely to one geomorphic process, even through commonly we are able to recognise the dominance of one.

Horberg in 1952 grouped the landscapes into five major categories:

1. *Simple Landscapes*: These are the landscapes which are formed by a single dominant geomorphic process.
2. *Compound Landscapes*: When two or more geomorphic processes operate for the development of existing topography. This type of landscapes are present in those areas which were glaciated before the Wisconsinian age. In such areas there may be found upland tracts which still retain many original glacial characteristics, yet along stream courses running water has formed the topography. A certain landscape is primarily the work of running water but weathering, gravity, and wind may have also contributed to its development.
3. *Monocyclic Landscapes*: These are those landscapes which bear the imprint of only one cycle newly created land surfaces as a newly uplifted part of ocean floor, volcanic area and area which is covered under glacial deposits of pleistocene.

4. *Multicyclic Landscapes*: These landscapes have been produced during more than one cycle of erosion. Much if not most of world's topography bears the imprint of more than one period of erosion. Aravallis in India bear the imprint of more than one cycle.

5. *Polyclimatic Landscapes*: These are those landscapes which have evolved under more accompanying variation in dominant geomorphic processes. These landscapes have experienced climatic changes leading to the development of various geomorphic processes. We have best example of polyclimatic landscape in Indiana (U.S.A.) where river Wabash flows across a buried preglacial topography, its valley is less than one and a half mile wide where the stream cuts into the bedrock of a buried ridge.

Concept 7

"Little of the earth's topography is older than tertiary and most of it not older than pleistocene."

The meaning of this concept is related with the age of the topography of the earth's crust. There is not an iota of doubt that present topography of the earth is older than pleistocene. In other words, the present topographic features are not so ancient. If there is any existence of these features they were formed during some past period of geologic time, then buried beneath a cover of mass of igneous or sedimentary origin, then still exposed through removal of the cover. Topographic features now being exhumed may date back as recent as pleistocene.

Prof. Ashley (1931) supported this concept and believed that "Most of the world's scenery as mountains, valleys, shores, lakes, rivers, waterfalls, cliff and canyons are post-miocene, that nearly all details have been carved since the emergence of man, and that few if any land surfaces today have any close relation to pre-miocene surface."

According to him, 90 per cent of our land surface has been developed in post-tertiary time and perhaps as much as 99 per cent is post-middle miocene in age. We can sum up this concept with this explanation that maximum topographic features of the earth's crust were formed after pleistocene period. Because before pleistocene features were carved or modified by after or during pleistocene geomorphic processes. The pre-pleistocene imprints or topographic features were sculptured by the pleistocene or after pleistocene geomorphic processes.

Concept 8

"A proper interpretation of present day landscape is impossible without a full appreciation of the manifold influence of the geologic and climatic changes during the pleistocene."

The main idea of concept reveals that for the study of present landscapes of the world you have to consider the effect of pleistocene or ice age on the topographic features of the earth's surface. The climatic and geologic changes during the pleistocene have also affected present day landforms. During the ice age more than 10,000,000 sq. miles area of the world was blanketed by glaciers. With the melting of these glaciers, glacial outwash and wind blown materials of glacial origin extended into areas which were not glaciated. The climatic and geologic changes of pleistocene period affected the courses of the rivers, lakes, canyons and sea levels of the oceans. Fluctuation in sea level of the oceans affected the coastal areas of the world. The structural changes of the pleistocene period affected the many mountains like, Himalayas and Rockies. Latest study of Nater Singh Raina reveals that the course of river Tawi of Jammu is affected by pleistocene glacier.

When the Shivaliks were rising the stream was getting its water supply continuously from the melting glacier of pleistocene periods existing in upper course of Tawi.

Concept 9

"An appreciation of world climates is necessary to the proper understanding of the varying importance of the different geomorphic processes."

The present concept is related with the influence of climate on the different geomorphic processes. Climatic variations affect the operation of geomorphic processes either directly or indirectly. Direct effect may be responsible for different intensity of weathering processes. Indirect effect may be related to how climate affects the amount, kind and distribution of the vegetal cover. Climatic effect includes the effect of precipitation, temperature, wind velocity and direction. In geomorphology basic concepts are evolved from humid temperature regions. There is a need to evolve basic geomorphic concept with respect to the humid, tropical, arctic and subarctic regions. For proper understanding of any geomorphic process we have to consider or study the effect of climate.

Concept 10

"Geomorphology although concerned primarily with present day landscape, attains its maximum usefulness by historical extension."

No doubt geomorphology concerns itself primarily with origin of the present landscape but most of the landscapes are in present forms that date back to previous geologic period. For studying previous geologic forms geomorphologists follow the principle of uniformitarianism. Previously there was no recognition of ancient erosion surfaces and the study of ancient topographies. Later on geomorphologists realised that the study of ancient landforms is a practical tool for geologic exploration. A new branch of geomorphology called Paleogeomorphology came into existence. Paleogeomorphology is the study of ancient landforms. Martin (1966) used paleogeomorphological study for exploration of oil in Canada. Prof. Ruhe in 1965 recognized three types of paleomorphic landforms. These are relict, buried and exhumed landforms. New paleogeomorphology is important field of geomorphology. The present day search for new mineral resources may apply this principle largely.

REFERENCES

Aalto, R. Montgomery, D.r. Hallet, Abbe, T.B. Buffington, J.M., Cuffey, K.M. and Sehmidt, K.M. (1997), A hill of beans *Science*, **277**, 1911-12.

Barry, R.G. (1981), Mountain weather and climate. London, Methuen, 313 p.

Chorely, R.J. (1967), Models in Geomorphology. In: Chorely, R.J. and Haggett, P.C. (Eds.), Model in Geography. London, Methuen, pp. 59-96.

Davis, W.M. (1930), Rock floor in aris and humid climates. *Journal of Gelogy*, **38** (1-27), 136-158.

Glock, W.W. (1931), the development of drainage systems: A synoptic view. *Geography Review*, **21**, 475-482.

Hallet, B. (1979), A theoretical models of glacial abrasion. *Journal of Glaciology*, **17**, 209-222.

Iverson, R.M. & Valance, J.W. (2001), New views of granular mass flows. *Geology* **29**, 115-118.

King, L.C. (1967), *The Morphology of the earth* (2nd ed.) Edinburth, Oliver and Boyd.

Lawrence, W.H. and Bettman, A.P., 1982. *Green Guide*. Northwest Working Press, Oregon.

Leonard, W.P., H.A. Brown, L.L.C. Jones, K.R. McAllister, and R.M. Storm. 1993. *Amphibians of Washington and Oregon*. Seattle Audubon Society, Seattle.

Loveland, T.R., J.W. Merchant, D.O. Ohlen, and J.F. Brown. 1991. Development of a land-cover characteristics database for the conterminous U.S. *Photogrammetric Engr. and Remote Sensing* 57(11):1453-1463.

Loy, W.G. 1976. *Atlas of Oregon*. University of Oregon Press, Oregon.

Mace, G.M., and R. Lande. 1991. Assessing extinction threats: toward a reevaluation of IUCN threatened species categories. *Conservation Biology* 5:148-157.

Marshall, D.B. 1992. *Sensitive Vertebrates of Oregon*, First Edition. Oregon Department of Fish and Wildlife.

Martin, Y. (2000), Modeling hill slope evolution: Linear and nonlinear transport relations. Geomorphology, **34**, 1-21.

Master, L.L. 1995. *Synoptic national assessment of comparative risks to biological diversity and landscape types.* Report to US EPA (Corvallis, Oregon). The Nature Conservancy, Arlington, VA.

McGarigal, K., and W.C. McComb. 1995. *Relationships between landscape structure and breeding birds in the Oregon Coast Range. Ecological Monographs* 65:235-260. (Note: habitat data were used from S. Garman's compilation of the original data used to write this paper, not from the published data).

Mellen, K., M. Huff, and R. Hagestedt. 1995. *HABSCAPES: reference manual and user's guide*. Unpublished manuscript, U.S. Forest Service.

Mumford, L., 1939. *Regional Planning in the Pacific Northwest : A Memorandum*. Northwest Regional Council, Portland, Oregon.

Myatt, W.G., 1958. *The Willamette Valley*, Doctoral dissertation., Clark University, 230 p.

Nash, D. (1980), Morphologic dealing of degraded normal fault scraps. *Journal of Geology*, **88**, 353-360.

Noss, R.F., and L.D. Harris. 1986. Nodes, networks, and MUMs: preserving diversity at all scales. *Environmental Management* 10:299-309.

Nussbaum, R.A., E.D. Brodie Jr., and R.M. Storm. 1983. *Amphibians and reptiles of the Pacific Northwest*. The University Press of Idaho, Moscow, Idaho.

Ollier, C.D. and Pain, C.F. (1994), Land scope evolution and tectonics in Southeastern Australia. *Journal of Australian Geology and Geophysics*, **15**, 333-345.

O'Neil, T.A., R.J. Steidl, W.D. Edge, and B. Csuti. 1995. Using wildlife communities to improve vegetation classification for conserving biodiversity. *Conservation Biology* 9:1482-1491. (Note: habitat data were used from O'Neil's unpublished wildlife—habitat cluster database).

Oregon Department of Environmental Quality (ODEQ). 1995. 1994/1996 Draft List of Water Quality Limited Bodies 303(d)(1). Oregon Department of Environmental Quality, Western Region, Eugene, OR. 97401, (541) 686-7838.

Oregon Department of Transportation (ODOT). 1993. Demographic and Economic Forecasts 1990.

Penck, W. (1924). Die Morphologische Analyse (Morphological Analysis of Landforms): J. Engelhorn's Nachfolger, Suttgart, 283 p. English translation by Czech, H. & Boswell, K.C., London, 1953, St. Martin's Press, New York, 429 pp.

Romey, W.D. (1982), Earth is my oatmeal EOS Transactions **63**, 162.

Ruxton, B.P. & McDougall, I. (1967). Denudation rates in northeast Paupa from potassium-argon dating of lavas. *American Journal of Science*, **265**, 545-561.

Small, E.E. & Anderson, R.S. (1998). Pleistocene relief production in Laramide mountain ranges, western United States. *Geology*, **26**, 123-126.

Strik, B.C., and D. deCalesta. 1992. Vertebrate pest management. Pages 179-180 in T. Casteel, editor. *Oregon winegrape grower's guide*, 4th ed. Oregon Winegrower's Association, Portland, OR.

Neil Ten-Eyck. 1996. *Personal communication*.

Tilman, D., R.M. May, C.L. Lehman, and M.A. Nowak. 1994. Habitat destruction and the extinction debt. *Nature* 371: 65-66.

Timm, R.M., editor. 1983. *Prevention and control of wildlife damage.* Nebraska Cooperative Extension Service, and Institute for Agriculture and Natural Resources: University of Nebraska Lincoln, Nebraska.

Towle, J.C., 1974. Woodland in the Willamette Valley: a historical geography. *Oregon Historical Quarterly.*

USDA and USDI. 1994. Record of Decision: for Amendments to Forest Service and Bureau of Land Management Planning Documents Within the Range of the Northern Spotted Owl. Washington, DC: U.S. Government Printing Office, Document 1994-589-111/00001, Region Number 10.

Vickery, P.D., M.L. Hunter, Jr., and S.M. Melvin. 1994. Effects of habitat area on the distribution of grassland birds in Maine. *Conservation Biology* 8:1087-1097.

Western, D. 1989. Conservation without parks: wildlife in the rural landscape. Pages 158-165 in D. Western and M. Pearl, editors. *Conservation for the Twenty-first Century.* Oxford University Press, New York, NY.

White D., P.G. Minotti, M.J. Barczak, J.C. Sifneos, K.E. Freemark, M.V. Santelmann, C.F. Steinitz, A.R. Kiester, E.M. Preston. 1997. Assessing risks to biodiversity from future landscape change. *Conservation Biology* 11(1).

Willamette Basin Task Force, 1969. Willamette Basin Comprehensive Study; Water and Related Land Resources; Appendix E; Flood Control.

Willamette Basin Task Force. 1969. Willamette Basin Comprehensive Study; Water and Related Land Resources; Appendix G. Land Measures and Watershed Protection.

Wilson, E.O., 1988, *Biodiversity: Washington, D.C.,* National Academy Press, 521 p.

ılson, M.V., Alverson, E.R., Clark, D.L., Hayes, R.H., Ingersoll, C.A., Naughton, M.B. The Willamette Valley Natural Areas Network. A partnership for the Oregon prairie. Restoration and Management Notes 1995, vol. 13, no. 1, pp. 26-2.

Wolff, J. O. 1995. On the limitations of species-habitat association studies. *Northwest Science* 69:72-76.

World Conservation Monitoring Centre. Brian Groombridge, editor; in collaboration with the Natural History Museum, London; and in association with IUCN—the World Conservation Union. *Global biodiversity : status of the earth's living resources.* London, New York, Chapman and Hall, 1992.

United States Department of Agriculture and United States Department of the Interior. 1994. Record of Decision: for Amendments to Forest Service and Bureau of Land Management Planning Documents Within the Range of the Northern Spotted Owl. Washington, DC: U.S. Government Printing Office, Document 1994-589-111/00001, Region Number 10.

United States Department of Agriculture. 1975. *Soil Survey of Benton Country*, OR.

United States Department of Agriculture (USDA). 1994. Expanding Horizons of Forest Ecosystem Management, Proceedings of the Third Habitat Futures Workshop, Vernon, British Colombia. General Technical Report no. PNW-GTR-336, September 1994, USDA Forest Service, PNW Research Station.

United States Census Office, 1854. Statistical View of the United States, embracing its territory, population—white, free coloured, and slave—moral and social condition, industry, property, and revenue; the detailed statistics of cities, towns, and counties; being a compendium of the seventh census, to which are added the results of every previous census, beginning with 1790, in comparative tables, with explanatory and illustrative notes, based upon the schedules and other sources of information. Compiled by J.D.B. DeBow, Washington, DC.

United States Environmental Protection Agency, 1995, Ecosystem Management Research in the Pacific Northwest Five Year Research Strategy, U.S. E.P.A. Office of Research and Development, EPA/600/R-95/069, May 1995.

Zybach, Bob and Angela Sondenna. Charlie Olson: Biographical sketch and early history of Sulphur Springs, Benton County, Oregon, 1900-1920/oral history interviews. Corvallis: Oregon State University Research Forests, College of Forestry, 1994.

6

ROCKS

The Earth's Crust

The Earth Sciences concern every part of our planet—from the centre of the metallic core, 6400 km below our feet, through its solid and liquid regions (Figure 6.1) to the outer reaches of the atmosphere. We will concentrate on the outermost, rocky part of the solid Earth, in other words the rocks of the Earth's crust that form the outermost layer of the lithospheric plates. On the scale of a human lifetime, these rocks and the landscapes of which they are part can seem static and immutable, at least in a country such as Britain where there are no active volcanoes, and earthquakes are infrequent and fairly small. Nonetheless, coastal erosion and the biological degradation of rocks to form soils are reminders that geological processes are occurring 'in our own backyard' and that the Earth's surface is continuously being reshaped.

Over hundreds of millions of years, the surface of the Earth has been rearranged by the slow grind of geological processes. The rigid outer layer of the Earth (some 100 km thick and known as the lithosphere) is divided into about a dozen plates, which change shape, size and position as their margins experience growth, destruction or collision (see Figure 6.1). So, although the processes can be very slow, given enough time the results include the generation and destruction of mountain ranges that stretch across continents, and the opening and closing of ocean basins. All these geological processes leave their mark by forming new rocks. And those rocks, whether sandstone formed from sand in a desert, or volcanic lava flows,

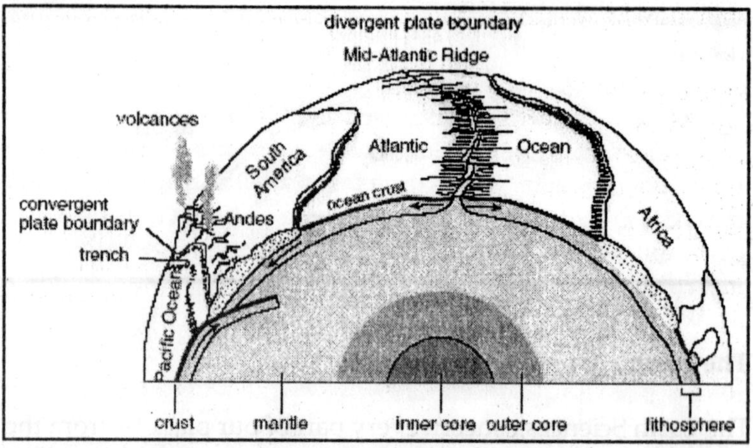

Fig. 6.1: A schematic cross-section through the Earth, showing its concentrically layered internal structure and the interacting lithospheric plates on the surface.

hold clues to the processes that formed them, and therefore, the conditions at the time of their formation. In some cases they contain information on past climatic conditions, and some rocks contain fossils that reveal the history of life and the story of evolution. Rocks can provide a narrative of the Earth's history, but we need to learn how to read their tales.

The aims of this section are to demonstrate how many of the features shown by rocks are inherited from the processes that formed the rock, to introduce you to the skills of sketching features seen in rock exposures, and to show how those features can be interpreted in a simple sequence of events through geological time.

Minerals and rocks

To begin with, it is necessary to explain the meanings of the two terms 'minerals' and 'rocks'.

Minerals

A mineral is a solid material, formed by natural processes and with a chemical composition that falls within certain narrow limits. Its constituent atoms are arranged in a regular three-dimensional array or pattern and because of this, minerals form crystals with characteristic shapes.

Although several thousand different kinds of minerals have been discovered, only a few are very common; for example, the mineral quartz, which forms many of the sand grains on a beach or in a desert. Because silicon and oxygen are the two most abundant elements in the Earth's crust, they are the main ingredients of common minerals, and these minerals are known as silicate minerals. They include quartz, whose chemical composition is silicon dioxide (formula SiO_2), and a range of others containing additional common chemical elements. A common mineral that is not a silicate is calcite; it has the chemical composition calcium carbonate ($CaCO_3$). Whereas most minerals are identified on the basis of physical characteristics (density, hardness, colour, etc.) calcite can also be readily identified with a chemical test. Calcite reacts with dilute hydrochloric acid, liberating bubbles of carbon dioxide (CO_2) gas in a vigorous fizzing froth. The chemical reaction is written as:

$$CaCO_3(s) + H^+(aq) + Cl^-(aq) \rightarrow Ca^{2+}(aq) + OH^-(aq) + Cl^-(aq) + CO_2(g)$$

In words, this equation can be expressed as: solid calcium carbonate (calcite) plus hydrogen ions in the acid plus chloride ions in the acid gives calcium ions, hydroxide ions and chloride ions dissolved in solution plus carbon dioxide gas.

Rocks

Any naturally formed solid assemblage of mineral grains can be described as a rock. The mineral grains may be fragments of crystals or intact crystals and their size can range from a few micrometres (1 micrometre = 10^{-6}m) to a few centimetres. A rock may consist of one type of mineral but more usually it consists of several minerals. Rocks can be classified according to the way in which the grains are arranged, although the identity of the minerals present (for example, the rock limestone is made mostly of calcite), the proportions of particular minerals, and the dominant size of mineral grains are also important. The shape of the grains in a rock, their size and the relationship between them (for example, whether or not the grains interlock with each other to form a mosaic) define the texture of a rock, and reflect the processes that formed it. A rock's texture has nothing to do with how the rock feels when you touch it. To 'read the rocks' and discover how any particular rock formed, we investigate its texture and work backwards to deduce the processes by which that texture was produced.

Because the mineral grains in most rocks are quite small it is often best to use a hand lens, typically with a magnification power of times 10, to get a clearer view. You may already have access to a small hand lens. If so, some notes on this are given in Box 6.1.

Box 6.1: Using the Hand Lens

The correct way to examine objects with a hand lens is shown in Figure 6.2a. Some additional hints are:

- Hold the hand lens 2–3 cm from your eye and bring the object up towards the lens until it is in focus.
- Make sure the surface of the object is well lit from the side.
- Keep the hand lens and the object parallel to each other and hold both steady.
- With rough surfaces you have to move the object back and forth to bring different parts into focus.

Rocks may be classified into three types, according to the three processes that form rocks:

1. *Igneous rocks*: These are formed from molten rock (magma) that becomes solid when it cools, either deep underground or after a volcanic eruption at the surface.
2. *Sedimentary rocks*: These are formed when eroded particles of pre-existing rocks (in other words sediment, such as sand on a beach or mud on the sea-bed) have been laid down in layers at the surface of the Earth and turned into solid rock by being buried and compacted under more layers of sediment.

Fig. 6.2: (a) The correct use of the hand lens.
(b) Incorrect use of the hand lens.

3. *Metamorphic rocks*: These are existing rocks that have 'changed form' by the action of high pressure or temperature causing new crystals to grow, for example after burial deep in the Earth.

The Formation of Igneous Rocks

Igneous rocks are defined as having solidified from a molten state, either inside the Earth or on the surface at volcanoes.

Igneous Rocks in the Landscape

The rocks that erupt from volcanoes are called *extrusive* igneous rocks, simply because they are formed by the extrusion of magma on to the Earth's surface. Igneous rocks can also form deep underground, and these are called *intrusive* igneous rocks, because the magmas were intruded into pre-existing rocks and then slowly cooled. The reason that intrusive igneous rocks are now visible at the surface is that over many millions of years erosion has stripped away the overlying rocks. In this way, bodies of igneous rock that were once subterranean pools of magma are revealed at the surface. Some of these intrusions can be up to several kilometres across. In other cases, magma had intruded pre-existing rocks to form long slab-shaped bodies of igneous rock whose longest dimensions are measured in kilometres but whose shortest dimension is no more than a few metres. These intrusions are called dykes in the case of vertical bodies, and sills in the case of horizontal bodies.

Texture of Igneous Rocks

What texture might we expect an igneous rock to have? An igneous rock will contain crystals that grew as the magma cooled. Each crystal will have started to grow unhindered by neighbouring crystals, so an igneous rock therefore has a crystalline texture in which the crystals are randomly oriented.

To picture this, consider a magma, at an initial temperature of perhaps 1000°C, as it slowly cools underground [Fig. 6.3, path (a) to (d)]. Initially, the magma is completely molten (Figure 6.3a) but, unlike water placed in the ice box of a freezer, magma doesn't turn from being totally liquid to totally solid at a single temperature when it is cooled. Instead, different minerals crystallize over a range of temperatures (in fact over one or two hundred degrees Celsius). So, when the temperature of a magma falls by a small amount, only a few mineral crystals will form (Figure 6.3b). On further cooling these crystals grow larger, and new minerals also start to crystallize (Figure 6.3c). Eventually, these crystals form an interlocking network, with the last crystals to grow filling the spaces between. When totally solidified, the rock has the crystalline texture shown in Figure 6.3d.

In contrast, very fast cooling allows crystallization to occur by the nucleation of many small crystals rather than the steady growth of a few crystals. The resultant igneous rock contains innumerable tiny crystals that may be so tiny as to be indistinguishable except under the high magnification of a microscope [Figure 6.3, path (a) to (e)]. In the most extreme case crystallization is totally inhibited and the starting liquid is quenched to form volcanic glass.

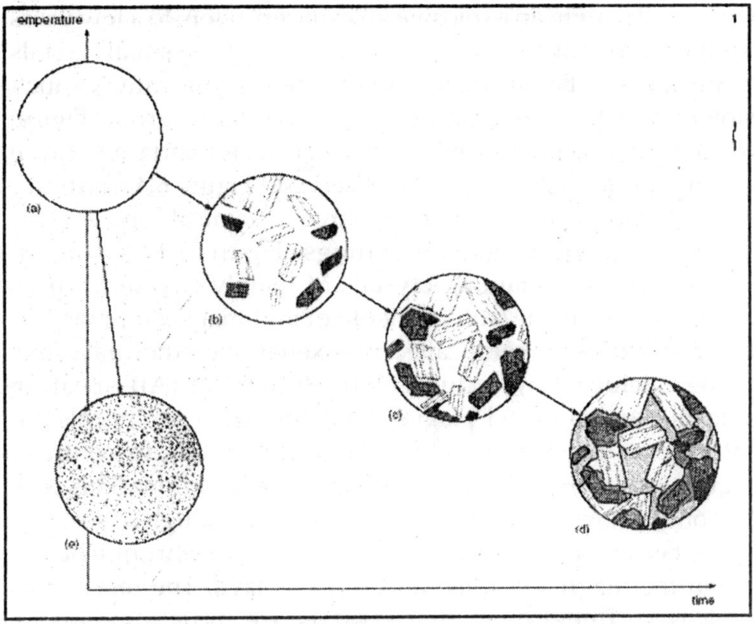

Fig. 6.3: The number and size of crystals that grow in a magma depend on the temperature and the cooling rate. Starting with a liquid (a) that undergoes very slow cooling, the magma solidifies by the gradual growth of large crystals of different minerals (indicated by different patterns of shading) to produce an igneous rock with large crystals with random orientations [cooling path (a) to (d)]. When cooling is fast, many very tiny crystals, rather than a few large crystals, grow with the result that a fine-grained igneous rock is formed [cooling path (a) to (e)].

Would you expect a fine-grained igneous rock to have formed deep below the Earth's surface or at the surface?

Now read the answer

A fine-grained igneous rock requires rapid cooling and this is more likely at the surface, where magma comes into contact with air or water, rather than in the hot interior of the Earth.

Generally speaking, the number and size of the crystals in an igneous rock depend on the amount of time available for their growth. The slower the cooling, the bigger the crystals. In the case of extrusive rocks, the amount of time is short—anything from a few seconds for droplets of magma flying through the air in an explosive volcanic eruption, to a few years for the interior of a thick lava flow. This results in small crystals (Figure 6.3e). For intrusive rocks, the cooling rate is much slower and there is time for larger crystals to grow (Figure 6.3d). (The times involved are not known for certain because the magma at depth cannot be observed.) Figure 6.4 illustrates this with two rocks of essentially similar chemical composition, but from different igneous settings. Figure 6.4a shows an intrusive rock containing crystals of mainly two minerals—one dark [this is the iron (symbol Fe) and magnesium (Mg)—bearing silicate mineral called pyroxene], the other pale [this is the calcium (Ca), sodium (Na), aluminium (Al)—bearing silicate mineral called plagioclase feldspar]. The crystals are intergrown (cf. Figure 6.3d) and easily visible. In contrast, Figure 6.4b shows a rock collected from a lava flow. This rock has only a few crystals (pale) that are large enough to see, and this is because cooling and crystal growth were abruptly halted when the magma erupted and froze as lava. The rest of this rock is so finegrained that crystals are indistinguishable without the help of a microscope. There are, however, a few small round dark areas (e.g. near the top left edge). These are gas bubbles formed when the magma came to the surface and gases that were dissolved in the magma came out of solution as the pressure on the magma decreased.

Chemical and Mineral Composition of Igneous Rocks

As well as varying in grain size (owing to different cooling rates), igneous rocks also vary in chemical composition and hence in the identity and proportions of minerals present. For instance, the common igneous rock granite contains (as part of the strict geological definition of the term 'granite') between 10% and 35% by volume of the mineral quartz (chemical composition silicon dioxide—SiO_2). On the other hand, the igneous rock gabbro (Figure 6.4a) does not usually contain any quartz. This difference is due simply to the fact that granite contains a much higher proportion of the element silicon (Si) than does gabbro. So, chemical composition determines which

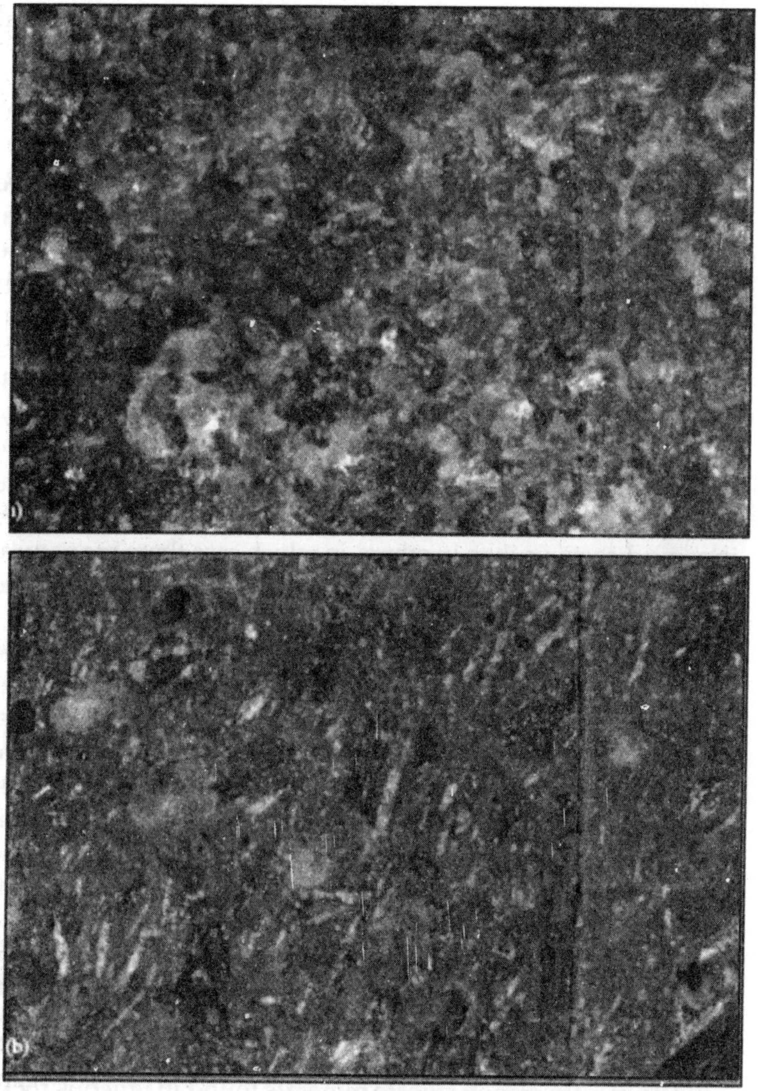

Fig. 6.4: Close-up views of two rocks formed by the cooling of magma of the same chemical composition. (a) Gabbro formed by slow cooling in a subterranean magma chamber. (b) Basalt formed by rapid cooling of magma at the Earth's surface in a lava flow. The field of view is about 2.4 cm across in each case.

minerals are present, and cooling rate determines the crystal size of those minerals.

Classifying Igneous Rocks

To classify (i.e. to name) igneous rocks, geologists use three pieces of information in combination—the grain size and the identity and proportions of the minerals present. The identification of minerals in a rock relies on recognising their particular distinguishing features. Such features include colour, lustre (the way in which light is reflected from the mineral's surface) and shape. The way in which certain minerals break apart along preferred planes, a property known as cleavage, can also be useful in identifying minerals because this property depends on the way that the mineral's constituent atoms are arranged. In some minerals the atoms are bound more strongly in some directions than in others, in which case there are natural planes of weakness present in the crystal. The mineral tends to break preferentially along these planes.

One mineral that shows this feature clearly is mica, a silicate mineral containing potassium (K), iron (Fe), magnesium (Mg) and aluminium (Al), together with hydroxyl (OH) groups. These compositional details need not distract us, but the important point is that the potassium ions occur in layers, which separate sheets of more tightly bound silicon, oxygen and other atoms. Mica therefore splits apart parallel to the sheets, so has just one set of cleavage surfaces parallel to each other (Figure 6.5a). This is why mica forms platy or flake-like crystals, rather like the pages of a book. Other minerals can have two or three sets of cleavages (which intersect at characteristic angles) whereas others, notably quartz, have no cleavage and break irregularly.

The most common silicate minerals found in igneous rocks (most of which are also common in sedimentary and metamorphic rocks) are listed in Table 6.1. Minerals rich in silicon appear towards the top, and those poorer in silicon but richer in magnesium and iron appear towards the bottom of Table 6.1.

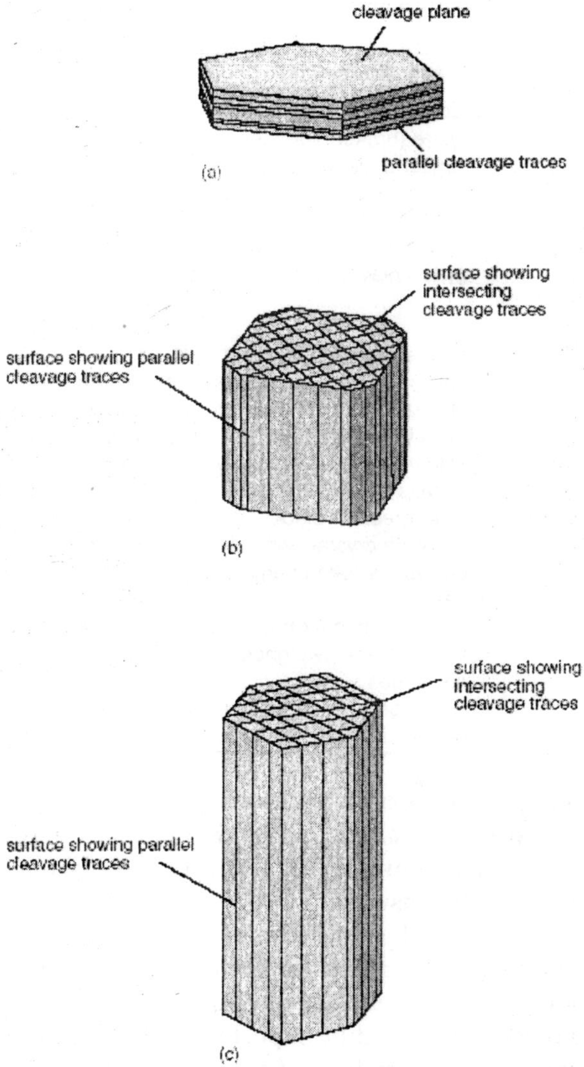

Fig. 6.5: Some examples of cleavage shown by different minerals. Note that in each case the number of cleavage traces with different directions depends on which face of the crystal is being looked at. (a) Mica, one set of cleavages when viewed edge on. (b) Pyroxene, two sets of cleavages intersecting at about 90° when viewed end on. (c) Amphibole, two sets of cleavages intersecting at about 120° when viewed end on. Table 6.1 gives additional information about these minerals, including their true colours.

Table 6.1: The Common Silicate Minerals of Igneous Rocks

Mineral	*Diagnostic features*	*Chemical composition*
Quartz	Colourless to pale; glassy lustre (broken surfaces similar to broken glass); curved fractures; no visible internal structure	SiO_2
Feldspar	Pale pink or white; two good cleavages	Ca, Na, Al silicates. Feldspar that is rich in Ca and Na is known as plagioclase, if rich in K and Na, alkali feldspar.
Mica	Forms platy crystals that split into flakes; one perfect cleavage. Brown or black mica is called biotite; colourless or silvery brown mica is called muscovite	K, Mg, Fe, Al silicate containing OH. Biotite is darker than muscovite because it contains more Mg and Fe.
Amphibole	Dark green or black; elongate crystals with two cleavages (often hard to see)	Na, Ca, Mg, Fe, Al silicate with hydroxyl group
Pyroxene	Dark green, dark brown or almost black; two good cleavages	Fe, Mg, Ca silicate
Olivine	Grass green; no cleavage, but has irregular fractures	Mg, Fe silicate

Would you expect an igneous rock that contained a lot of magnesium (Mg) and iron (Fe), but not much silicon (Si), to contain all of the minerals in Table 6.1 in equal abundance?

Now read the answer

No. The type and proportions of the minerals present must reflect the chemical composition of the rock. So an igneous rock that contains a relatively high proportion of Mg and Fe and a low amount of Si would be expected to contain more olivine and pyroxene and very little, if any, quartz.

The variation in the mineral content of igneous rocks is shown in Figure 6.6, and this diagram provides the means to classify igneous rocks. The percentage of each mineral present is represented on the vertical scale, and the range of rock type (effectively the chemical composition) is given on the horizontal scale. Coarse-grained rocks (typical grain size greater than

2 mm) are named separately from fine-grained rocks (typical grain size less than 0.25 mm). Notice that the proportion of pale-coloured minerals (felsic minerals) increases from left to right, reflecting the increase in the concentration of silicon in the chemical composition of the rocks. According to the diagram, granite is defined as a coarse-grained igneous rock containing quartz, feldspar (both alkali and plagioclase feldspar), mica and sometimes amphibole. The relative proportions of these different minerals are given by the width of the appropriate band on the diagram. For instance the quartz content of granite can range from about 10% to 35%, and the mica content from about 5% to 15%.

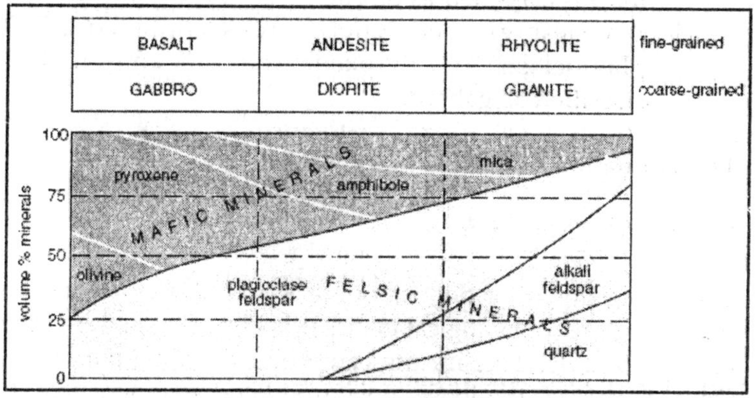

Fig. 6.6: Classification of common coarse- and fine-grained igneous rocks according to their mineral content. (Medium-grained rocks are omitted for simplicity.) Dark-coloured minerals are known as mafic minerals [they are rich in magnesium and iron (Fe)]. Pale minerals are known as felsic minerals (from feldspar and silicon).

Question 1

Do igneous rocks with small crystals form at the surface of the Earth?

Give reasons for your answer.

Now read the answer.

A fine grained igneous rocks form at surface of the earth with rapid cooling.

Question 2

Use Figure 6.6 to decide which of the following statements are true :

1. Olivine is not present in granite.
2. The proportion of pyroxene in gabbro can vary from about 25% to 50%.
3. Alkali feldspar is less abundant than plagioclase feldspar in coarse-grained igneous rocks.
4. A coarse-grained igneous rock containing 60% plagioclase, 15% pyroxene, 15% amphibole and 10% mica is classified as diorite.
5. A coarse-grained igneous rock containing 50% plagioclase, 35% pyroxene and 15% amphibole is classified as gabbro.

The Formation of Sedimentary Rocks

Sedimentary Material

The laying down, or deposition, of layers of rock fragments, mineral grains, or biological material, such as the shells or other hard parts of dead organisms, can produce sedimentary rocks. Once deposited, the loose, unconsolidated sediment may be converted into a solid rock by compaction and cementing of the grains together by chemical action deep below the surface. These rocks consist, therefore, of fragments of sedimentary material, bound together by even smaller fragments, or some sort of cementing material, and so display a fragmental texture, as shown in Figure 6.7.

Sedimentary Processes

Sedimentary grains are formed when the rocks at the Earth's surface are slowly broken up physically by exposure to wind and frost, and decomposed (chemically) by rainwater or biological action. These processes are collectively termed weathering. Once a rock has been broken up by weathering,

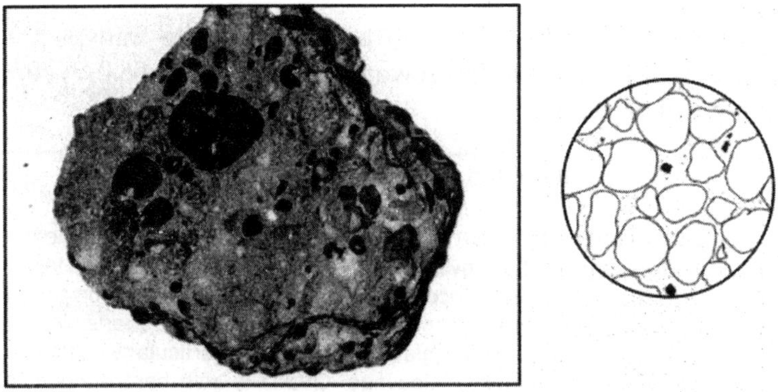

Fig. 6.7: The fragmental texture of sedimentary rocks illustrated by (a) a sample of conglomerate, about 20 cm across, comprising rounded pebbles and enclosing sand grains, and (b) a generalised sketch showing round grains bound together by much smaller grains.

the small rock fragments and individual mineral grains can be eroded from their place of origin by water, wind or glaciers and transported to be deposited elsewhere as roughly horizontal layers of sediment. The resulting sediment reflects the original rock types that were weathered, the efficiency of erosion and transport, the extents of chemical and physical degradation of the sediment grains during transport, and the conditions under which the grains were deposited from the transporting water, wind or ice. For example, sand-sized grains of quartz are one of the main constituents of sandstone, but those grains may have been transported by water in a river, carried by waves on a sea-shore, or blown around in hot desert sandstorms (to give just three possibilities). How we might distinguish which of the many possibilities is the most likely in any given case?

One approach is to use the size and shape of the grains in a sediment or sedimentary rock to reveal quite a lot about the origin of the sediment. For example, a vigorous river transports much larger grains than a gentle current in a lake, so the size of the grains gives an indication of the strength of the currents that could have transported and deposited the grains. In other words, the grain size depends on the energy of the environment in which the sediment was deposited. The general shape of the grains will tell you about the nature of the transporting medium; for example, was it water or air? (See Box 6.2, A story in a grain of sand).

Box 6.2: A Story in a Grain of Sand

Quartz is a hard mineral that is very common in the Earth's continental crust. It is resistant to attack by chemicals and physically strong, so it tends to survive the weathering process that disaggregates and decomposes pre-existing rocks. Many sedimentary rocks contain grains of quartz. Quartz grains are recognisable by their glassy appearance (particularly on freshly broken surfaces) and lack of cleavage. Quartz is also hard enough to scratch steel.

Whether quartz grains are transported and deposited by moving air (by being blown around by the winds in a desert) or by moving water (in a river or in ocean currents) determines how rounded they become (Figure 6.8). The degree of rounding of quartz grains depends on the intensity and frequency with which grains collide with each other, and these factors depend on the environmental conditions. Air is less viscous than water, so windblown quartz grains collide more violently than quartz grains carried in water, which has a cushioning effect. Also, the wind speeds needed to move a sand grain of given size are higher than the speeds for flowing water. This means that collisions between grains will be much more energetic in air than in water, so the corners of windblown grains are readily knocked off, and the grains are usually very much more rounded (Figure 6.8b) than water-transported grains (Figure 6.8a).

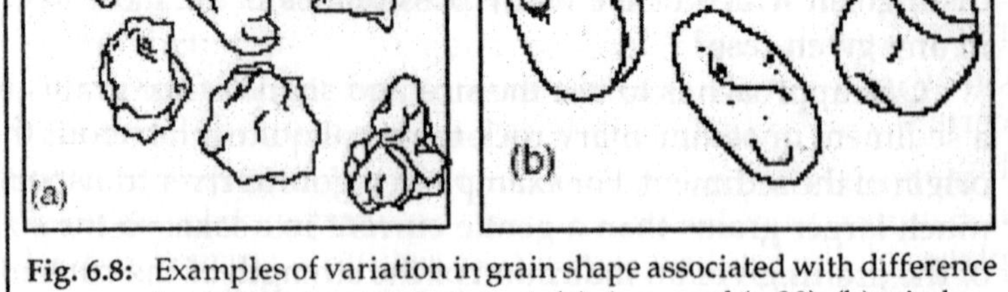

Fig. 6.8: Examples of variation in grain shape associated with difference in sedimentary environment: (a) river sand (× 20); (b) wind-blown sand (× 20)

(contd.)

Box 6.2 *(contd.)*

Wind-blown sand deposited in desert environments differs from water-lain sand in another way. Quartz sand grains in a desert often have a coating of red or orange iron oxide. This red-orange colouration is typical of desert landscapes and it is derived from the insoluble 'rusty' residue from weathering of iron-rich minerals. Water-lain sand grains lack such an obvious coating.

The degree of sorting in a sediment is another useful method for distinguishing different types of depositional situation. Sorting is a measure of the range of grain sizes present in a sediment or sedimentary rock. A poorly-sorted sediment (Figure 6.9a) has a wide range of grain sizes as a result of rapid deposition, such as occurs during a storm. On the other hand, a well-sorted sediment has a narrow range of grain sizes (Figure 6.9c), and is the result of extensive reworking of a sediment by wind action in deserts, or wave action on beaches and in shallow shelf seas.

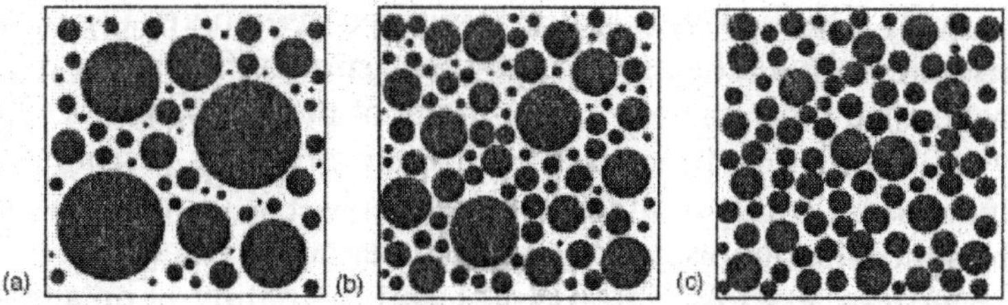

Fig. 6.9: (a) Poorly-sorted sediment. (b) Moderately-sorted sediment. (c) Well-sorted sediment.

Are all sediments composed of fragments of rock and minerals eroded from pre-existing rocks?

Now read the answer

Any record of ancient life preserved in a rock is known as a fossil—sometimes fossils are rare, whereas a few rocks are composed of virtually nothing else but fossils. In particular, many limestones were formed by the accumulation of the calcite (calcium carbonate—$CaCO_3$) shells and skeletons of certain marine organisms. Chalk is a well-known type of limestone that outcrops extensively across southern England; it is almost pure calcite, and consists largely of minute calcite plates of countless planktonic algae (phytoplankton) fossils. Other limestones, such as those found in the Peak District of northern England, contain abundant fossils of reef-building corals. Another example of a biologically formed sedimentary rock is coal, which is formed from compressed layers of woody plants.

Fossils are important when reconstructing the geological past because they are records of the environment at the time and place the fossil organisms were living. For instance, limestones rich in corals typically indicate warm shallow seas—the conditions needed for a coral reef ecosystem to thrive.

It is important that as many lines of evidence as possible are used to give a consistent interpretation of a rock's origin. No single feature should be taken as unequivocally diagnostic. For instance, think of a desert sand, composed of well-rounded, red oxide-coated, well-sorted quartz grains. Now imagine that climatic conditions change and these sand grains are swept away by flowing rivers and re-deposited elsewhere. The new sand deposit will be produced by the action of flowing water, but the sand grains may retain many of the characteristics of wind-deposition inherited from their previous history. A misleading interpretation can be reached if other lines of evidence are ignored. Such supplementary evidence could come from any fossils in the rock and the nature of adjacent sedimentary layers. Getting all of the necessary information involves a mixture of making observations and measurements at rock exposures in the field as well as examination and analysis of samples in the laboratory.

Sedimentary Strata

We have seen that the detective work of piecing together a part of Earth's history from sedimentary rocks involves detailed investigation of rock samples, but this can give only a partial picture. On the larger scale of a rock exposure, there can be plenty for us to see and to interpret. Sedimentary rocks are usually found as layers referred to as strata (Fig. 6.10), with each stratum (layer) recording the particular conditions at the time of its deposition. [*Note*: sedimentary layers are often referred to as beds, but strictly speaking the term 'bed' is reserved for those strata thicker than 1 cm; thinner layers are known as laminae (singular lamina).] Over time, conditions may have changed, either gradually or quickly, causing the

Fig. 6.10: A Succession of Sedimentary Strata (layers) Exposed in a Cliff on the Dorset Coast.

nature of the sediment being deposited to change. In this way, a vertical stack of sedimentary strata is a record of changing conditions during a segment of geological time. The oldest sediments will be at the bottom, with progressively younger strata laid down on top. Geologists refer to this as the principle of superposition—older rocks are overlain by younger rocks; an individual layer is younger than the one beneath it and older than the one above it; the oldest layer lies at the bottom. This provides a relative time-scale. Changes in sedimentary rocks (or in the types of fossil they may contain) up through a sequence of strata provide a record of changing conditions over the passage of time throughout which the rocks were deposited.

Stratigraphy and Geological Time

Stratigraphy is the study of how the types of strata have varied over time and also considers how they are distributed geographically. One of the most useful results of stratigraphy is a generalised geological succession—the stratigraphic column—that defines the divisions of geological time. Fig. 6.11 shows the geological time-scale. The broadest division of Earth

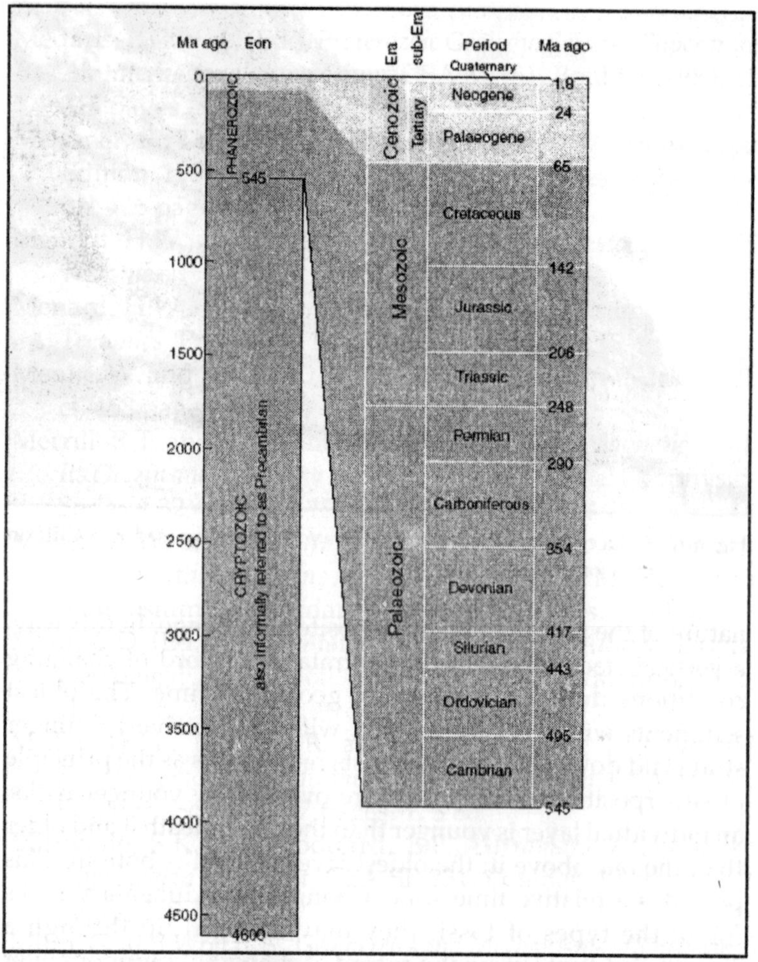

Fig. 6.11: The Geological Time-Scale.

history is into two intervals (called eons) of very different length: the Cryptozoic Eon and the Phanerozoic Eon. The Cryptozoic is a vast amount of time—from the origin of the Earth, 4,600 million years (Ma, short for mega-annum) ago, to the start of the Phanerozoic, 545 Ma ago. Cryptozoic is derived from Greek words meaning 'hidden life'. This is the period of time when organisms did not possess hard parts such as shells, so their remains are difficult to find in sedimentary rocks. In contrast, Phanerozoic is derived from Greek words meaning 'visible life', reflecting the great abundance of fossils derived from the hard parts of organisms throughout this eon. The Phanerozoic Eon is divided into three eras—the Palaeozoic, Mesozoic and Cenozoic Eras (meaning 'ancient life', 'middle life', and 'recent life', respectively). Each of these eras is divided into a number of periods of unequal length. (Fig. 6.11) The current period, which started 1.8 Ma ago, is the Quaternary Period.

Looking at this column, you might think that there is a complete sequence of rocks everywhere, but there is not, in the same way that there are no historical records for certain times from certain areas. In the field, the Earth scientist might find rocks of, say, the Triassic Period lying above and in direct contact with rocks of the Carboniferous Period, so that rock evidence for the Permian Period is not present. This means that either there was no deposition of sediment during the Permian Period, or that these rocks were deposited, and then eroded, before rocks of the Triassic Period were laid down. This type of relationship in the geological rock record, where at a contact there are beds of the intervening age missing, is called an unconformity. The recognition of unconformities in the field is a key part of unravelling the geological history of an area because they represent some sort of hiatus in the conditions at the Earth's surface where the sediments were deposited.

Fossils and Ancient Environments

An essential component of any environment is the plant and animal life that is adapted to the prevailing conditions. Fossil plants and animals are therefore wonderful sources of information about ancient environments. Plants can leave behind remains ranging from roots, leaves and twigs to seeds and pollen. Leaves and twigs are relatively fragile, and require a comparatively low energy environment (e.g. the mudflats of an estuary) for their preservation. Seeds, pollen and spores are surprisingly robust, and are often the only parts of a plant to survive. Animals can be preserved in one of two ways—either by some part of their body remaining as a fossil, or by some trace, such as the animal's footprints preserved in a muddy sediment, becoming preserved as sedimentary rock (a trace fossil). Even dung can end up fossilised, the resultant fossils being known as coprolites.

Body fossils of animals include shells, skeletal frameworks, bones and teeth. We do not have room here to describe the most common fossil groups. Figure 6.12 illustrates some of them, with their characteristic features labelled. It is more important that here we consider how an organism gets fossilised, and what they can tell us about the environment in which they lived.

First of all, we have to consider the organism itself. Does it have any hard parts? If not, e.g. a jellyfish, then its chances of fossilisation are very low indeed. However, some soft-bodied

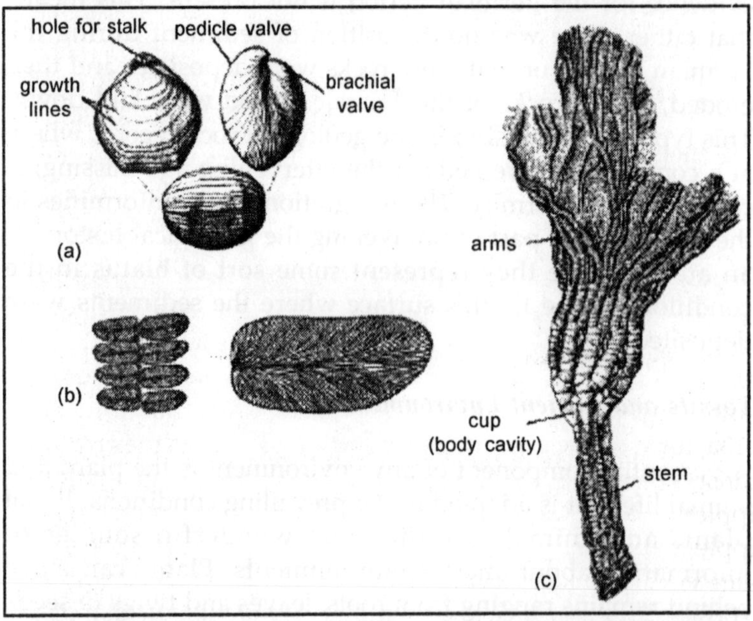

Fig. 6.12: Drawings of three types of fossil. (a) The shell of a member of the group of marine animals known as brachiopods (*Obovothris*, from the Jurassic Period), life size; (b) Fragment of a land plant (*Neuropteris*, from the Carboniferous Period) (half life size) with detail shown at 2.5 times life size; (c) Upper part of a crinoid (*Pentacrinites*, from the Jurassic Period), a member of the group of marine animals that includes sea urchins.

organisms burrow into sediment, e.g. the lugworm that leaves the familiar worm casts on the shore, and at least there is the chance that their burrow becomes preserved as a trace fossil. An organism with a one-piece shell, such as a periwinkle or garden snail, stands a better chance of being preserved intact, than if its skeleton is made up of lots of pieces, like a sea urchin or crinoid. (Fig. 6.12c)

Which of the following is indicative of deposition in a high-energy environment, and which in a low-energy environment? (a) A sedimentary layer containing complete examples of thin fragile shells. (b) A sedimentary layer containing broken shell fragments.

Now read the answer

Fossils of land-dwelling organisms (Fig. 6.12b) indicate deposition in or at least near a terrestrial environment, whereas marine fossils indicate a marine environment suitable for the fossil organisms to have lived in (e.g. correct temperature range, light levels, salinity, water depth). This evidence can then be added to evidence from the features of sedimentary grains to help reconstruct the environment.

Sedimentary Structures

Consider some of the places where sedimentary materials are moved and deposited. Are the sediments always laid down in perfectly horizontal, perfectly flat layers? No; as often as not, the depositing surface is not perfectly flat. Instead, a system of parallel ridges, or ripple marks, like the ones shown in Fig. 6.13a, form by the action that flowing water has on the erosion, transport and deposition of sand grains. In sedimentary strata, ripples and dunes (which are effectively larger versions of ripple marks) are normally seen in cross-section rather than in plan view, because pristine depositional surfaces are rarely exposed. In cross-section, the thin sedimentary layers that built up each ripple or dune can be visible, inclined at an angle to the top and bottom of the bed as a whole. (Fig. 6.13b) These structures are called cross-stratification and are a clue to the strength, direction and setting of the currents that produced them.

Fig. 6.13: (a) Ripple marks on a sandy beach. (b) Cross-stratification in sandstone formed by underwater dunes advancing from right to left; the exposure is about 3 m high.

Many features associated with either erosion or deposition in a sedimentary environment can be preserved. For instance desert dunes can yield cross-stratification on a scale measured in many metres. All such structures are referred to as sedimentary structures and, as we have just seen, provide evidence with which to interpret ancient sedimentary environments.

Question 3

Explain whether the energy of the environment of deposition of the following sediments is (i) high, (ii) medium or (iii) low.

(a) A sediment made up of pebbles and boulders (High).
(b) A sediment made up of sand grains (Medium).
(c) A sediment made up of fine clay particles (Low).

The Formation of Metamorphic Rocks

Any type of rock can become a metamorphic rock if it is heated to temperatures of several hundred degrees Celsius, and/or if subjected to high pressure (because of the weight of overlying rocks). During metamorphism, the minerals making up the rock become chemically unstable, meaning that their constituent ions are redistributed. The result is that either large crystals grow at the expense of existing smaller ones, or a new set of minerals is formed. Generally speaking, the overall chemical composition of the rock remains about the same. Although igneous and metamorphic rocks both form at high temperatures, an important distinction is that metamorphism occurs in the solid state, whereas igneous activity involves liquid rock (magma).

Causes of Metamorphism

What natural process could cause a rock to be heated?

Now read the answer

On the other hand, an increase in both pressure and temperature will come about if the rock becomes more deeply buried as a result of Earth movements, particularly at convergent plate boundaries where continents collide, or is covered by a deepening layer of sedimentary deposits.

There are, therefore, two settings where metamorphic rocks can be found. In the first, and simplest situation, a narrow zone around the edge of an igneous intrusion becomes heated by the magma and undergoes metamorphic recrystallization. This is known as contact metamorphism because it is caused by hot magma coming into contact with cold rocks; contact metamorphism is due to heating alone.

The second common setting of metamorphism is far more extensive, and is caused by the deep burial of crust at continental collision zones. In this setting, huge volumes of rock experience increases in temperature and pressure, causing metamorphism on a regional scale; this is known as regional metamorphism.

Metamorphic Recrystallization

To consider metamorphic recrystallization at its simplest, let's begin by imagining a sedimentary rock composed entirely of quartz grains—a quartz sandstone. Sandstone is a sedimentary rock and so has a fragmental texture (see Fig. 6.7b). When it is subjected to high temperature and high pressure no new minerals can form because there are no other minerals present with which the quartz grains could react. All that can happen is that the quartz grains recrystallize and the rock known as quartzite is formed. The original fragmental texture is obliterated and replaced by a crystalline texture. Likewise, when a pure limestone, comprising calcite (i.e. $CaCO_3$) is metamorphosed, the calcite recrystallizes, and marble is produced. In these cases the rock adjusts to a high pressure or temperature by slowly recrystallizing in a denser, more compact, form.

In most cases, however, we start with a rock containing several different minerals, giving a richer chemical mix for metamorphic reactions to work with. At sufficiently high temperature or pressure, the original minerals react with each other, and new mineral crystals grow. In order to do this the constituent atoms must diffuse at different rates through the rock, but diffusion is extremely slow so atoms can move only very small distances in a given amount of time. The chemical rearrangement of the rock therefore entails the growth of small crystals unless the temperature is particularly high, in which case larger crystals can grow. Whenever metamorphism occurs in the compressional environment of continental collision zones (regional metamorphism), the rock is also subject to directed pressure and this also has an effect on the way the minerals crystallize. In the case of the metamorphism of mudrocks, mica crystals are formed during metamorphism. Mica crystals are characteristically platy in shape, reflecting the fact that the atoms in mica are arranged in layers or sheets. When platy minerals grow during metamorphism the energetically most favourable pattern of growth is one in which their flat surfaces lie more or less parallel, and at right angles to the main direction of imposed pressure (Fig. 6.14). Likewise, any elongate crystals grow aligned parallel with each other.

Fig. 6.14: A sketch of the interlocking crystalline texture in a metamorphic rock formed by regional metamorphism; mineral banding develops at right angles to the direction of compression.

In what way is the metamorphic texture shown in Figure 6.14 the same as the texture found in igneous rocks (Figure 6.3) and in what way does it differ?

Now read the answer

The alignment of platy minerals means that the rock has a series of closely spaced planes of weakness running through it, so it tends to split apart into fairly flat pieces. The rock is said to have a foliation, because its mineral grains are aligned like a stack of leaves (foliage) lying one on top of the other. A classic example is slate, a metamorphic rock with an extremely fine grain size that can be split into thin sheets used for roofing. Slate, and the coarser-grained metamorphic rock schist (pronounced 'shh-ist', to rhyme with 'mist') (Fig. 6.15) have marked foliation because they contain a lot of mica. But not all metamorphic rocks are as rich in mica. In such cases, metamorphism can still produce a banding effect, but here the minerals grow segregated into alternating bands a few millimetres to centimetres thick. Bands of light-coloured minerals alternate with bands of dark-coloured minerals, and the rock is known as gneiss (pronounced 'nice'). In general, the coarser the grain size of a metamorphic rock, the higher the temperature and/or pressure.

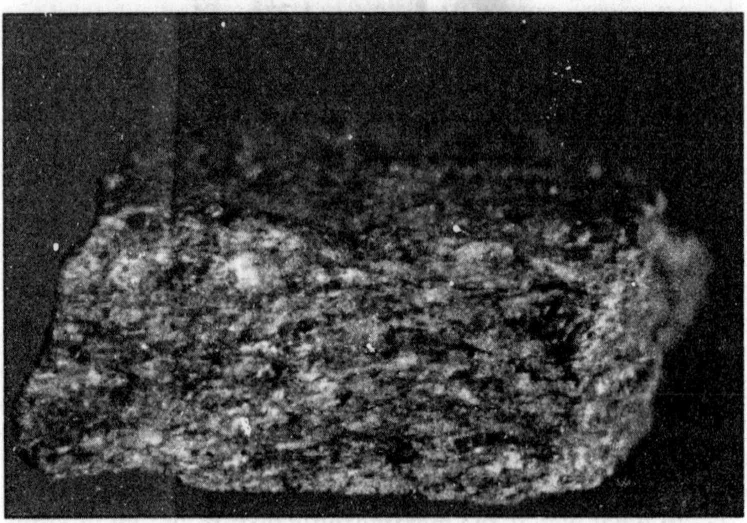

Fig. 6.15: The metamorphic rock schist. The platy mica crystals (dark) are arranged in undulating layers, defining a foliation. The sample is about 6 cm long.

Is the control on grain size in metamorphic rocks is same as in igneous rocks?

Now read the answer

The end product of metamorphism depends on two main variables—the chemical composition of the starting rock, and the pressure and temperature conditions under which metamorphism occurred. A useful analogy is with cooking—the product of baking in an oven depends on the ingredients that went in (chemical composition) and the temperature that the ingredients were subjected to in the oven. Just as the shape, colour and taste of food from the oven gives us clues about the ingredients and the baking conditions, so the texture and mineralogy of a metamorphic rock allow us to say something about the original rock type and the temperature and pressure conditions in the crust where metamorphism occurred.

Question 4

Compare and contrast contact metamorphism and regional metamorphism in terms of (a) the presence of foliation and (b) the distribution of the affected rocks.

Now read the answer

Interlude

Now that we have covered the features found in igneous, sedimentary and metamorphic rocks, and seen how these features can be explained by the processes that formed the rocks, here is a useful point at which to have a break before continuing with the next section. Before returning, you might like to see for yourself what types of rock you can find in your area. Can you identify their texture, or spot any fossils? Surfaces that haven't been obscured by grime or lichens are by far the best, as it is the attributes of the rock itself, not any later weathering processes that we are interested in. Polished work surfaces, ornamental stones, grave stones and shop fronts are interesting possibilities to start with. And avoid rocks whose grain size is so small that you can't recognise individual grains; a microscope is needed for studying those rocks!

The Rock Cycle

As you are reading this, rocks are being formed and destroyed on the Earth. Rocks are being heated and squeezed to form new metamorphic rocks; other rocks are melting to form magmas, which eventually cool and solidify as new igneous rocks; and the processes of weathering, erosion, transport and deposition are generating new sediments. The continuous action of rock-forming processes means that (given time) any rock in the Earth's crust will become transformed into new types of rock and that these too may in turn be transformed into yet other rocks. We call this recycling of rock materials the rock cycle.

Moving Around the Rock Cycle

One way of illustrating the possible ways of moving material around the rock cycle is to draw a diagram that places the processes into their geological contexts. Since the rock cycle involves processes occurring on the Earth's surface and also within its interior, we use a cross-section through the Earth's

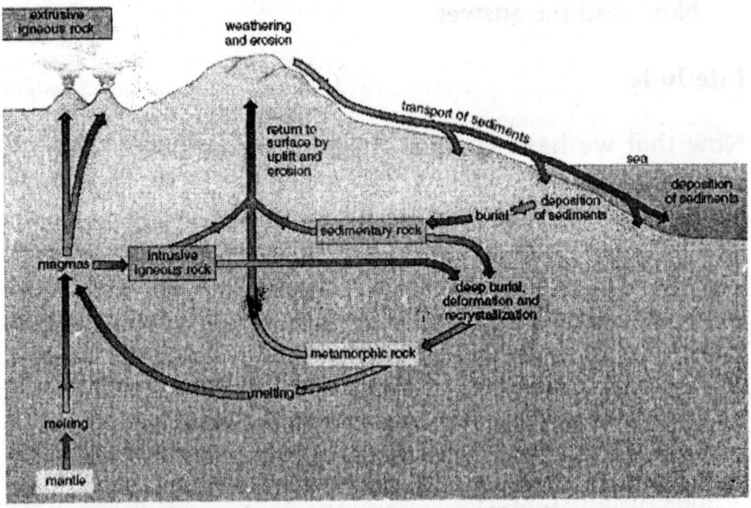

Fig. 6.16a: The rock cycle. The arrows show the paths and processes taken in transforming one type of rock (contained in a box) to another type.

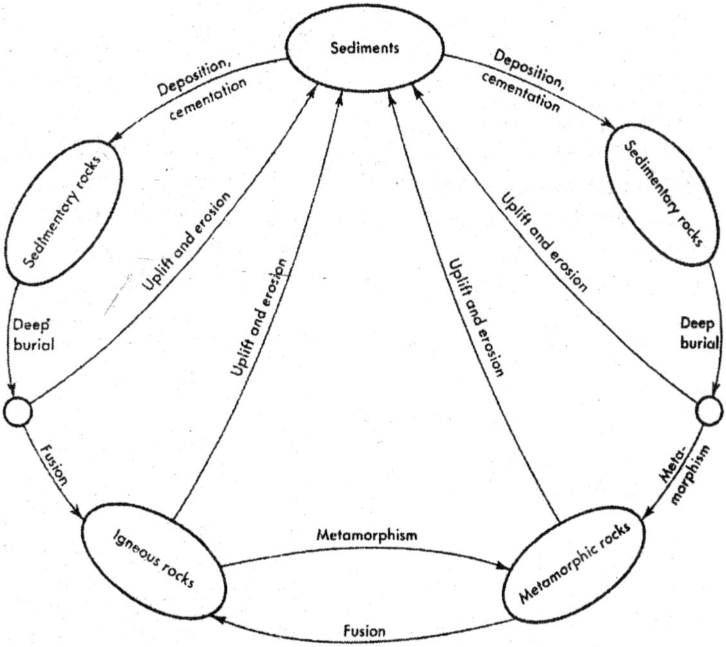

Fig. 6.16b: Diagrammatic representation of "rock cycles," indicating that materials in the Earth's crust may pass through the igneous, sedimentary, and metamorphic states repeatedly and in several possible sequences. The diagram might show additional variations; for example, magma might form either volcanic or intrusive igneous rocks, and rocks in either group might be metamorphosed, reduced to sediments, or fused again to start new generations of igneous bodies. Rocks in an active mountain zone are subject to most frequent and radical changes.

crust and uppermost mantle to do this, as shown in Fig. 6.16a. In this diagram we have concentrated on the most common processes within the rock cycle (Fig. 6.16b).

On the grand scale of the Earth, many rock cycle processes occur at plate boundaries, where plates are created or destroyed. For example, the increases in pressure and temperature needed to cause metamorphism in the crust occur in continental collision zones. Metamorphism also occurs in the crust at subduction zones, where rocks get dragged down to great depths and hence experience great pressures. Igneous processes take place wherever melting occurs in the mantle—

beneath divergent plate boundaries, above the subducting plate at convergent plate boundaries, and far from plate boundaries at hot spots.

The sedimentary processes of erosion and deposition that are important elements of the rock cycle occur in a great variety of surface environments, such as glacial environments, deserts, and continental shelves. Great rivers, such as the Nile and Mississippi, transport sedimentary material many thousands of kilometres across plate interiors, laying down fertile muddy sediments and modifying the coastline by forming deltas. Other great rivers, such as the Ganges, Indus and Brahmaputra, rise in the high mountain belts where plates are colliding. In these collision zones, plate convergence thrusts mountains into the sky, but the steep mountain slopes and high rainfall encourage weathering and erosion. Consequently, sites of mountain building generate prodigious amounts of sediment, which get washed away to be deposited elsewhere. For example, the Bay of Bengal contains several million cubic kilometres of sediment derived from the Himalayan mountains. The fate of sediments is to be eroded and redeposited, or to be turned into sedimentary rocks. Those rocks can then be caught up in subduction zones, either entering the mantle, being forced onto the edge of a continental plate, or being trapped and metamorphosed in a continental collision.

Although the rock cycle moulds the geology of the crust, it does not operate in isolation. The weathering of rock is influenced by climatic effects. The transport and deposition of sediment varies according to the strength of water currents, wind speed or glacial action. The rock cycle is also greatly influenced by organisms. An obvious example is the formation of limestones from corals and other organisms, which form calcium carbonate skeletons or protective structures. Another way in which organisms affect the rock cycle is through weathering. Rocks are partly broken down by the physical action of wind and frost, and by chemical decomposition by rainwater containing dissolved carbon dioxide, but this is greatly accelerated by plant roots, micro-organisms, and other living things in the soil.

READING FOR EARTH SCIENCES

How the Earth Works

Davidson, J.P., Reed, W.E. and Davis, P.M. *Exploring Earth*, Prentice Hall, New Jersey 0-13-018372-5 2001 second edition.

Brown, G.C. and Mussett, A.E. Chapman and Hall, *The Inaccessible Earth*, New Jersey 0-412-48160-X 1993 second edition.

Kearey, P. and Vine F.J. Blackwell, *Global Tectonics*, Oxford 0-86542-924-3 1996 second edition.

Decker, R. and Decker, B. Freeman, *Volcanoes*, New York 0-7167-3174-6 1998 third edition.

H. Sigurdsson (ed.) *Encyclopedia of Volcanoes*, Academic Press, San Diego 0-12-643140-X 2000.

Tatsumi, Y. and Eggins, S. Blackwell, *Subduction Zone Magmatism*, Oxford 0-86542-361-X 1995.

J.P. Davidson, *Continental and Island Arcs*, Academic Press, San Diego, 1991.

Davis, G.J. and Reynolds, S.J., *Structural Geology of Rocks and Regions*, John Wiley and Sons 0-471526215 1996 second Yes No.

Earth Materials

P.F. Kerr, *Optical Mineralogy*, McGraw Hill, Gribble and Hall, Chapman Hall.

Best, M.G., *Igneous and Metamorphic Petrology*, Blackwell Publishing, 2001.

MacKenzie, W.S. and Adams, A.E., *Rocks and Minerals in Thin Section*, Manson Publishing, London 1-874545-17-0, 1994.

Tucker, M., *Sedimentary Petrology*, Blackwell 0-632-05735-1, 2001 any edition.

Earth History and Life

Cowen, R., Blackwell, *History of Life*, 1-4051-1756-7, 2005, 4th edition.

Stanley, S., *Earth System History*, Freeman, 1996.

Benton and Harper, *Basic Palaeontology*, Addison Wesley Longmn, 0-582-22857-3, 1997.

7

DESCRIPTIVE CLASSIFICATION OF MASS WASTING

The essence of mass wasting is that it bridges the gap between weathering, which is defined as occurring in place and erosion which requires as one element of its definition transport by some agent or medium. Mass wasting combines elements of weathering and erosion, yet separates those two large categories of processes. In every sense of cause and effect, mass wasting, is a category of transitional phenomena. There is broad international agreement that the definitive criteria of mass wasting should include:

1. The type of material in motion, including its coherence and dimension.
2. The type and rate of movement, whether falling, toppling, sliding, spreading or flowing. An abbreviated classification of land slides (slope movements) are given in Table 7.1.

Table 7.1: Abbreviated classification of slope movements (landslides)

Types of Movement	*Bed Rock*	*Engineering Soil Predominantly Coarse*	*Predominantly Fine*
Fall	Rock fall	Debris fall	Earth fall
Topple	Rock topple	Debris topple	Earth topple
Slide	Rock slide	Debris slide	Earth slide
Rotational (slump)			
Translational			
Spread	Rock spread	Debris spread	Earth spread
Flow	Rock flow	Debris flow	Earth flow

Source: Cruden and Varnes 1996. (Fig. 7.4)

Creep

Creep is barely detectable and unaccelerating downslope movement. It is most common, in fact ever present in regolith on slopes, but the term is also applied to slow movement of otherwise unweathered joint blocks. The material moving down slopes is called *colluvium,* whether it be derived from *in situ* weathering or from transported sediment such as glacial drift.

Particle Creep: Individual surface pebbles or clods of soil are constantly shifting because of wetting and drying, heating and cooling and freezing and thawing. In frost climates, shale fragments and pebbles are lifted several centimetres above the surrounding soil surface by ice crystals. Volumetric expansion by any cause displaces towards the free face of the expanding mass, or perpendicular to the ground surface. On contraction, however, the particles are not pulled back into their former position but settle down slope with the gravitational component. Only rarely are the cohesive forces of soil and water strong enough to draw particles back into the ground during contraction with net down slope motion. The result is *particle creep.* Wind, raindrop impact and disturbance by plants and animals complicate any simple isolation of particle creep as a distinct process (Fig. 7.1).

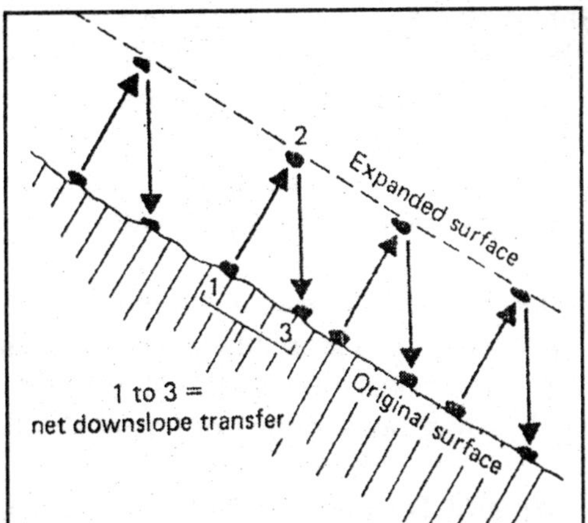

Fig. 7.1: Path of a surface particle during expansion and contraction of soil on a slope. See Figure 7.2 for more complex motions.

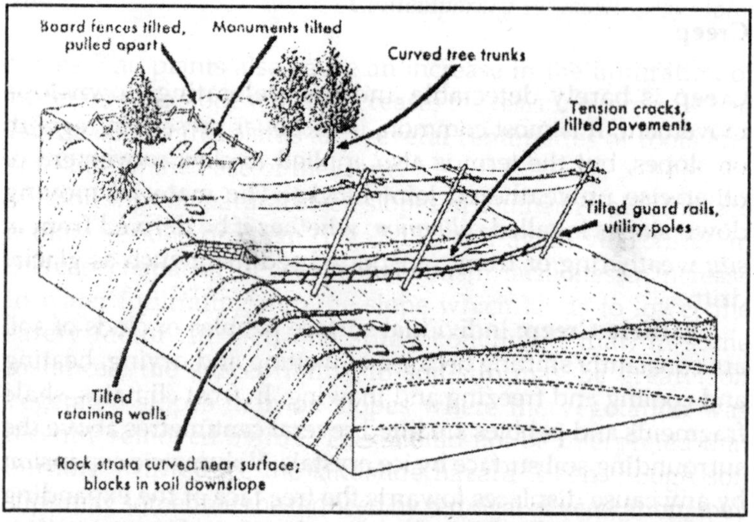

Fig. 7.2: Common effects of creep. Not all will be present in one place.

Soil Creep: Defined as imperceptible except over a period of a year, soil creep is more difficult to measure than particles creep. In theory soil creep should essentially be laminar with each layer of soil carried down hill by the motion of the layer beneath it. The effect should be cumulative decreasing to zero with depth. As a result soil creep should not shear across an immobile substrata at depth and should not abrade a buried surface. Measured profile of creeping soil show many variations, however *Depth Creep* is plug like with the entire surface layer, perhaps bound by plant roots, moving over a narrow shear zone at depth.

A variety of surface phenomena are usually attributed to creep, such as downslope titled trees, walls and utility poles (Fig. 7.2).

Tree trunks on steep slopes are commonly concave uphill although their curvature is more likely because of initial light-seeking growth perpendicular to the slope and later more nearly vertical growth as saplings outgrow shading ground cover.

Rock Creep and Topple: According to Geophysicists and Engineers rock creep means slow, permanent internal rock deformation (strain) under low stress. As a process of mass wasting, rock creep is less precisely used to describe slow movement of large coherent rock masses over a discrete basal surface. Most often, rock creep is observed on cliff face, whereas massive rock such as sandstone overlies shales. Large joint bounded sheets or blocks, slowly creep or lean outward until they become stable and topple over or slide downhill.

Flow: Incoherent rock debris may be mobilized sufficiently so that it flows like a viscous fluid. The criteria for defining of flow are evidence of internal turbulence and either discrete boundaries or narrow marginal zones of shear. In both criteria, the flow is distinct from creep. Sub-division of flow can be based on the kind of material, the degree of saturation, or the speed of advance. The factors of sediment concentration in water and the flow rate can be combined in a rheologic classification, ranging from fluid flow with low sediment concentration, through plastic flow with a finite yield stress but essentially fluid behaviour above that stress level, to granular flows that are dry or nearly dry. Flow rates for all rheologic behaviour can range through many orders of magnitudes.

Solifluction: If soil is saturated with water, the soggy mass may flow downhill a few mm or a centimetres per day or per year. This type of movement is called solifluction. It is a form of mass wasting common whenever water cannot escape from a saturated surface layer of soil by percolation into deeper levels. A clay hardpan in a soil or impermeable bedrock layer can promote solifluction as effectively as a frozen substratum.

Gelifluction and Frost Creep: According to Washburn, solifluction associated with frozen ground is called gelifluction. He was able to separate the components of gelifluction and frost creep (soil creep due to the ratchetlike motion of particles during alternating freeze-thaw cycles (Fig. 7.3). His periodic resurveys proved that frost creep was quantitatively slightly more important on most slopes than gelifluction. An unexpectedly large component of retrograde motion due to cohesion caused its targets to move uphill during the summer thaw season, with respect to the vertical plane through the

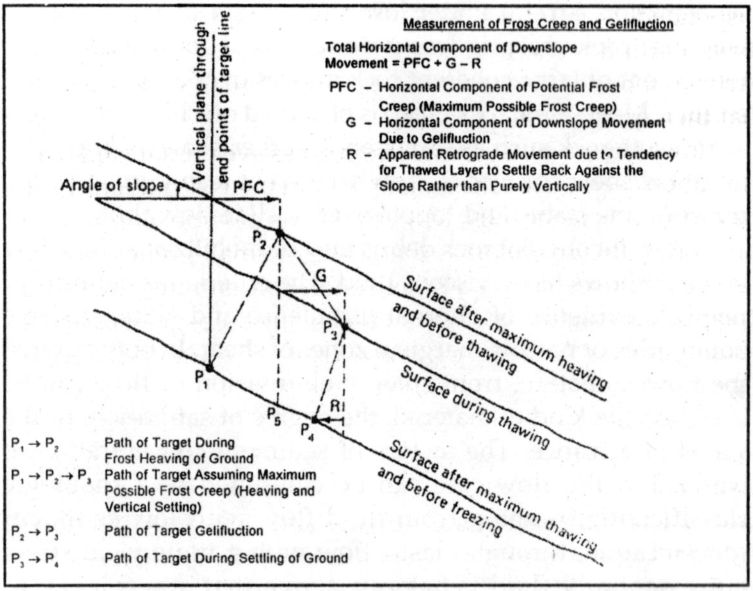

Fig. 7.3: Diagram of frost creep, gelifluction, and retrograde components of movement by a survey target on a perennially frozen hillside in northeast Greenland (Washburn, 1967, Fig. 5).

targets at the time of maximum frost heave. The retrograde motion partly or entirely cancels down slope movement by gelifluction.

Debris Flow, Earth Flow, Mud Flow: These three terms are applied to type of mass wasting very similar to solifluction. They are somewhat more rapid and they commonly flow along valleys, where solifluction sheets or lobes cover an entire hillside with moving debris. Some geomorphologists lump all three terms under the general debris flow to stress the similarity of their mechanisms despite differences in their flow rate or percentage of fine particles. The transition between a debris flow and a muddy river is illustrated by measurement of densities of 2.0 to 2.4 g/cm^3 in debris flow (dense enough to buoyantly support large blocks), densities of 1.3 to 1.8 g/cm^3 in more fluid, hyper concentrated flows and densities of only slightly greater than 1.0 g/cm^3 in a stream with a heavy suspended load (Hooke, 1987). The distinction is that in debris flows, earth flows, and mud flows, gravity acts directly on the particles that drag along interstitial fluid, rather than on a fluid that then entrains the particles.

Flows almost always result from excessive rainfall and over saturation seems to be the major factor in their formation. They may form on planer hillsides or at the toe of large slumps, but most often they develop at the heads of small gullies, where groundwater seepage or surface runoff concentrates to initiate streams. Unusually deep weathering along seepage lines at the head of incipient drainage networks may be the reason for the high frequency of flow initiation there.

Case history of flows from around the world emphasize the combination of steep slopes, weak or easily weathered rock or surfical cover and abnormal precipitation in triggering them. Artificial alteration, especially excavation at the foot of a slope and vegetation clearing are very often additional factors.

Spread: A spread is defined as a lateral extension of a cohesive rock or soil mass combined with general subsidence of the fractured cohesive material into a softer substrate. The cause is usually liquefaction of the substrate while the surface retain cohesive strength. Rock spreads are typically extremely slow, as when brittle layers are rafted along by slow deformation of underlying ductile material such as shale or slat. Spread are common form of landslide on low gradient terrains during earthquakes, when rapid shaking liquefiers sediments below the water table and displace overlying blocks of drier material.

Slide: Mass wasting wherein a mass of rock or weathered debris moves downhill along discrete shear surfaces is defined as a slide. Slides are subdivided into **rotational** and **translational** categories based on shape of the shear surface. The distinction is important for engineering analytical and control procedures. Subcategories of slides in various classifications include *slump, rock slide, block glide, mud slide, debris and avalanche* (Fig. 7.4).

Slump: It is the form of slide most common in thick, homogenous, cohesive sediments. The surface of failure

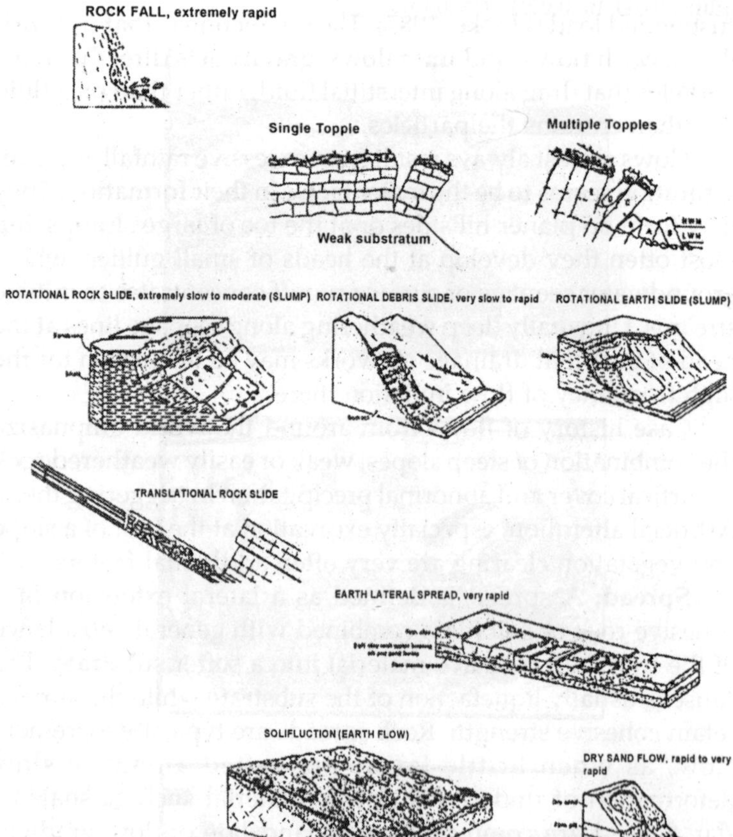

Fig. 7.4: Main types of landslides, illustrated.
(Compare Table 7.1 based on Cruden and Varnes, 1996)

beneath a slump block is spoon shaped concave upward or outward. The upper surface of a slump block commonly is tilted backward because the entire mass rotates as the lower part moves outward and downhill. Vegetation or even houses may be carried intact on the surface of the large slump block. Slumps may be caused by water erosion undercutting the foot of the slope. They are also a common result of faulty engineering design of cut embankments. The most diagnostic feature of an

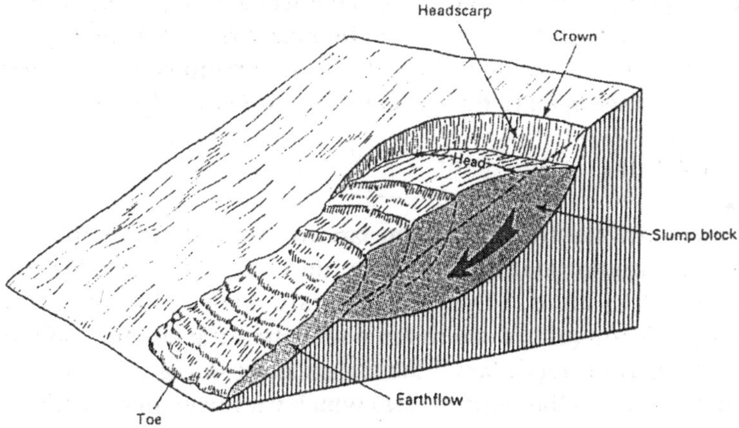

Fig. 7.5: A slump-earth-flow with principal parts labelled.

ancient slump area is the hummocky or chaotic landforms on it (Fig. 7.5).

Mud Slide: Depending on the slope gradient, degree of saturation, grain size of sediment emerging from the toe of a slump, it can be called a mud flow or mud slide. A mud slide moves over or between discrete shear surface in a lobate shape, somewhat slower than a more turbulent mudflow. At its terminus it may form a bulbous lobe.

Rock Slide, Block Glide: The simplest form of translational slide is a rock slide or block glide. The movement is relatively rapid and most commonly occurs where steeply dipping strata or sheeting nearly parallels the surface slope. Rock slides are generally shallow. A heavy rain or freezing and thawing provides fluid pressure or vibration breaks off obstruction and reduces the coefficient on the glide plane and a detached slab or block slides down. The dip strata or sheeting is an obvious factor in rock slides. Other inherent strength of massive rocks, the controlling factor in such rock slides is the presence, spacing and orientation of joints and other discontinuities in the rock mass.

Avalanches and Debris Slides: Any large catastrophic landslide may be called an avalanche. They can include broken rock, ice, and snow usually so mixed that the material term debris is most appropriate. The general cause for humid-climate debris avalanches is saturation by heavy rain, but the precise trigger may never be known. In diverse regions of the world various factors play their role for triggering avalanches.

Landslides

Slopes are the most common landforms, and although most slopes appear stable and static, they are actually dynamic, evolving systems. Slope processes are one significant reason why stream valleys are much wider than the stream channel and adjacent floodplain. Material on most slopes is constantly moving down the slope at rates which vary from imperceptible creep of soil and rock to thundering avalanches and rockfalls which move at tremendous speeds. As with floods, it may not be the largest (rare and infrequent) nor the smallest (most common) event that moves the greatest amount of material down slopes as valleys develop. Rather, events of moderate magnitude and frequency may play the most important role.

Landslides and related phenomena such as mudflows, earthflows, rockfalls, snow or debris avalanches, and subsidence are natural events that would occur with or without human activity. However, human use and interest has led to both an *increase* in some of these events such as landslides on hillside developments due to oversteeping of slopes and a *reduction* of landslides by stabilizing structures or techniques on naturally sensitive slopes likely to experience landslides.

Earth materials on slopes may fall and move or deform in several ways including flowage, sliding, falling, and subsidence (Fig. 7.6); and typical downslope movement event (landslide) will often involve more than one type of failure (Fig. 7.7). A brief classification of the common types of downslope movements is given in Table 7.2, and reflects the terminology used in this discussion and in other chapters. The important variables in classifying downslope movements are type of movement, slope material type, amount of water present, and rate of movement. In general, a landslide is considered to move rapidly if the motion can be discerned with the naked eye,

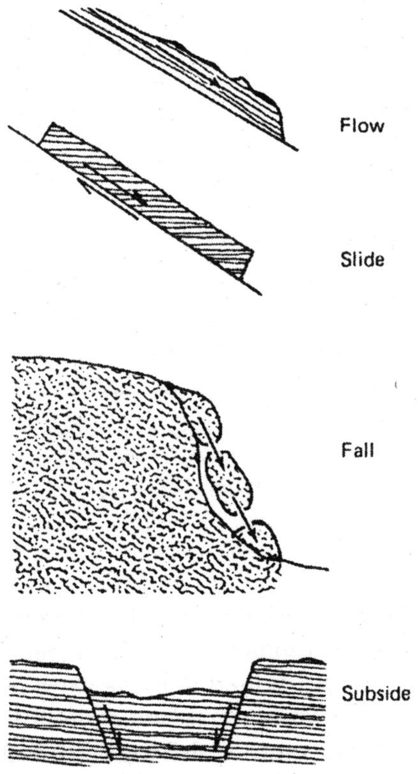

Fig. 7.6: Common ways that earth materials fall and move in landslides and related phenomena.

otherwise it is classified as slow. Actual rates vary from a slow creep at rates of a few millimetres or centimetres per year, to very rapid at 1.5 metres per day, to extremely rapid at several tens of metres per second.

Slope Stability

The nature and extent of slope stability for the various types of downslope movements can generally be explained by examining forces on slopes in terms of the role of these significant, and somewhat interrelated, variables: type of earth materials, slope (topography), climate, vegetation, water and time.

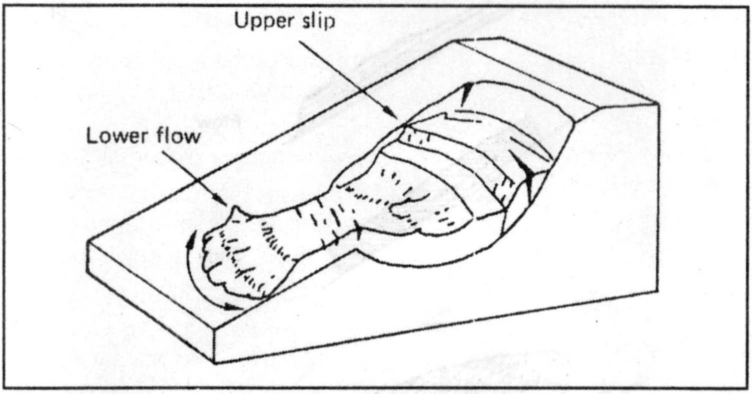

Fig. 7.7: Block diagram of a common type of landslide consisting of an upper slip motion and a lower flows. (After R. Pestrong, *Slope Stability*, American Gelogical Institute, 1974).

Forces on Slopes

The causes of many landslides and related types of downslope movement can be examined by studying relations between *driving forces*, those which tend to move earth materials down a slope, and *resisting forces*, those which tend to oppose such movement. The most common driving force is the downslope component of the weight of the slope material, including

Table 7.2: Classification of Landslides and Other Downslope Movements

Type of Movement		*Materials*	
		Rock	*Soil*
Slides (variable water content and rate of movement)		Slump blocks Translation slide	Slump blocks Rotational slide Soil slip (planer)
Falls		Rock fall	Soil
		Rock creep	Soil Creep
Flows	Slow	Unconsolidated materials (saturated)	
		Earthflow Mudflow	
	Rapid	Debris avalanche	
Complex		Combinations of slides and flows	

anything superimposed on the slope, such as vegetation, fill material, or buildings. The most common resisting force is the shear strength of the slope material acting along potential slip planes.

Slope stability is evaluated by computing a safety factor (SF) defined as the ratio of the resisting forces to the driving forces. If the safety factor is greater than one, the resisting forces must exceed the driving forces and the slope is considered stable. If on the other hand, the safety factor is less than one, then the driving forces exceed the resisting forces and a slope failure can be expected. Of course, anything that reduces the resisting forces or increases the driving forces will lower the safety factor and thus increase the chances of a landslide or other type of downslope movement. Driving and resisting forces are not static, rather they tend to change with time. Thus, depending on changes in local conditions, the safety factor may increase or decrease. For example, consider construction of a road cut in the toe of a slope with a potential slip plane (Fig. 7.8). The effect of the road cut in the critical toe of the slope will

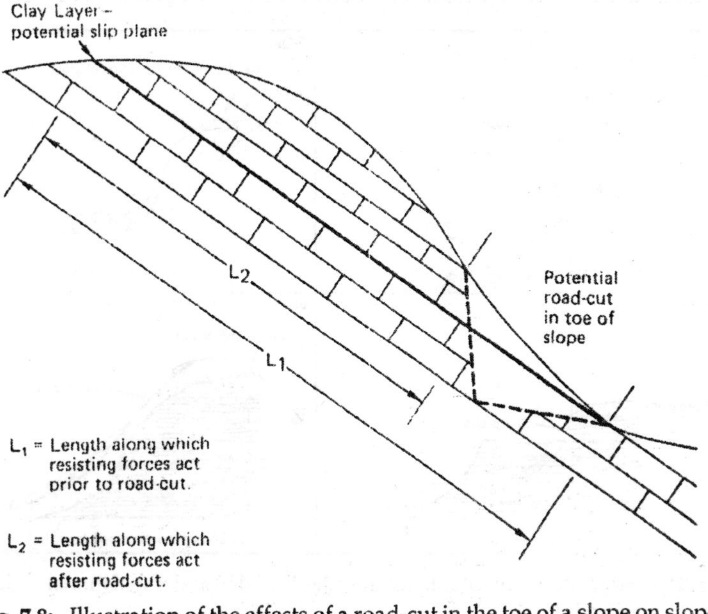

Fig. 7.8: Illustration of the effects of a road-cut in the toe of a slope on slope stability (See text for explanation.)

(1) reduce the driving forces because some of the slope material is removed, and (2) reduce the resisting forces because the length of the slip plane is reduced, and it is this plane along which the resisting force (shear strength) acts. The overall effect of the road cut will be to lower the safety factor. This results because the reduction of the driving force is small compared to the reduction of the resisting force. That is, only a small portion of the potential slide mass is removed, but a relatively large portion of the total length along which resisting forces act is removed. Thus, the safety factor must be reduced.

Role of Earth Material Type

The type of downslope movement is partly a function of slope material type. For landslides, two basic patterns of movement are recognized, *rotational* and *translational* (Fig. 7.9). Rotational slides are most common for soil slopes, but are also associated with slumping in rock slopes, especially where a permeable rock such as sandstone overlies a weak rock such as shale. However, if a very resistant rock overlies a rock of very low

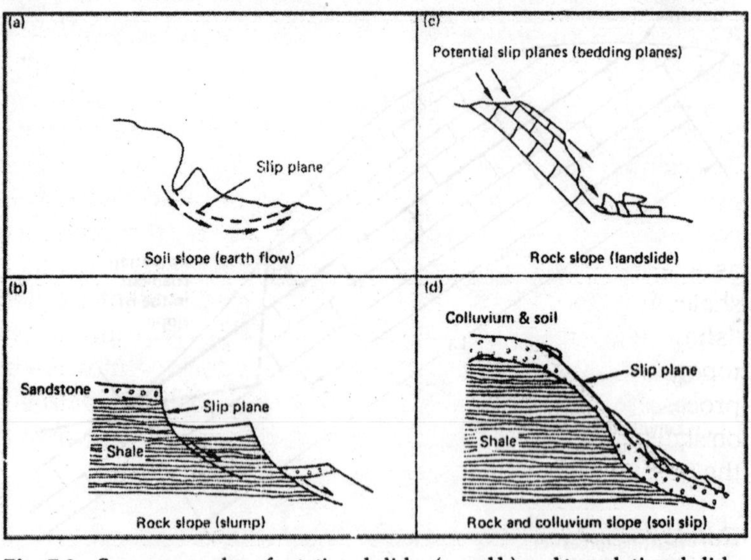

Fig. 7.9: Some examples of rotational slides (a and b) and translational slides (c and d).

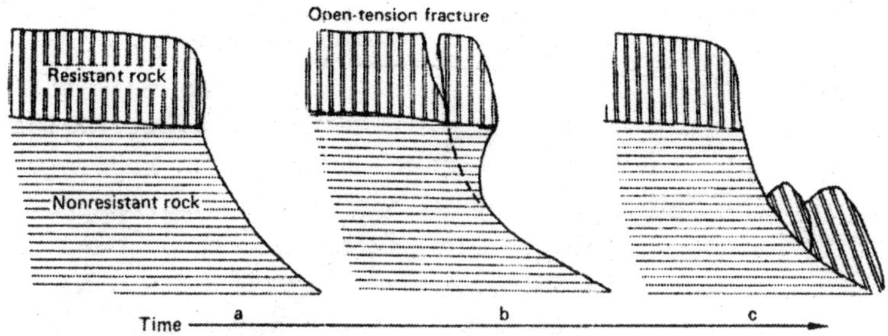

Fig. 7.10: Development of a slab slide. Undercutting of resistant "cap" rock at time b causes the development of tension fractures and eventual failure by rockfall at time c.

resistance, then rapid undercutting of the resistant rock may cause a slab failure (rock fall) (Fig. 7.10).

Translation slides occur along geologic planes of weakness within rock slopes. Common translation slip planes include fractures, bedding planes, clay partings, and foliation planes. However, very shallow slides parallel to the slope, known as *soil slips,* also occur in some areas. For soil slips, the slip plane is usually above the bedrock but below the soil within slope material known as *colluvium,* which is a mixture of weathered rock and other material.

The strength of the slope materials may greatly influence the magnitude and frequency of landslides and related events. For example, creep, earthflows, slumps, and soil slips are much more common on shale slopes or slopes on weak pyroclastic (volcanic) materials than on slopes on more resistant rock such as well cemented sandstone, limestone, or granite. In fact, shales are so notorious for landslide activity that what is called "shale terrain" is characterized by irregular, hummocky topography produced by a variety of downslope movement processes. For all types of land use, from agricultural to urban, on shale or other weak rock slopes, one must carefully consider the possible landslide hazard prior to development.

Role of Slope and Topography

Angle greatly affects the driving forces on slopes. Fig. 7.11 illustrates that, as the angle of a potential slip plane increases,

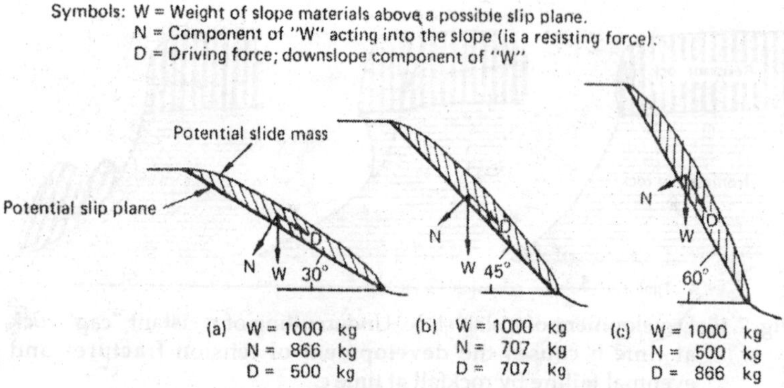

Fig. 7.11: Effect of slope on the driving force. Notice that, for hypothetical slide mass weighing 1000 kilograms (force), the driving force increases as the slope increases. (100 kg = 220.5 lb).

the driving force also increases. Thus, everything else being equal, landslides should be more frequent on steep slopes. In a general way, this is true. For example, a study of landslides that occurred during two rainy seasons in California's San Francisco Bay Area established that 75 to 85 per cent of landslide activity is closely associated with urban areas on slopes greater than 15 per cent, or 8.5 degrees. (Fig. 7.12)

To some extent, the type of landslide activity is also a function of slope and topography. For example, rockfalls and debris avalanches are associated with very steep slopes, and shallow soil slips in southern California are common on steep saturated slopes. These soil slips often are transformed downslope into debris flows that are extremely hazardous (Fig. 7.13). At the other extreme, earthflows may occur on moderate slopes, and the effects of creep may be observed on slopes of only a few degrees.

Role of Climate and Vegetation

Climate and vegetation can influence the type of landslide or other downslope movement that occurs on a particular slope. Climate is significant in influencing downslope movement because it controls the nature and extent of the precipitation and thus the moisture content of slope materials. For example,

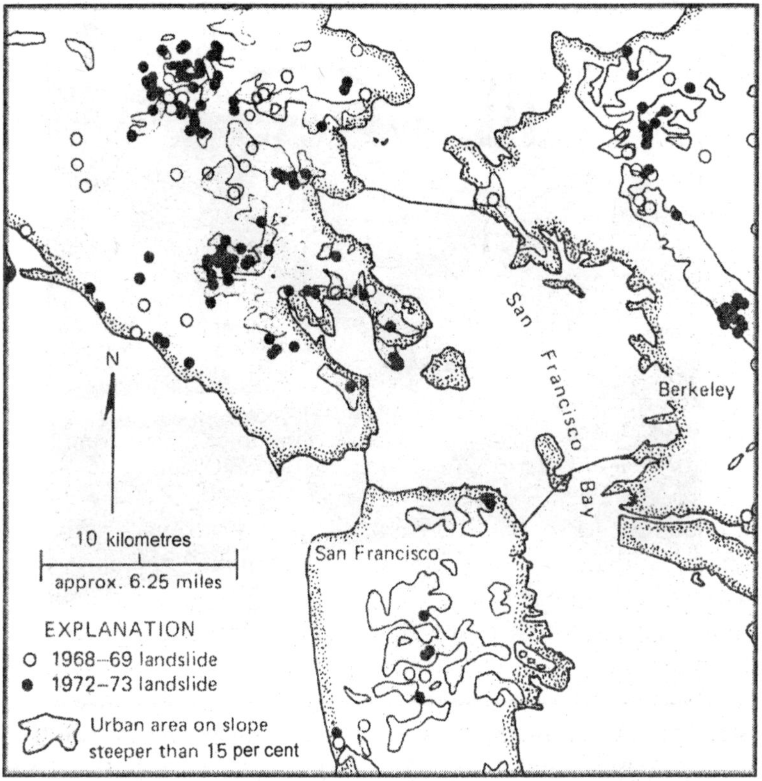

Fig. 7.12: Relationship between landslide frequency and urban areas on slopes exceeding 15 per cent. (After T.H. Nilsen, F.A. Tayler, and R.M. Dean, U.S. Geological Survey Bulletin 1424, 1976).

both earthflows and mudflows involve downslope movement of water-saturated earth materials. However, earthflows are common and mudflows relatively rare in humid regions. This results because a good deal of water infiltrates into slopes in the humid areas, facilitating earth-flows. Furthermore, humid regions have many perennial streams which continuously transport the materials delivered from slopes out of the drainage basin. Thus, there is little built-up of debris in the basin and, therefore, little material available to be mobilized in a mudflow.

The role of vegetation in landslides and related phenomena is complex. This results because the vegetation in an area is a

Fig. 7.13: Home in southern California destroyed by debris flow that originated as a soil slip. This 1969 event claimed two lives. (Photo courtesy of Los Angeles Department of Building and Safety).

function of several factors including climate, soil type, topography, and fire history, each of which also influences what happens on slopes. Vegetation is a significant factor in slope stability for three reasons. First, vegetation provides a cover that cushions the impact of rain falling on slopes, thus facilitating infiltration of water into the slope while retarding grain-by-grain erosion on the surface. Second, vegetation has root systems that tend to provide an apparent cohesion to the slope materials. And third, vegetation adds weight to the slope.

Most problems concerning slope stability and vegetation result from disturbance or removal of vegetation from slopes. However, in some cases vegetation can increase the probability of a landslide. This is especially true for specific types of shallow

Fig. 7.14: Shallow soil slips on ice plant near Santa Barbara, California.

soil slips on steep slopes. For example, one type of soil slip in southern California coastal areas occurred on steep-cut slopes covered with ice plant (Fig. 7.14). During especially wet winter months, the shallow-rooted ice plants take up water, adding considerable weight to steep slopes and increasing the driving forces. The plants also cause an increase in the infiltration of water into the slope which decreases the resisting forces. When failure occurs the plants and several centimetres of roots and soil slide to the base of the slope.

Soil slips on natural steep slopes in southern California are a more serious problem (Fig. 7.13), and again vegetation, in this case chaparral (dense shrubs or bush), facilitates an increase in water infiltrating into the slope which tends to lower the safety factor. In fact, one study concluded that, in some instances, the susceptibility to soil-slip may be greater on vegetated slopes than on slopes where the vegetation was recently removed by fire. This should not be interpreted that burning will reduce the landslide hazard. Even though soil slips may sometimes be reduced by removal of vegetation, they are not eliminated and, in addition, the grain-by-grain erosion caused by rain splash and sheet wash on the surface is greatly increased. The eroded sediment tends to fill up the ravines (steep stream valleys) with a metre or more of debris which may be mobilized into mud-flows during wet winters.

The type of vegetation also affects the frequency and morphology of shallow soil slips in southern California. For example, in comparing grass-covered with chaparral-covered slopes, soil slips on grassy slopes are 3 to 5 times more frequent, have a lower minimum angle at which failure occurs, and tend to be shorter and wider than those on chaparral-covered slopes.

Fire may facilitate the development of hydrophobic soil horizons that retard the infiltration of water and the establishment of vegetation after burning. The "nonwettability" or hydrophobic properties are produced when waxy organic coatings on soil particles are mobilized by the fire and accumulate in a soil horizon. Hydrophobic soils have been recognized in areas with chaparral vegetation as well as forest cover, and the problem is more serious than generally realized because the behaviour of water in the near surface environment is critical to the establishment of vegetation and retardation of soil erosion. Unfortunately, our (often necessary) practice of not allowing periodic natural burns has increased the chances of hydrophobic soils forming because the less frequent fires have more fuel and burn very hot, increasing the likelihood of the hydrophobic properties developing in the soils.

The disturbance or removal of vegetation by logging has also been associated with an increase in landslides. Clearcutting, or the removal of all trees, has caused several problems. First, the rate of transpiration (loss of soil water through the trees) is greatly reduced; thus soil moisture increases, tending to reduce the stability of the slopes. Second, in specific instances infiltration of water into a slope may be increased. This is especially likely if a permeable soil on relatively low-gradient slopes is covered in winter with a thick snow pack that slowly melts in the spring; as the slope becomes saturated, resisting forces are lowered. Third, with the exception of redwood trees which regenerate after logging, the roots of cut trees decay with time, reducing their strength and thus the apparent cohesion of the soil. This tends to reduce the resisting forces within the slope, helping to explain the increased frequency in landslides several years following timber harvesting.

Role of Water

Water is almost always directly or indirectly involved with landslides, and thus its role is of particular importance. Water is the universal solvent, and much weathering of rocks, which slowly reduces their shear strength, takes place near the surface of the earth. The weathering contributes to the instability of rocks and is especially significant in areas with limestone, which is very susceptible to chemical weathering. It has been argued that water has a lubricating effect on individual soil grains and potential slide planes. This notion is probably incorrect. Water may actually act as an antilubricant, and very little water is needed to produce nearly saturated conditions, implying that all slopes in a humid climate are fully lubricated. The amount of water in a soil is very significant. For example, dry sand will form a cone when dumped, whereas moist sand will stand nearly vertical because surface tension in the water

Fig. 7.15: Idealized diagram showing the development of a perched water table in colluvial material during a heavy rainfall. (After R.H. Campbell, U.S. Geological Survey Professional Paper 851, 1975).

holds the grains together, and saturated sand with all of its pore spaces filled will flow as mud.

The effects of water on slopes and landslides are quite variable. First, saturation of earth materials causes a rise in pore water pressure. In general, as the water pressure in slopes increases, the shear strength (resisting force) of the slope decreases and the weight (driving force) increases; thus, the net effect is to lower the safety factor. This is thought to be a significant factor in the development of soil slips and debris avalanches, as well as other types of landslides. Soil slips generally occur during heavy rainfall when near-surface temporary or *perched* water table conditions may be present. During the rainstorm the rate of surface infiltration in the unsaturated zone of the soil or colluvium exceeds the rate of deep percolation in the rock below the colluvium, and even though some water moves as seepage parallel to the slope, a temporary (perched) water table develops (Fig. 7.15). Failure of the slope occurs when the resisting forces are reduced sufficiently, that is, when the safety factor becomes less than one. The safety factor will be a minimum when the perched water table rises to the surface, indicating that the potential slide mass is entirely saturated.

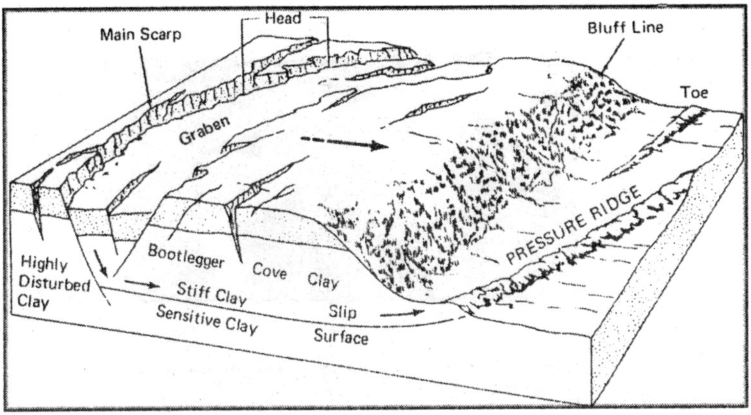

Fig. 7.16: Block diagram showing how landslides developed in response to the 1964 Alaskan earthquake in Anchorage. The graben is a down-drop block of earth material. (From W.R. Hansen, U.S. Geological Survey Professional Paper 542A, 1965).

A rise in water pressure is present prior to many landslides, and it is safe to say that most landslides are due to an abnormal increase in the water pressure in the slope-forming materials.

Rapid draw-down, when the water level in a reservoir or river is lowered quickly at a rate of at least a metre per day, is a second way that water may affect the stability of slopes. When the water is at a relatively high level, a large amount may enter the banks, producing a bank storage phenomenon. Then, when the water level suddenly drops, the water stored in the banks is left unsupported. This produces an abnormal pore water pressure which reduces the resisting forces; simultaneously, the wieght of the stored water increases the driving forces. This explains the tendency for bank failures (slumps) to occur along streams after the flood water has receded.

A third way that water affects landslides is by contributing to spontaneous liquefaction of clay-rich sediment or *quick clay*. When disturbed, some clays lose their shear strength, behave as a liquid, and flow. The shaking of clay below Anchorage, Alaska, during the 1964 earthquake produced this effect and was very destructive (Figs. 7.16 and 7.17). Other examples of slides associated with sensitive clays are found in Quebec,

Fig. 7.17: Landslides in Anchorage, Alaska, caused by the 1964 earthquake. The slides are similar to those diagrammed in Figure 7.16: Government Hill School slide (a), aerial view of slide near the hospital on Fourth Avenue (b) and slides that caused the collapse of Fourth Avenue near C Street (c). [Photos (a) and (b) by W.R. Hansen, courtesy of U.S. Geological Survey. Photo (c) courtesy of U.S. Army].

Canada, where several large earth movements have destroyed numerous homes and killed about seventy people in recent years. The slides occur on river valley slopes in initially solid material that is converted into a liquid mud as the movement begins. These slides are often initiated by river erosion at the toe of the slope and, although they start in a small area, may develop into large events. Since they often involve the reactivation of an older slide, future problems may be avoided by restricting development. Fortunately, the older slides, even though masked from ground view by vegetation, are often visible on aerial photographs. The Quebec slides/flows are especially interesting because the liquefaction of clays occurs without earthquake shaking, unlike those in Anchorage.

Seepage of water from artificial sources, such as reservoirs and unlined canals, into adjacent slopes may also affect slope stability. This can lower the cohesion and thus the shear strength by removing the cementing materials, as happened in the St. Francis Dam failure. Seepage may also cause an increase of water pressure in adjacent slopes, causing a reduction in the resisting force.

The ability of water to erode also affects the stability of slopes. Stream or wave erosion may remove and steepen the toe area of a slope, thus reducing the safety factor. This problem is particularly critical if the toe of the slope is an old inactive landslide that is likely to move again if the stability is reduced. Therefore, it is important to recognize old landslides along potential road cuts and other excavations prior to construction so that potential problems can be isolated and corrected.

Role of Time

The forces on slopes often change with time. For example, both driving and resisting forces may change seasonally as the moisture content or water table position alters. Furthermore, these changes are larger in especially wet years. This is reflected by the increased frequency of landslides in wet years. In other slopes there may be a continuous reduction in the resisting forces with time. This might be due to weathering, which

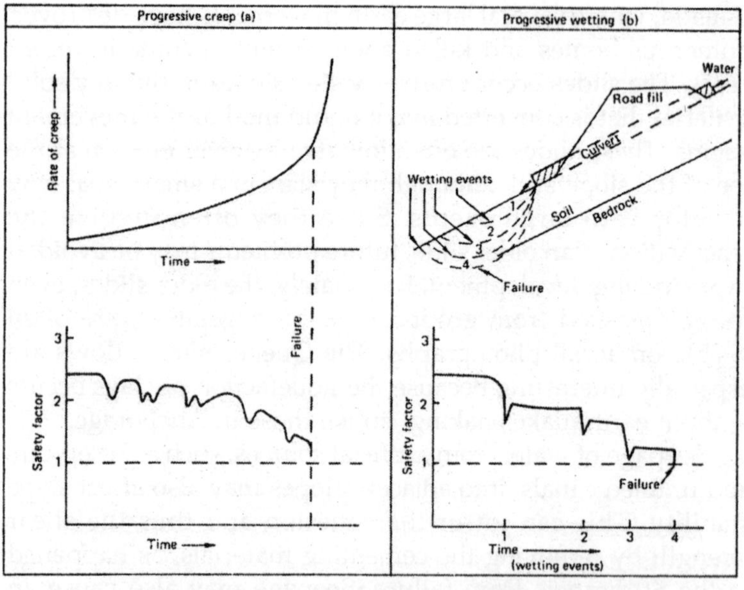

Fig. 7.18: Idealized diagrams showing the influence of time on the development of a landslide: progressive creep (a) and progressive wetting (b). Progressive creep is symptomatic of a decreasing safety factor, whereas progressive wetting may cause a reduction in the resisting forces and thus produce a lower safety factor. [Diagram (b) after C.S. Yee and D.R. Harr, Environmental Geology, Vol. 1, p. 374, 1977].

reduces the cohesion in slope materials, or to a regular increase in water pressure due to natural or artificial conditions. A slope that is becoming less stable with time may have an increasing rate of creep until failure occurs (Fig. 7.18a). The safety factor of a slope might also be decreased with time due to progressive wetting which causes a disarrangement of the soil particles in the slope, lowering the internal friction and thus the strength of the slope materials. Fig. 7.18b illustrates this phenomenon as a result of road building and culvert installation that periodically dumps water downslope. This emphasizes an important point: we may design slopes that are stable when constructed, but stability has a way of changing with time. Therefore, slope design should include provisions to minimize processes that might progressively weaken slope materials.

Causes of Landslides

In many cases the real causes of landslides, increase in driving force or decrease in resisting force, are marked by immediate causes such as earthquake shocks, vibrations, or a sudden increase in the amount of moisture in slope material. The distinction between real and immediate causes is very important in court when a landslide case is heard and a definitive statement by an earth scientist concerning the cause of a landslide is expected. For example, a translation slide may have as an *immediate* cause heavy rains which saturated the earth material; but the *real* cause is the potential to slide upon long, weak, clay layers. A similar hypothetical example might be given for an artificial slope in a housing development in which the immediate cause is an earthquake shock, but the real cause is a poorly designed slope.

Causes of landslides may also be grouped according to whether they are external or internal. External causes produce an increase in the shear stress (driving force per unit area) at relatively constant shear strength (resisting force per unit area). Examples of external cause include loading of a slope, steepening of a slope by erosion or excavation, and earthquake shocks. Internal causes produce landslides without any recognized external changes, and include processes that reduce the shear strength. This would include changes such as increase in water pressure or decrease in cohesion of the slope materials. In addition, some causes of landslides are intermediate, having some attributes of both external and internal causes. For example, rapid draw-down involves an increase in the shear stress accompanied by decrease in shear strength. Other intermediate causes include spontaneous liquefaction, and subsurface weathering and erosion.

Human Use and Landslides

Effects of human use and interest on the magnitude and frequency of landslides vary from insignificant or nearly so to very significant. In cases where our activities have little to do with the magnitude and frequency of landslides, we need to learn all we can about where, when, and why they occur to

Fig. 7.19: Floresta Creek Valley mudflow at the base of the Sierra Das Araras Escarpment. Before the mudflow, the valley bottom contained a village and a highway construction camp. Several hundred people lost their lives when the mudflow, which was about 4 metres deep, moved through the area. (Photo by F.O. Jones, courtesy of U.S. Geological Survey).

avoid hazardous areas and minimize damage. In cases where human use has increased the number and severity of landslides, we need to learn how to recognize, control, and minimize their occurrence wherever possible.

Mixtures of adverse geologic conditions such as weak soil or rock and potential slip plains on steep slopes with torrential rains, heavy snowfall, or seasonably frozen ground (permafrost) will continue to produce landslides, mudflows, and avalanches regardless of human activities. These are natural processes reacting to natural conditions. For example, the night of January 22, 1967, produced a landslide disaster of tremendous magnitude in Brazil. Following an electrical storm and cloudburst lasting 3 and ½ hours, an area of about 194 square kilometres was devastated by landslides and erosion (floods) which claimed the lives of abut 1700 people. Damage to property and effects on industry were inestimable as tens of thousands of landslides turned the green hills into wastelands and valleys into mud baths (Fig. 7.19). Virtually, all the slides

Fig. 7.20: Debris avalanche in thin, residual soils. (Photo by F.O. Jones, courtesy of U.S. Geological Survey)

were debris avalanches and mudflows. The slopes in disaster areas characteristically have only a thin veneer of residual soil over hard rock, resulting in a contact of relatively easy parting (Fig. 7.20). It is interesting to note that more avalanches cut areas of high vegetation than grass-covered slopes, and many destroyed forests left in their natural state for many years. The combination of the high-frequency vibrations of the electrical storm with the pressure of the falling water is considered a primary cause of the avalanches. However, rising water pressure, groundwater movement, vegetation, and erosion undoubtedly play a part also. This magnitude of landslide activity, while unique in recorded history, has probably been relatively frequent along the mid-southern coast of Brazil during the geologic evolution of the slopes.

The above example is similar to a widespread episode of debris avalanches that occurred in August of 1969. The remnants of hurricane Camille moving eastward from Kentucky and the Appalachian Mountain ridges mixed with a mass of saturated air to provide thunderstorms of catastrophic proportions. These storms locally produced 71 centimetres of rain in 8 hours and triggered a great many debris avalanches in central Virginia. The storms claimed 150 lives, most to flooding. However, most people died from broken bones and blunt-force injuries rather than drowning. It is impossible to estimate how many died as a result of the avalanches, but the debris delivered to channels in floods certainly was significant. The avalanches generally followed pre-existing depressions, moved a layer of soil and vegetation 0.3 to 1 metre thick, and left a pronounced linear scar of exposed bedrock.

8

DEVELOPMENT OF LANDSCAPES-I

Base Level: It is that stage or temporary level towards which the land surface may be reduced and below which the land surface cannot be reduced by sub-aerial agents of erosion. With this as the lower limit of effective degradation we can deduce that the ultimate or last stage in the evolution of landforms, regardless of what combination of agents act, will be the production of a broad, nearly flat erosion surface. Further evolution of the region which has been reduced to such a surface must await uplift or igneous activity to build the area up again.

Diastrophism or Deformation: Uplift or igneous activity initially brings units of land into position above the sea. Then the cycle of erosion may commence to sculpture and reduce the land. Because diastrophic movements may be very slow they may continue over a longer period of the erosional history of the region.

Interruptions: Diastrophism can create interruption in the sequence of development of landforms. This interruption may be in a number of forms: (a) It may be very slow up warping or a down drag. (b) It may be a very rapid up warping or down drag. (c) It may be an intense deformation, folding faulting of the crust.

Structure: The structure of the rock also plays an important role in determination of what type of landforms are possible. The surface of an uplifted part of the crust has certain unique physiographic forms before the new cycle begins. Consider for example the initial differences in surfacial features of the each of the following:

(a) Sections of the marginal sea floors,
(b) Folded mountain structure or faulted mountains,
(c) Volcanic mountains and Lava Plateaus,
(d) Glaciated plains,
(e) Lake beds, and
(f) Deserts.

Time: Each stage in the evolution of the land is characterized by a particular group of landforms.

Polygenetic Nature of Geomorphological Landscape

The word polygenetic comprises of two words: 'Poly' means many and 'Genetic' means origin or evolution. The polygenetic landscape consists of complex landforms which are formed by different sources. A landscape having geomorphic features that developed under different types of climates and processes at various times during its formation. A landscape modified by glaciations is an example. A polygenetic landscape genetically displays more than one erosional surface with a complexity of landforms. There are areas in the world where complex landforms exist. At present in such areas we find many topographic features formed by stream action. Along some streams glacial features especially outwash plain may be found. This outwash plain may have the features created by wind action.

Triccart and Cailleux (1965) observed that many landscapes give evidence of having developed under changing morphogenetic systems associated with the changing climate of Pleistocene period. In India there are many areas where the imprints of polygenetic landform are preserved.

Himalayas are most complex landforms on the earth. The endogenetic and exogenetic processes have created complexity in the landforms. The endogenetic processes were responsible for upliftment of uplands above the glacial limit and exogenetic processes modified landscape by subaerial environments related to arid, humid and glacial geomorphic system. All these factors are responsible for complexities of Himalayan geomorphology.

The geomorphic study of Aravalli region reveals that the Aravalli range had reached the old age some time in the late Cretaceous period. The present ranges are the stumps or roots of the synclines and are good example of polygenetic landforms. There are three erosional surfaces identified at elevations of:

(a) 1200 mts. which is marked by the highest peak.
(b) 600-680 mts. around Udaipur in the valleys, and plain towards the east of the range.
(c) 300-430 mts.

The following are the main evidences further support the polygenetic landscape:

(1) The uniform skyline of the ranges.
(2) Numerous superimposed drainage in Udaipur region and of the Chambal in the Bungli hills.
(3) Erosional scarps facing the northwest bank of the Chambal river.
(4) The ravine formation of Chambal and its tributaries.
(5) The used meanders identified in Chambal and its tributaries.

Another example of Barakar basin displays four erosional surfaces; each surface characterized by minor landforms. These erosional surfaces are of the polygenetic nature. They still preserve form of ancient surface in modern relief, although the surfaces appear to be uplifted epeirogenetically.

Indo-Gangetic plains present best example of complex landscape.

The complex plains are formed by the different rivers. These rivers originate from entirely different sources and from different directions. The material of Khadar, Banger, Babar and

Terai genetically is formed by different sources and in different directions.

The polygenetic landscapes are also preserved along the coast where sea embraces the land through geomorphic effect of land-sea-air contact. The coastal plains of Kerala and Tamil Nadu show the evidences of polygenetic landscapes.

Fluvial Landscape

Fluvial Landscape refers to the topography of normal cycle of erosion. Although the normal cycle of erosion is that which is controlled by river action, the erosional cycle may as well be applied to the work done by glacier, winds and waves.

The concept of cycle of erosion was given by William Morris Davis (1850-1934). In the cycle of erosion the surface forms undergo changes as a result of the processes acting upon them. These changes are not erratic but follow each other in regular sequence. According to Davis every land is a result of structure, process and stage. He has given four stages: Initial, Youth, Maturity and Old. In the initial stage landform is generally even and is raised high above the sea level, slow and small change occur. However, vast changes take place by the time the landscape attains maturity. In old stage again landscape evolution slows down.

In the initial stage the geomorphic agents work slowly and remove small material. The most of the landforms of this stage are erosional in origin. Maturity stage is associated with bolder relief features. These are marked with contrast in hill and valley forms.

In the old age topography is much subdued and tends to be lowered to the base level of erosion. This is a level to which the initial relief is lowered. High relief features are worn down and low relief features are filled up. (Fig. 8.1)

Peneplains and *monadnocks or inselberges* are the characteristic features of this stage. Peneplains is a low and gently undulating plain. In other words, peneplain is broad low, generally smooth outlined plain which represents the completion of cycle of erosion.

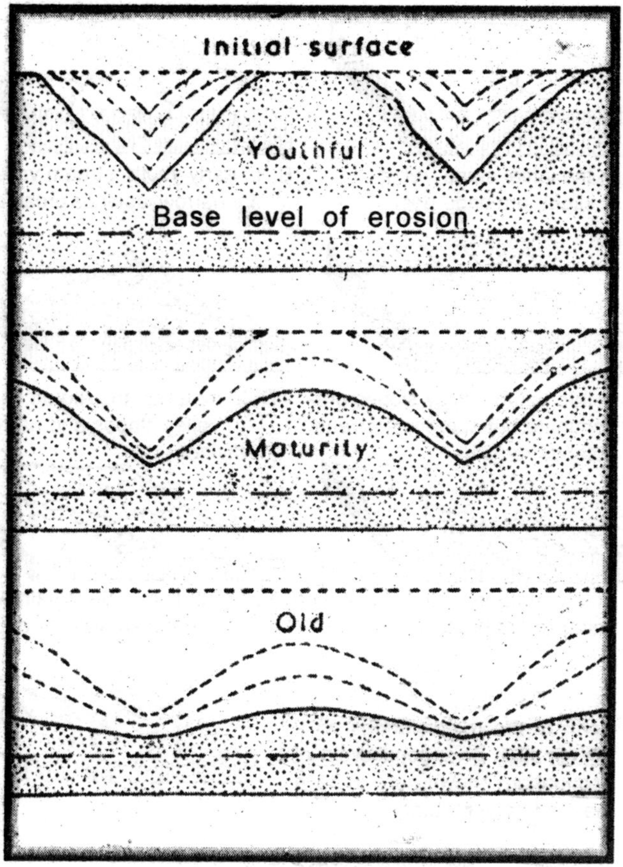

Fig. 8.1: Cycle of erosion.

Inselberges or Monadnocks are isolated features; these are more resistant to weathering and erosion, glacial cycle, Karsts cycle and marine of erosion.

Fluvial Landscape in hills, plains and coasts are formed by running water in different stages. More than 80% landforms of the world are modified by the running water hence it is called normal cycle of erosion. There are three stages of a river:

1. Youthful Stage: In general, a Youthful Topography is associated with a recently uplifted landmass. In the youthful stage the elevation of the landmass is highest. (Figs. 8.2-8.6)

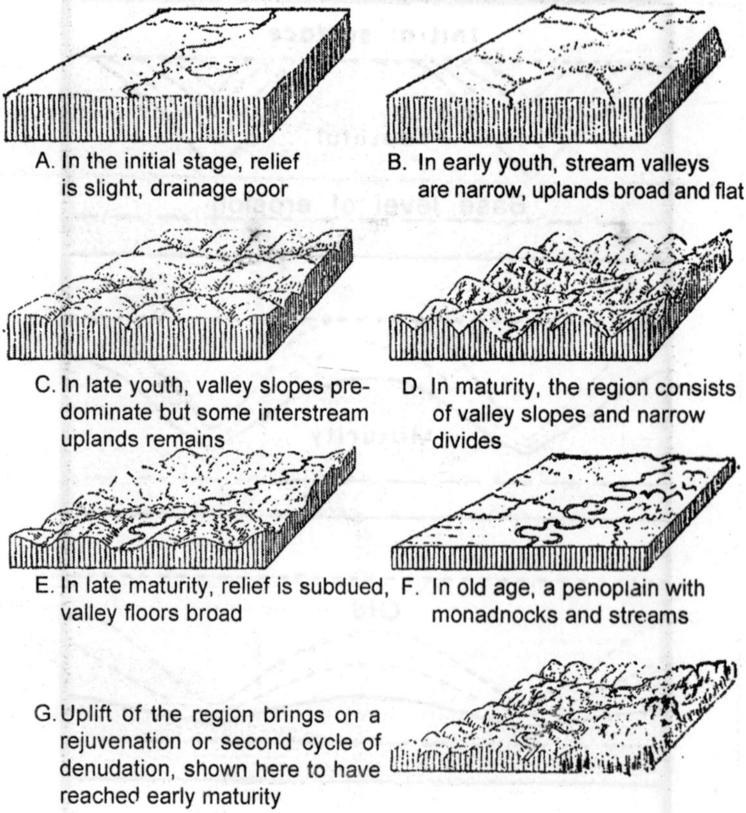

A. In the initial stage, relief is slight, drainage poor

B. In early youth, stream valleys are narrow, uplands broad and flat

C. In late youth, valley slopes predominate but some interstream uplands remains

D. In maturity, the region consists of valley slopes and narrow divides

E. In late maturity, relief is subdued, valley floors broad

F. In old age, a penoplain with monadnocks and streams

G. Uplift of the region brings on a rejuvenation or second cycle of denudation, shown here to have reached early maturity

Fig. 8.2

Some of the characteristic features in this stage are channel deepening due to bed scurrying and pot-hole drilling, formation of V-Shaped valleys. The inter-fluvial divides are wide and generally simple in their outlines. *Finger tip channels, rills, extended gullies, headwater amphitheatres, semi-circular crests, waterfalls, rapids, valley headwalls and inter-locking are other main features of this stage.*

Spurs are some of the other features developed in this stage. Rills are small furrows. These rills develop into small stream-channels, and are important in removing weathered material and wearing back the slope when more than one rills join together a deep stream channel known as gully is formed.

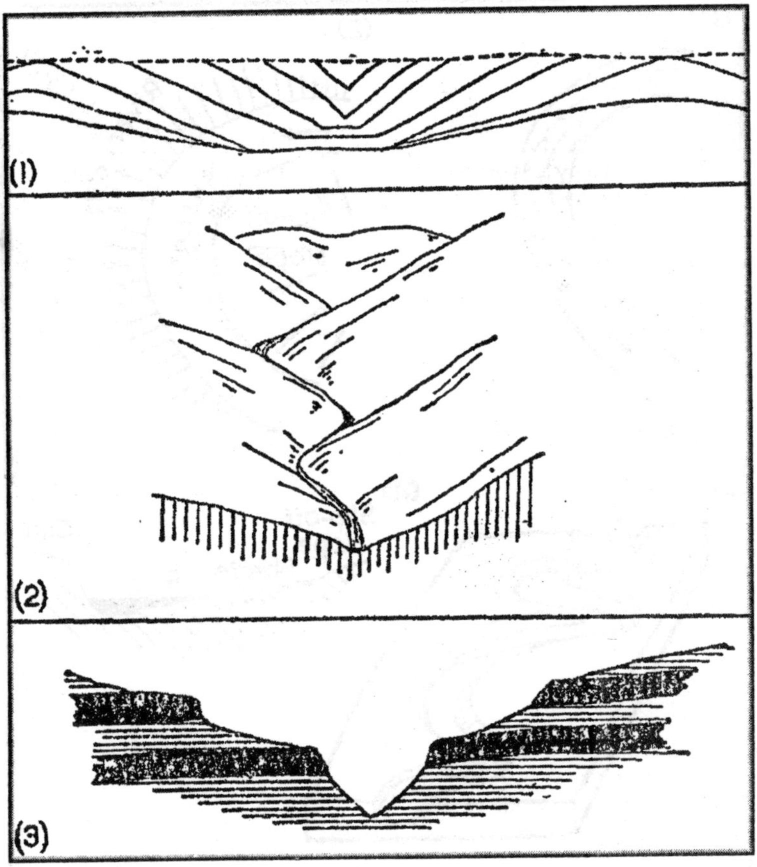

Fig. 8.3: Features of the cross-profile of a river valley. (1) Successive stages in the evolution of a cross-profile, (2) Interlocking spurs, (3) The effects of almost horizontal resistant strata (in black) upon the form of the profile.

Waterfalls own their existence to some sudden drops in the rivers longitudinal profile. Waterfalls are formed under different conditions. The variation in resistance or rocks may lead to the formation of a waterfall. Waterfalls are also found where there is a sharp well defined edge of the plateau. Some waterfalls are formed due to faulting. The commonest type is formed where the fault line scarp brings a less resistant rock on the downstream side against a more resistant rock.

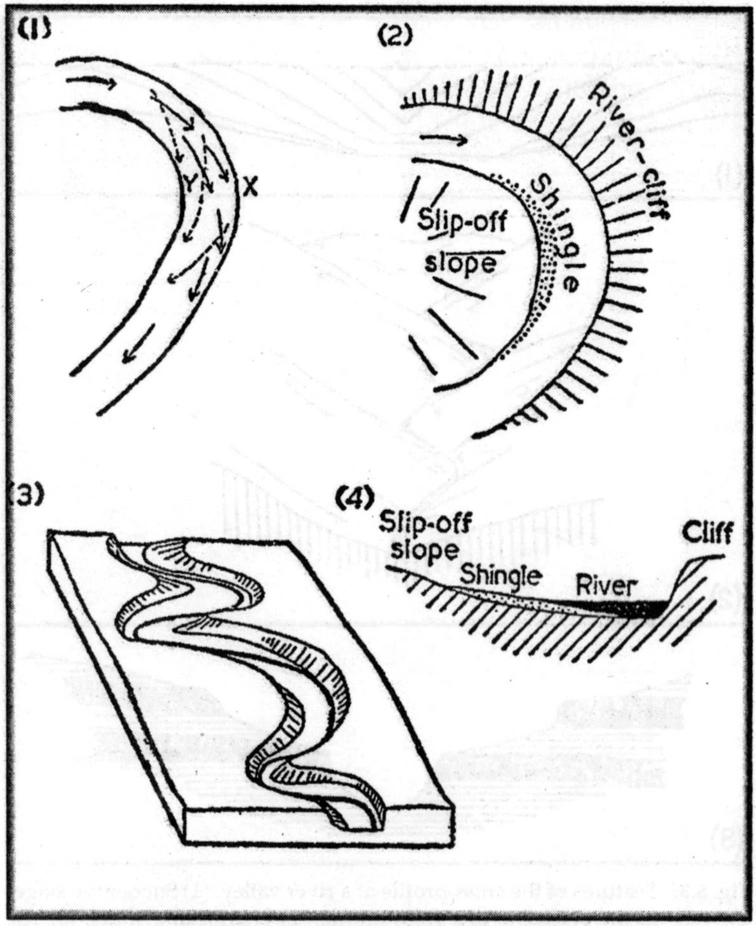

Fig. 8.4: Meanders: (1) The direction of the main current is shown by solid arrows; as it swings round the bend, it impinges on the outside of it, thus eroding a steep river-cliff or bluff at X. As the water heaps up on the outside of the bend by centrifugal force, causing powerful turbidity, a return bottom-current (as shown by the pecked arrows) flows towards the inside bank (helicoidal flow), depositing a shingle or sand-bank at Y. (2) The developing meander. (3) Block-diagram of meander development; each bend is enlarged, both sideways and downstream. The valley floor is therefore gradually widened. (4) Section of meander.

Rapids like waterfalls occur at a sudden increase in the gradient of the streams channels. When a stream flows over a

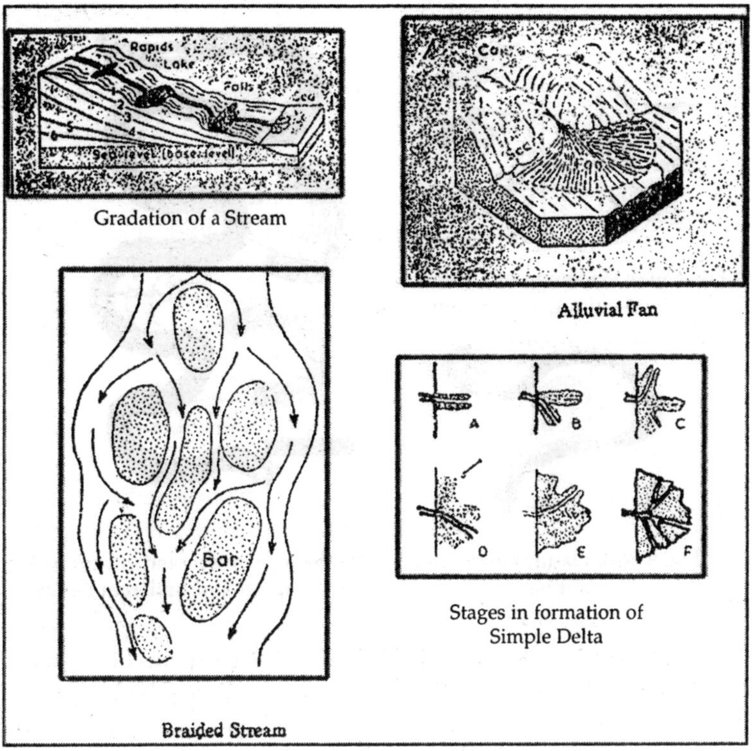

Gradation of a Stream

Alluvial Fan

Stages in formation of Simple Delta

Braided Stream

Fig. 8.5

slope alternating hard and soft layers the softer layer is removed forming depression resulting in a waterfall. Although rapids do not plunge straight down as waterfalls do, the underlying cause of their formation is often the same. In fact many rapids have developed directly from pre-existing waterfalls. (Figs. 8.2 to 8.6)

At the base of the waterfalls the churning of water containing boulders produces pot-holes, the process being known as pot-hole drilling. *Pot-holes* are rounded depressions and deep holes carved into solid rock by sand grains and pebbles swirling in the fast eddy.

An upland stream follows a winding course, since it tends to flow around obstacles formed by more resistant rocks. These bends are gradually emphasized since the current tends to be

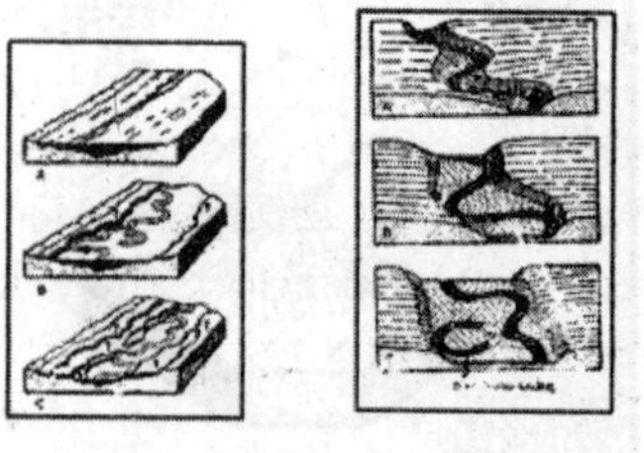

Fig. 8.6(a): Widening of a valley by lateral cutting of a Stream.

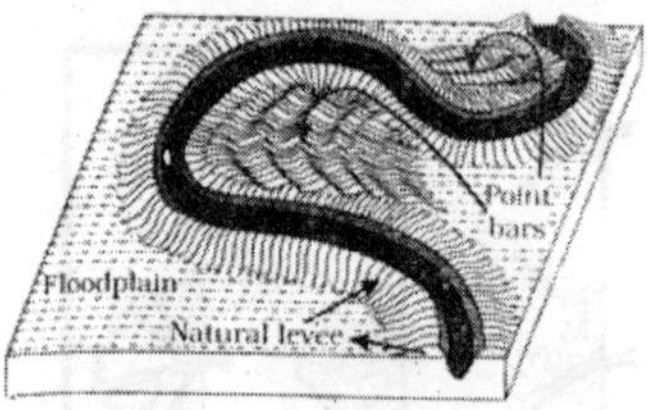

Fig. 8.6(b): Natural levee and point bars.

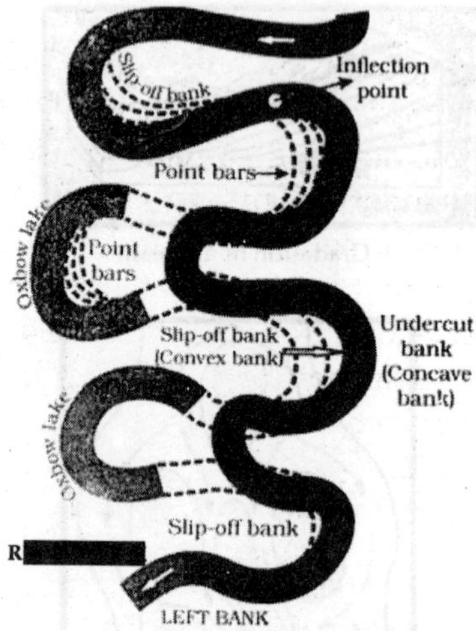

Fig. 8.6(c): Growth and cut-off loops and slip-off and undercut banks.

strongest on the outside of the bend and so spurs alternating on either side of the river 'Interlock' each other. The bed of the stream consists of pools, pot-holes, boulders and small rapids or falls caused by locally resistant rocks.

Gorges and *Canyons* are also common in the youthful stage. A gorge is an elongated steep-sided valley, deep in proportion to its width. A river gorge is found where fluvial erosion cuts down more rapidly than the forces of weathering can wear back and open up the sides.

A gorge may also be formed in arid climates, where an exotic river deriving its volume mainly from snow melt on mountains beyond the desert is able to maintain its flow and its erosive power. Such gorges are known as canyon in America.

1. Fluvial Landscape in Plains (Mature Stage)

In the plain area river develops mature features and valley becomes wider. As the water flows round a bend the river tends to accentuate the curve. (Figs. 8.2 to 8.6)

Since the current impinges most strongly on the concave side or outside of the curve causing maximum erosion, even undercutting takes place there. There is little erosion and even some deposition occurs in slack of the current on the inside of the bend. Thus an initial 'swing' in the river may be transformed into meander with a river cliff hanging over the undercut bank and a sloping spur known as a *Slip-Off Slope,* projecting from the opposite side (Figs. 8.2 to 8.6).

Drainage lines now get integrated in a density network. The river flow with a graded profile. During mature stage both erosional and depositional activities are at their maximum. As a result of heavy lateral erosion and deposition a large number of landform features are developed in this stage, such as *Hog Backs, Cuestas, Plateau Tables, Esplanade, Meander Core, Meander Cut off, Ox-Bow Lakes, Meander Scar, Natural Bridge, Arch, Flood Plain, Flood Plain Carp, River Bluff, Terraces, Alluvial Fans and Linear Lakes.* (Figs. 8.6A—8.6C)

Hogback is a long narrow ridge in which both dip slope and scarp slope are steep, due to the steep inclination the beds are thus differing from cuesta. *Cuesta* is a Spanish word. It is a ridge, or belt of hilly land formed on gently dipping rock strata. It has a gentle dip slope on one side, and a relatively steep scarp on the other.

As we have already discussed that both depositional and erosional processes are involved in the formation of meander. In their down valley migration different meanders as well as different parts of the same meander may move at different rates. The result is that the narrow neck of a meander may be cut through and stream takes a new course. The narrow channel is called Neck Cut Off.

The abandoned meander—the meander scar is called an Ox's collar. Usually both ends of the depression are gradually silted and the old meander becomes completely isolated from the new channel. If the abandoned meander fills up with water an Ox-*Bow lake* is formed.

The flood plain is formed by repetitive deposition, beyond the banks of material brought by the flood of the rivers. It is wide and long and is parallel to the river. On the flood plain the characteristic features are natural *levees* which are long depositional ridges extending parallel to the river, abandoned channels, ox-bow lakes and linear lakes.

In many flood plains the water surface of the streams is held above. The level of the valley floor is raised above the bank of sand and silt known Natural levees a name derived from the French lever means "to raise".

A tributary stream entering river valley with high levees may be unable to find its way directly into main channel, so it runs parallel before finding an entrance. It is called a *Yazoo.* Rivers following similar course are known as Yazoo river.

When a river enters into the plain its velocity is reduced as a result its carrying capacity is also reduced. At this point the river deposits a large amount of load which assumes a fan or conical shape termed as *Alluvial Fans.* In the lower parts many fans join laterally to produce a bajada or piedmont plain. A stream terrace is a relatively flat surface running along the valley, with a steep bank separating it either from the flood plain or from a lower terrace. It is a remnant of the former channel that now has succeeded in cutting its way down to a lower level. Terraces that face each other across the stream at the same elevation are known as paired terraces. Sometimes, the downward erosion by stream created unpaired terraces.

In the mature stage the main stream have well developed valleys separated by inter-fluvial divides. The valleys are bounded by river bluff which overlooks the flood plains. Near the base of the bluff are located linear lakes which represent the inactive flood plain scour routes through which the channels had flown at one time.

In the flood plain two types of material is deposited:

(1) Fine silt clays, and sands which may be spread across the plain by a river that overflows its banks during flood. These are called over bank deposits.
(2) The other group of deposits is made of coarse material, e.g. gravel and sand and is related directly to the channel of the stream.

These deposits include chiefly the material that are found inside of the bends of the winding or meandering river. These bars of sand are called *Point Bars*. Along various sections during this stage the main channels disintegrate into many small channels separated by *lenticular sand bars*. These disintegrated stream channels again meet the main channel in the downstream. This channel is termed as *braided stream*.

Old Stage Topography

In the third stage rivers are flowing very near to their mouth. Because of gentle slope and low velocity the energy of the river flow is small. So aggradational processes are predominant in this stage. (Figs. 8.2 to 8.6)

The relief of the topography is small and outline of the land is gently rolling with wide valleys and low inter-valley divides.

Some of the main characteristic features of this stage are delta distributaries, Delta Flank depressions, linear depressions, lenticular sand bar and saucer shaped, inter-fluvial.

The word *Delta* was first used by Herodotus in the fifth century B.C. while he was walking in the plains of Nile. He used term delta for triangular plain because of the similarity between this plain and outline of the Greek letter.

Whenever a stream flows into a body of standing water, its velocity and transporting power are quickly stemmed. If it carries enough material and if conditions in the body of standing water are favourable a delta will be formed gradually. If the drainage across a delta is carried by a series of channels diverging from the apex, the branching channels or distributaries channels build a delta.'

At the outer end of distributary's channels sediments may be arranged in a definite pattern. The coarse material is dumped first, forming a series of dipping beds called forest beds. But the finer material is swept further along the channel and settles across the sea or lake floor as bottom beds.

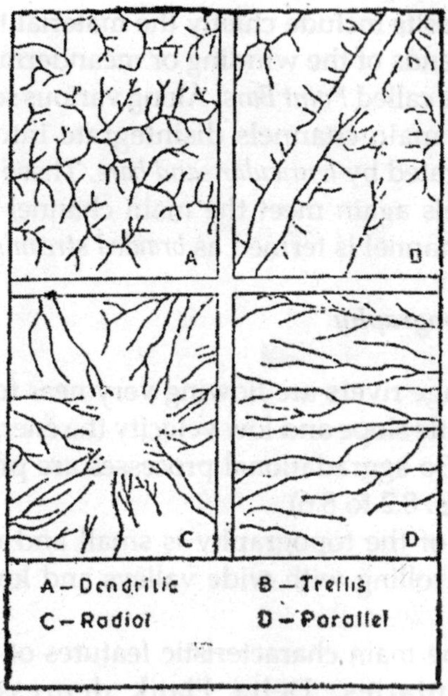

Drainage Patterns

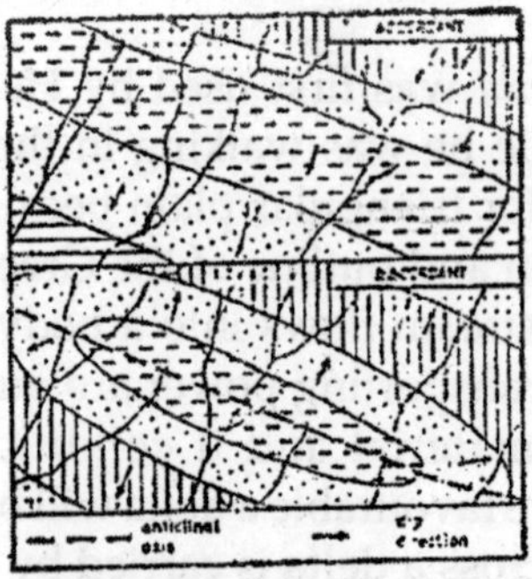

An example of accordant and discordant drainage

Fig. 8.7.

Depressions filled with water and forming lakes occur on the flanks of the deltas. These are termed delta flank depression, lake Kelleru of Andhra Pradesh is best example. River Nile, Mississippi and Rhine also have many examples of this feature.

Drainage Patterns

Fluvial topography develops well in a drainage basin. In the drainage basin the streams join each other to form a mainstream. The joining of the tributaries with the master stream produces a pattern termed as drainage pattern. The common drainage patterns are dendritic, parallel, Trellis, Annular and Radial. (Fig. 8.7)

A dendritic pattern resembles a tree in which angles of confluence of the streams are small. A parallel pattern develops on the steep slopes where the tributaries and the master stream flow parallel to each other. A Trellis pattern develops in a topography created on a fold structure. A radial pattern consist of drainage lines radiating from a common water parting in different directions.

GENETIC CLASSIFICATION OF DRAINAGE PATTERNS

Consequent and subsequent rivers and their valleys. River systems that develop as a direct result of and in harmony with preexisting land surface are called to be **Consequent.** One may considered that the flow of water was a direct consequence of whatever slopes were initially present (A less common term for small valleys or gullies in **insequent** shortened from initial consequent). Table 8.9a shows the significance of basic and modified basic drainage patterns. (Figs. 8.8 and 8.9)

Antecedent drainage: In orogenic regions, the valleys that are first eroded over a landscape may later have new tectonic ridges rise across their trends. The river may be defeated (force to abandon their valleys) or they may be able to maintain gorges across the landmass rising in their paths. Streams that have maintained their valleys across tectonic ridges are said to be antecedent because the river is older than or antecedent to deformation. River Tawi, Narmada, Tapi and Satluj are the best examples of antecedent drainage.

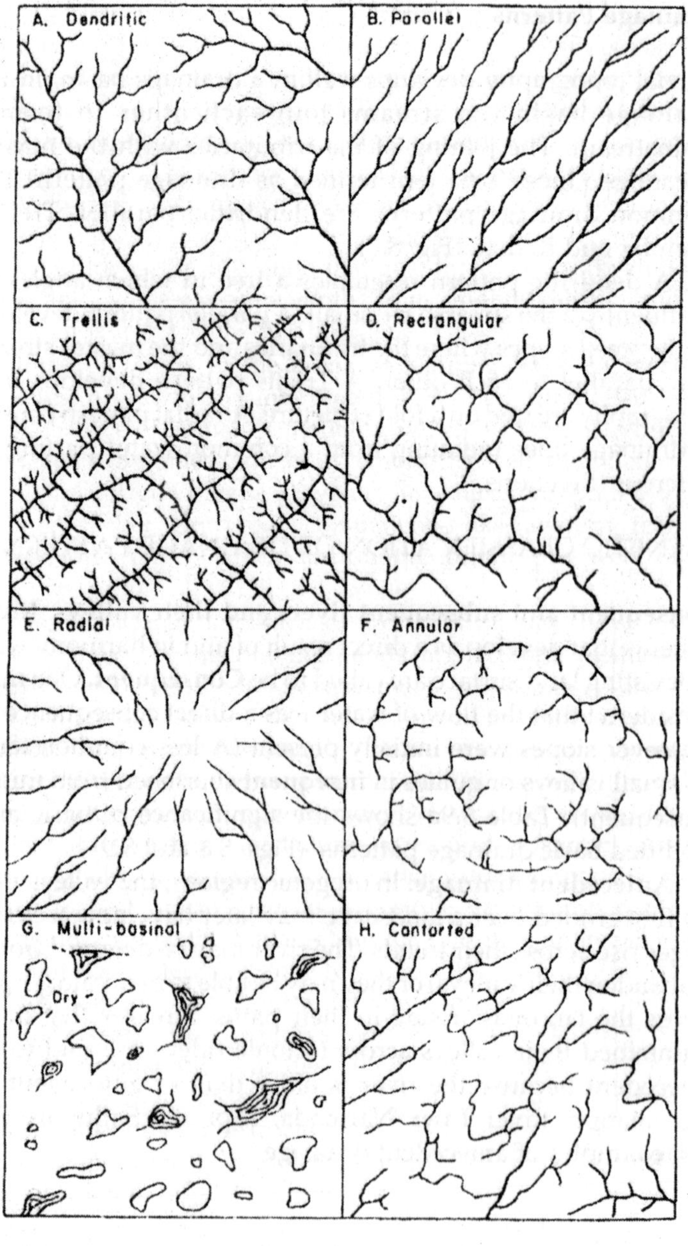

Fig. 8.8: Basic drainage pattern. See Table 12-1 (Howard, 1967, Figure 1).

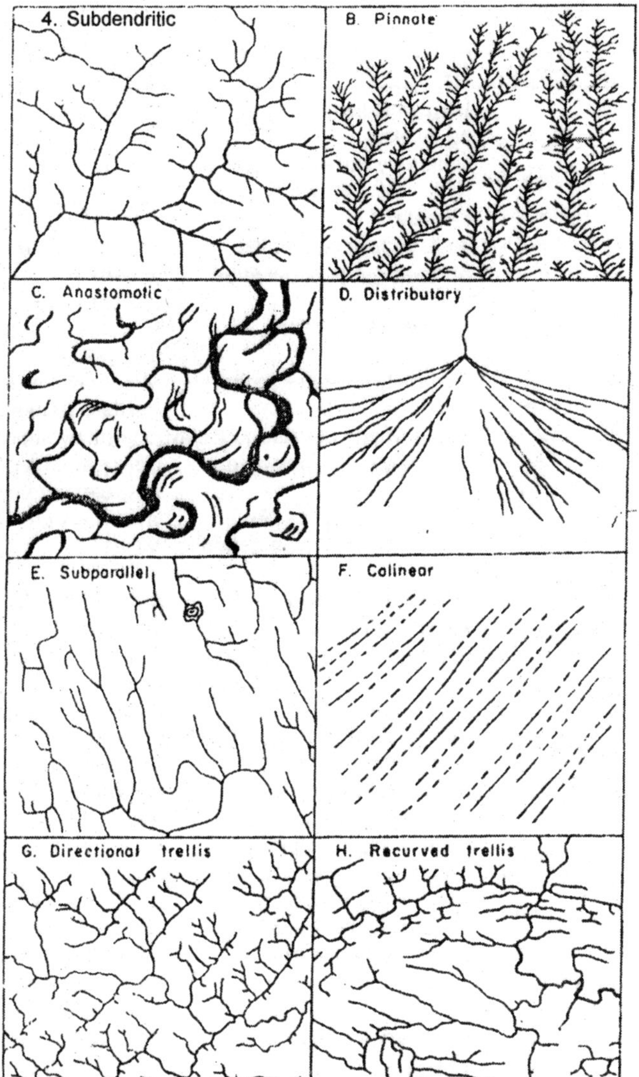

Fig. 8.9: Modified basic patterns. Each pattern occurs on a wide range of scales. See Table (Howard, 1967, Figure 2).

Table 8.9a: Basic and Modified Basic Drainage Patterns

Basic Significance	*Modified Basic*	*Added Locale*
Dendritic	horizontal sediments or be levelled Uniformly resistant, crystalline rocks. Gentle regional Pinnate Type pattern resembles Spreading oak or chestnut tree. Mashes. Alluvial fans	Subdendritic Minor secondary con- trol, generally structu- ral. Fine texture easily slope at inception. Anastomotic erodible materials. (Dichotomic) Floodplains and tidal and deltas.
Parallel	Generally indicates moderate Steep slopes but also found in area of parallel, elongate All transitions possible this pattern and type dendritic and trellis.	Subparallel Inter- mediate slopes or control by subparallel Collinear landforms. Between landforms. linear loss and sand ridges.
Trellis	Dipping and folded sedimentary Volcanic or low grade meta- sedimentary rocks; area of trellis Lake or seafloor ribbed trellis Branching converging by beach ridges. All transitions to parallel pattern. Type pattern is regarded here as one in which small tributaries joints. are essentially same size Subsequent streams.	Subtrellis Parallel elongated land Directional Gentle homoclines. Gentle slopes with parallel fractures; exposed Recurved Plunging folds. Fault trellis diverging, roughly parallel faults. Joint trellis Straight parallel faults and/or on opposite sides of long parallel
Rectangular	Joints and/or faults at right angles. Lacks orderly repetitive quality of trellis patterns; streams and divides lack regional continuity.	Angulate Joints and/ or fault at other than right angles. A compound rect- angular-angulate Pattern is common.

Basic Significance	*Modified Basic*	*Added Locale*
Radial	Volcanoes, domes, and erosion Residuals. A complex of radial patterns in a volcanic Field might be called multi-Radial.	Centripetal Crater, calderas, and other depressions. A complex of centripetal patterns in area of Multiple depressions might be called multi-centripetal.
Annular	Structural domes and basin diatremes and possibility Stocks. And permit distinction	Longer tributaries to annular subsequent Streams generally indicate direction of dip Between dome and Basin.
Multibasinal	Hummocky surficial deposits Differentially scoured or deflated bedrock ; area of recent Volcanism, limestone solution and permafrost. This descriptive is suggested for all multiple-depression patterns whose exact origin are unknown.	Glacially Glacial erosion and/or disturbed deposition. Limestone Karst Permafrost. Themokarst Coastal plains and elongate bay deltas.
Contorted	Contorted, coarsely layered Metamorphic rocks. Dykes, veins and magmatized bands provide the resistant layers in some areas. Patterns differs from recurved trellis in lack of regional orderliness, Discontinuity of ridges and valleys and generally smaller scale.	The longer tributaries to curved subsequent streams generally indicate dip of metamorphic layers and permit distinction between plunging anti-clines and synclines.

Superposed: The word superposed is a contraction of superimposed and implies that a river system has been let down or laid over a landscape. Suppose a land newly emerged from the sea has a thin cover of sediments that bury an old terrene of complexly folded rocks. On emergence a consequent drainage system perhaps with a dendritic pattern might form. As the valleys deepen, they are eventually eroded into the old

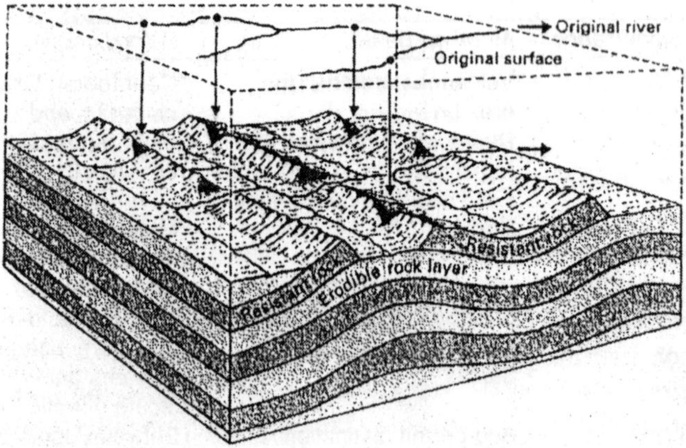

Fig. 8.10: Expansion of a consequent valley system over an initial landscape, with a later stage of erosion when subsequent tributaries have become dominant. Arrows project downward from valley segments of the consequent system that were defeated by resistant layers and stream capture to become wind gaps.

land under covering strata. The dendritic pattern becomes superposed on the older rocks without regard to structural control. When the covering beds have been entirely removed by erosion, the only clue to superposition is the anomalous position of the river valleys. An even more complex case of superposition would be if the former cover were a deformed, folded thrust sheet or nappe. One can only guess at the possibilities of drainage anomalies superposed on the rocks below the thrust. Superposition is extremely common in glaciated regions, where interglacial and preglacial drainage network are filled with glacial drift and the present streams almost randomly occupy alternating segments of old re-excavated valleys and new preglacial valleys.

River Capture: Piracy has been a popular concept of geomorphologists for generations. Either by chance or aided by structural advantage, one stream may erode into the drainage basin of another and either progressively or abruptly divert or capture new tributaries. The trunk stream that is thereby deprived of part or its entire former watershed is said to have been beheaded. Quite commonly subsequent drainage systems captured segments of either consequent drainage.

Figure 8.10 suggest how the process could occur. After capture, the former course across the resistant ridges might be marked by **wind gaps**, notches that mark the former line. Any sharp change in direction of a stream, downstream from an anomalously steep section suggests an **elbow of capture.**

STRUCTURAL CONTROL OF FLUVIAL EROSION

During the progress of erosion it is inevitable that some structures will be more resistant to destruction than others. Relatively erodibility or rock mass resistance is a subjective property, not simply defined. A loose distinction can be made between the sedimentary *cover strata* of continents and the igneous and metamorphic *basement* rocks that underline them. Some cover strata, for example, the thick bedded quartz-rich sandstones, are significantly more resistant to erosion than many igneous and metamorphic rocks. More commonly, the covered strata are more erodible and easily stripped off the basement rocks by fluvial, glacial and eolian processes.

FRACTURES AS STRUCTURAL CONTROLS

Joints

Systematic or non-systematic joints sets cut most indurated rock masses. Joints are probably the most common rock structures breaking igneous, metamorphic and sedimentary rocks in all kinds of orogenic and epeirogenic setting. Whatever their origin joints provide opening for groundwater movement weathering and mass wasting are intensified. Joints that opens to parallel gorges and steep sided valleys probably aid in enlarging and perpetuating the valley. Masses of jointed bare rock stacked like well worn eggs blocks are called **tors.** They are not transported but result from differential weathering and erosion along joints at least one set of which is horizontal. Tors may formed at depth by joint controlled weathering within the regolith or they may formed subaerially by differential weathering and erosion of structure within the rock mass.

Faults

In a faulted terrane of massive igneous rocks or metamorphic rock, the only crushed zone along faults may be subject to more rapid weathering and erosion. A confused or chaotic landscape is likely to result from structurally controlled erosion of tectonic mélanges.

The mélange structure may extend to a depth of 10 km. A unique aspect of mélange terrane is that the hills of resistant rock have no roots and no prediction of sequential landscape evolution can be made. Although not strictly an example of fault controlled scarps, the abrupt structural boundaries of mélange hills seem to fit into any other available descriptive category.

DIFFERENTIAL FLUVIAL EROSION ON LAYERED ROCKS

Most sedimentary rocks and many igneous and metamorphic rocks have layers of contrasting erodibility. The layering may be the result of sedimentary stratification, igneous intrusion, metamorphic differentiation or a host of other causes. Where the layers are horizontal their exposed edges or *outcrops* are parallel to contours and form attitudinally zoned structural landforms where layers are tilted much more complex geometry develops. Much of the geomorphic diversity of landscapes is due to structural control of layered rocks and an extensive descripitive and explanatory terminology has evolved.

Layered Igneous and Metamorphic Rocks

Volcanic plateaus are built of piles of lava and tephra sheets and intrusive sills crossed by numerous dikes. Either a single flow or a series of rapidly sequential eruptions may cover a very large area with fresh flows of lava or ignimbrite which than cool as a unit. The interior of cooling unit looses heat upward and downward as it cools it contracts and fractures into a crudely hexagonal pattern of columnar joints. However, the upper and marginal zones usually cool more rapidly and shatter into unsystematic blocks of glassy or vesicular rock. Columnar jointing provides high angle sets of closely spaced joints that promote rockfalls, cliffs and taluses. The glassy or shattered marginal zones commonly taluses separating joint controlled cliffs.

Differential fluvial erosion on volcanic structures other than basalt or ignimbrite plateaus also produce distinctive structurally controlled landforms although they are rarely regional extent. Metamorphic terranes have a dominant regional structural grain. Hills have their long axes on strike. Streams follow erodible belts or cross ridges at fracture zones. Entire regional landscapes such as the New England geomorphic province of the northeastern US become a series of structurally controlled subparallel ridges and valleys.

Folded Sedimentary Rocks

Most of the present day subaerial landscape is eroded on stratified sedimentary rocks. Sedimentary rocks cover 66% of the total continental area (excluding ice-covered Antarctica) although the coverage varies greatly from 87 % of Europe to only 52 % of North America. Epeirogeny has, however, uplifted large continental areas with only slight warping of the cover strata which are only 1 to 2 km in thickness. Thus epeirogeny and orogeny have created regionally extensive subaerial landscape that is largely eroded from sedimentary structures (Fig. 8.13). The differential erodibility of shale, limestone, and sandstone is the dominant structural control of landscapes over two-thirds of the continental area. Flat-lying sedimentary rocks of contracting erodibility form *structural benches* on valley walls. Many plateau surfaces are structurally controlled by resistant *caprock* that persists at a high elevation, while the terrain around it is being lowered. The caprocks is usually a sedimentary rock, but it can also be lava flow.

Generally, the outcropping edge of resistant stratum will form an escarpment or cliff, commonly joint controlled. The dip slope, away from the crest of the escarpment, is parallel to the stratification and tends to resist dissection as weaker over

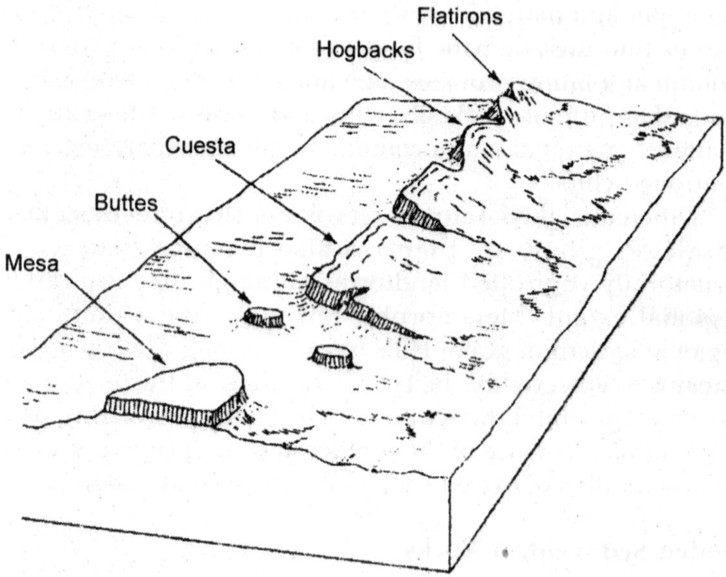

Fig. 8.11: Terminology of structurally controlled ridges.

lying strata are eroded. A single thin layer of sandstone or limestone may armour or protect an extensive shale dip slope.

Popular terminology has evolved to differentiate structurally controlled ridges on tilted resistant layers (Fig. 8.11). If the more resistant unit is nearly horizontal and caps a broad flat topped hill is called a **Mesa.** If the diameter of the cap rock is less than the height of the hill above the surrounding terrain, the term **Butte** is more often applied. If the resistant beds dips gently, an asymmetric **Cuesta** results with a steep escarpment and a gentle dip slope. When the dip of the layer is comparable to the angle of repose of the sliderock on the escarpment face, the ridge becomes approximately symmetric and is known as a **Hogback**. " Flatirons" are interfluve remnant of dissected hogback. The structural control of tilted strata imposes a powerful asymmetry on drainage networks. Escarpment streams are steep, short and have high gradients. Dip slope streams are likely to have more gentle gradients, larger watersheds, more tributaries and more sustained flow. (Fig. 8.12)

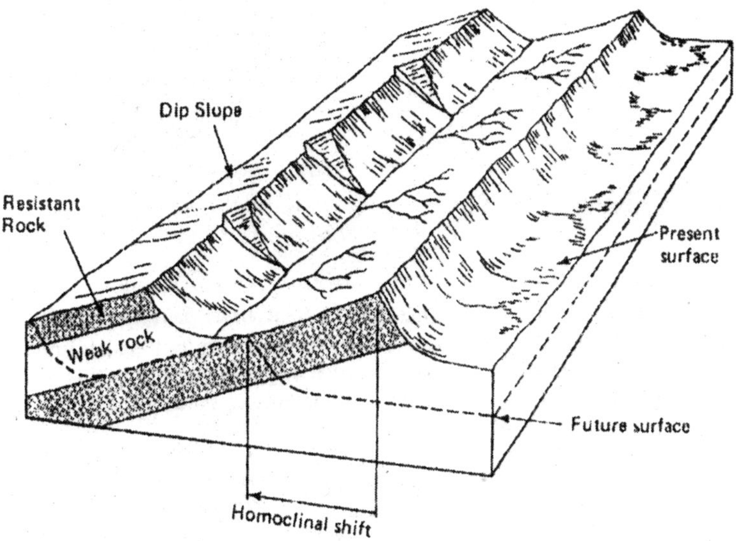

Fig. 8.12: Asymmetric drainage networks on homoclinal ridges. Steep-gradient escarpment streams erode faster than gentle-gradient dip slope streams. Ridge crests and valleys will migrate laterally downdip as the landscape is lowered by erosion.

As erosion progresses, a trunk stream flowing along a weak-rock low land between two cuestas will migrate laterally in the down-dip direction, as it strips away the less resistant strata. The entire ridge and valley system migrates laterally as well as downward with time in a process termed **Homoclinal shifting**. By this process structurally controlled ridges and valleys must migrate laterally many times their own width as a region of dipping strata is lowered by erosion.

Beyond homoclinal dips, the next degree of structural complexity in stratified rocks is a series of parallel anticlines and synclines with horizontal fold axes. (Fig. 8.13)

When structural depression such as syncline axes become ridges or hills led to the concept of **topographic inversion.**

Structurally controlled domes and basins are the result of erosion on double-plunging fold. Although common in fold belts of orogenic regions broad domes and basins are important features of epeirogeny as well.

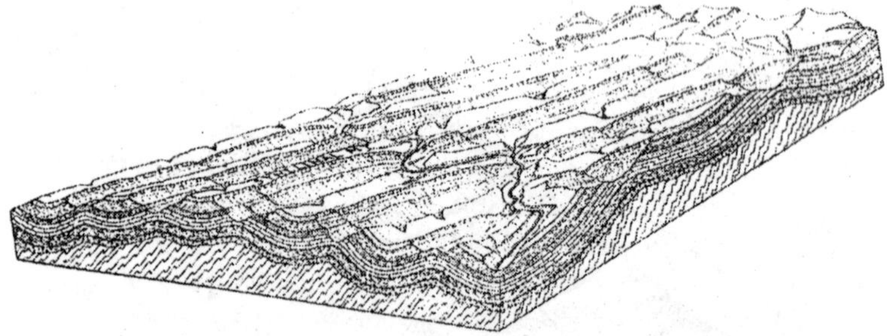

Fig. 8.13: Block diagram of dominantly synclinal ridges in a terrane of folded sedimentary rocks (from von Engeln, 1942, Figure 183). It is unlikely that basement rocks would be folded, as suggested in this figure. More likely a *décollement*, or detachment zone, would separate basement blocks from the folded cover strata.

KARST TOPOGRAPHY

Introduction

The word karst denotes to the solution topography. The term connotes both an assemblage of landforms and a set of processes. Karst is "terrain with distinctive hydrology and land-forms arising from a combination of high rock solubility and well developed secondary porosity." Soluble rocks with well-developed secondary porosity are involved. Solution becomes a process of both weathering and erosion inasmuch as the weathered products are removed from the surface and carried away in solution. The first systematic studies of karst were carried out in Dinaric Alps of Slovenia and Croatia in a region traditionally known as the Karst. Karst is the most strongly developed in humid uplands where very thick strongly jointed limestones occurs. The term karstic is used for karst scenery. Karstification is the formation of drainage systems with the associated development of passages and caves and with protracted surface solution.

Karst Processes and Hydrogeology

The erosional work of groundwater is performed through the mechanism of corrosion or solution, corrosion or abrasion, attrition and hydraulic action but the last three types of erosion are not effective because of exceedingly slow movement of groundwater and thus corrosion of chemical erosion is the only effective method of denudation of carbonate rocks such as limestones, dolomites, chalk etc. by surface and subsurface water (groundwater).

Rainwater mixed readily absorbs CO_2 from the atmosphere and especially from soil air to form a weak acid Carbonic acid which disassociates to hydrogen and bicarbonate ion in solution and dissolves carbonate rocks to form numerous tvpes of solutional landforms. According to R.M. Garrels, there are seven variables which control limestone solution:

1. Partial pressure of CO_2
2. H_2CO_3 (carbonic acid)
3. $H_2CO_3^-$ (bicarbonate ion)
4. CO_3^{2-} (carbonate anion)
5. H^+ (hydrogen ion)
6. OH^- (hydroxyl ion)
7. Ca^{2+} (calcium cat ion)

Biologic Effects

In addition to providing the decaying humus that enriches the soil air in CO_2, plants and animals may corrode limestone directly, it is called **phytokarst.** The surface is intricately pitted with pits intersection to form a sharp-edged, spongy network of pinnacles more than 2mm in height. Cyan bacteria (blue-green algae) are the principal micro-organism that live on the surface (epilithic) within pre-existing cavities (chasmolithic) and within the outer millimetre (endolithic) of limestone in tropical regions.

On certain tropical limestone coasts flocks of sea birds have nested for many generation and accumulated thick deposits of dropping. This guano is rich in phosphate from the scales and skeleton of the small fish that are the bird's primary food. Decay and leaching of the guano makes phosphoric acid, which strongly corrodes the limestone and react with it to form commercial deposits of **rock phosphate.**

Minor Karst Landforms

Limestone and dolomite and to a lesser degree other soluble rocks, commonly develop distinctive surface and near surface weathering features sometimes termed **epikarst. Phytokarst** is independent of gravity, with surface pits corroding equidimensionally beneath an algal film, whereas forms created by flowing water tend to elongated vertically or downslope.

Solution Pits, Facets, Runnels, Flutes and Karren

On exposed rock (bare karst, pavement karst) horizontal surfaces develop shallow **solution pits** (rain pits, makatea) on surface inclined few degrees from horizontal, sheet runoff produces flat **solution facets** (solution bevels). **Solution runnels** (Rinnenkarren) are branching channels in network on gently inclined bare surface. Many runnels probably first form under soil cover but have formed and perpetuated by run off on bare rock surface. On steeply inclined karst surfaces elongated **solution flutes** (Rillenkarren, Lapies) are common landforms. These are straight, parallel-sided grooves with sharp crested separating them. They follow the steepest slope down a rock face.

Covered Karst: Grikes, Clints, Solution Pipes and Solution Notches

If soil or an organic mat covers a limestone surface, the effectiveness of solution increases manifold. Solution pits are even more common under soil cover than bare ledges, but the divides between pits are likely to be more rounded and smooth than the knife edged ridges between intersection pits on bare karst. For vertical opening such as joints that have been widened by solution the British term is **grike** (Kluftkarren, solution slot). Grikes intersect as circular, vertical **solution**

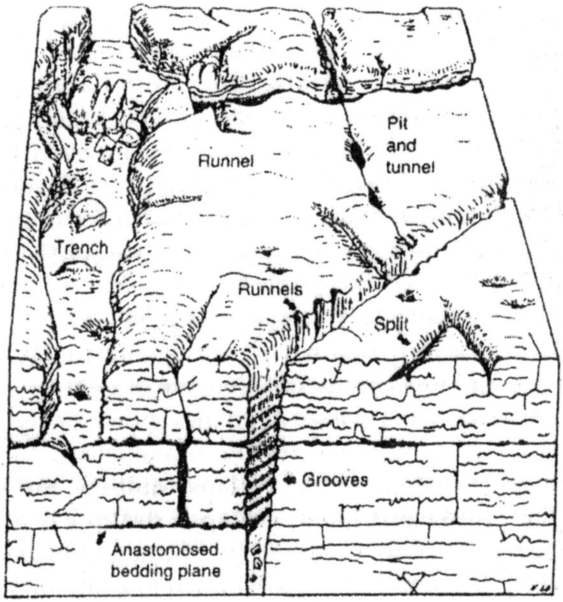

Fig. 8.14: Labelled sketch of minor karst forms on Silurian dolomite near Hamilton, Ontario (after Pluhar and Ford, 1970, Figure 2).

pipes (karst wells, potholes). The remnant pavement surfaces between open grikes are called clints, on which minor solution features develop. (Figure 8.14) At the level where bare rock projects above the soil level, a **solution notch** may develop. A distinct undercut measures the greater effectiveness of karst processes below ground level. An extreme example of a solution notch has been named a **swamp slot.** It is probably restricted to the humid tropics.

Major Karst Landforms

Dolines (Sinkholes, Sinks, Swallow Holes, Cockpits, Blue Holes, Cenotes)

Karst is essentially a near-surface process in vegetated and soil-covered landscapes, where most of the solution takes place in or just beneath the soil profile. On the bare rocks solution is concentrated at the water table in the zone of mixing.

The fundamental geomorphic component of karst topography is the **doline** or limestone sink. Whether due to joint control differential solubility, or random events such as cave collapse, localized areas of karst terrains are lowered more rapidly than the surrounding area and form closed depressions. Their sides slope inward at the angle of repose of the adjacent material typically 20° to 30°. Some have much more gentle side slopes and others especially those forms known as *collapse sinks*, have vertical or over-hanging cliffs. They are usually circular in plan view and less commonly elongate or oval.

There are five major classes of dolines shown in Figure 8.15. Two most contrasting types are the funnel-shaped **solution dolines** and the steep or cliffed **collapse dolines. Subsidence dolines** and **subjacent karst, collapse dolines** are surface forms in non-soluble rock, caused by solution of buried karst. Small dolines (*swallow holes ponors*) form in a stream bed and although alluvium completely conceals the buried karst relief, the stream abruptly disappears and abandons the downstream segment of its former channel. These have been called **alluvial stream sink dolines.**

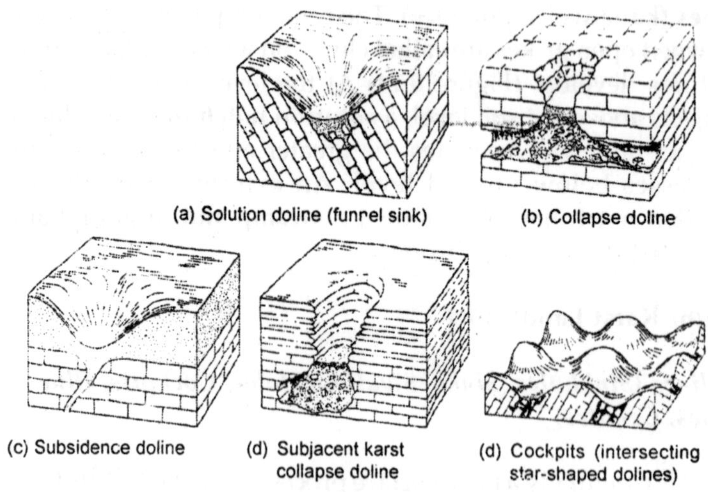

Fig. 8.15: Five major classes of dolines (from Jennings, 1985, Figures 37 and 43).

Dolines in tropical regions develop rapidly and grow very large. Adjacent dolines intersect, fuse, or engulf smaller ones to form compound, reticulate honey comb, or even star-shaped depressions. If they are of smaller size and closely spaced, the intersection ridges will form a polygonal pattern. Their floor are permanent or seasonal swamps at the water table. The **cockpit country** of Jamaica is the classic locality for tropical dolines.

Uvalas (Compound Sinkholes, Karst Windows, Blind Valleys)

A genetic progression can be visualized in which dolines progressively abstract surface runoff into groundwater circulation and leave network of dry valleys as relict surface forms. Term **Uvala** was given to such a compound dolines or chain of intersection dolines. A **karst window** is similar in that an unrooted segment of an underground stream channels becomes a surface valley in the window and passes underground again downstream.

Allogenic rivers originate in a terrane or climate unlike the one through which they flow, for example, the Colorado or Nile. In the karst context, an allogenic river may have a network of valleys on insoluble rocks, then flow into karst prone terrane. If the exposed segment has normal stream valley form and an alluvial floor but disappears downstream into a swallow hole, the term **blind valley** is often used.

A **karst valley** is a special form of blind valley that occurs where a surface drainage network first developed on insoluble rock and later eroded into soluble rock, where after subsurface drainage totally defeated the surface flow.

Poljes (Plans, Wangs, Hojos)

Large blind valleys or karst valleys enclosed by either soluble or insoluble rocks, usually enlongated along tectonic axes are called **Poljes**. Many are grabens or half grabens bu: others are simple large scale examples of structurally controlled karst processes. Their floors truncate folded limestone formation or deep sedimentary infills at a local base level determined by a narrow gorge through insoluble rock or a series of swallow holes at the downstream end of the poljes. (Fig. 8.16) Poljes have broad, flat floor that about sharply against steep enclosing walls. On occasion, during floods, the swallow holes by which some valley floors are drained becomes choked with debris and the entire poljes floor becomes a shallow lake.

Many geomorphologists tend to consider karst landforms as a genetic sequence dolines → Uvala→ Poljes. Dolines grow and merge into uvala through time in a genetic sequence.

Speleothem and other cave deposits

Dripstone and **Flowstone** are descriptive terms for cave deposits from water, the equivalent term **Speleothem** is gaining popularity. Banded calcareous deposits are called **travertines** whereas the calcareous deposits, softer than travertine at the mouth of the caves are called **tufa** or **cale-tufa.** The columns of dripstones hanging from the cave ceiling are called **Stalactites** while the calcareous columns of dripstones growing upward from the cave floor are known as **Stalagmites. Cave pillars** are formed when stalactites and stalagmites meet together. Numerous needle-shaped dripstone hanging from the cave ceiling are called **drapes** or **curtains.** The dripstones growing sideward from stalactites and stalagmites are called **helictites** and **heligmites** respectively. The helictites of globular structure are called **globulites.** Floor deposits caused by seepage water and water flowing out of stalagmites are called **flowstones.**

Caves that have flowing streams contain a variety of alluvial deposits. Many limestone caves have thick layers of red clay that was formerly thought to be the insoluble residue of subsurface solution. (Fig. 8.17)

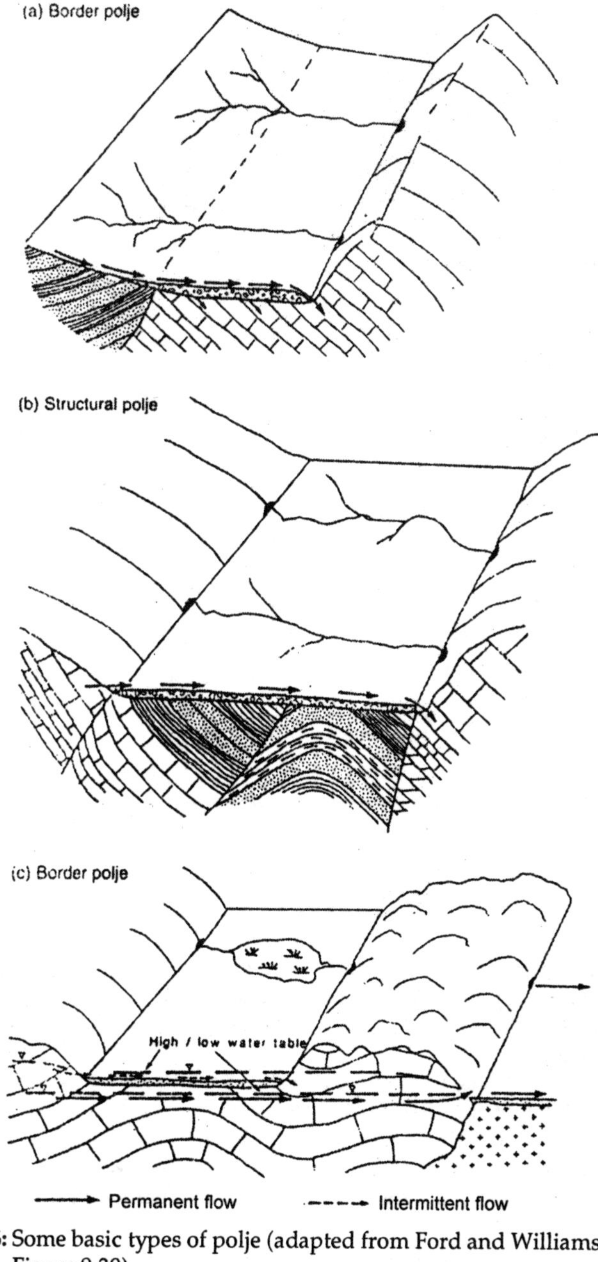

Fig. 8.16: Some basic types of polje (adapted from Ford and Williams, 1989, Figure 9.30).

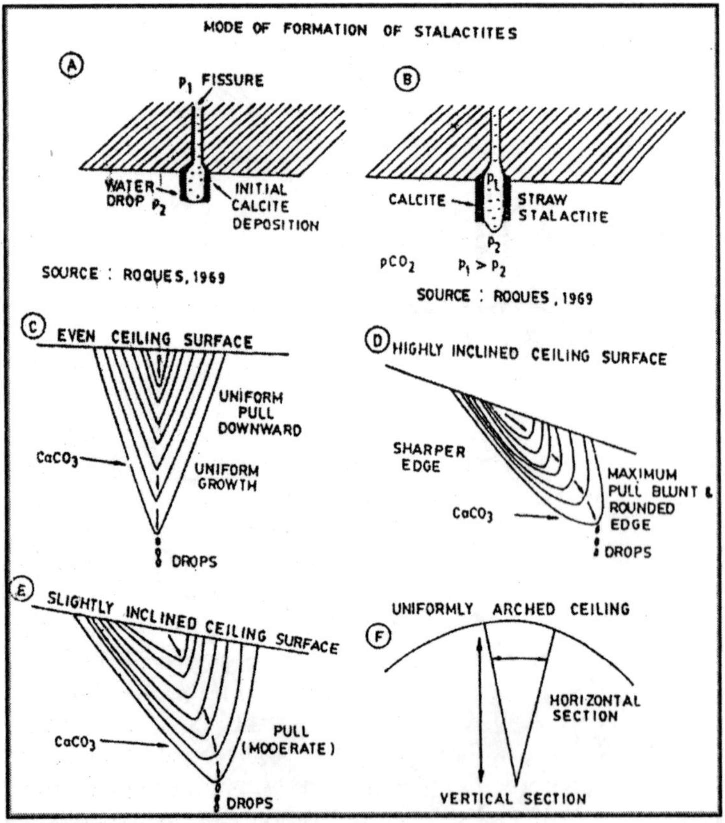

Fig. 8.17: Nature of Cave Ceiling and Formation of Stalactites. A and B after H. Roques, 1969.

REFERENCES

Chorley, Richard J., Stanley Alfred Schumm and David E. Sugden (1985). *Geomorphology*. London: Methuen. ISBN 0-416-32590-4.

Edmaier, Bernhard (2004). *Earthsong*. London: Phaidon Press. ISBN 0-7148-4451-9.

Kondolf, G. Mathias, Heré Piégay (2003). *Tools in fluvial geomorphology*. New York: Wiley. ISBN 0-471-49142-X.

Needham, Joseph (1954). *Science and Civilisation in China*. Cambridge, UK: Cambridge University Press. ISBN 0-521-05801-5.

Scheidegger, Adrian E. (2004). *Morphotectonics*. Berlin: Springer. ISBN 3-540-20017-7.

Selby, Michael John (1985). *Earth's changing surface: an introduction to geomorphology*. Oxford: Clarendon Press. ISBN 0-19-823252-7.

9

DEVELOPMENT OF LANDSCAPES-II

Glacial Landscape

Glacial Landscape means the landform formed by the glacier. A glacier is a tongue of ice, sliding from the mountain with gravity. On the back of a glacier there are large snow fields. The thickness of a glacier varies from a few feet to a few hundreds of metres. Its width also varies from tens of feet to tens of miles. The length of the glacier also varies from areas to areas from a few hundreds feet to thousands of miles.

In areas where the winter snowfall exceeds the amount of snow that melts away during the summer, stretches of perennial snow known as snowfields, cover the landscape. At the lower limit of a snowfield lies the snow line. The exact position of snow line varies from one climatic region to another.

Classification of Glaciers

There are four main types of glaciers:

I. Valley Glaciers

These are like streams of ice that slide down the valley in mountainous area. They vary in width, depth and length. Valley Glaciers are nourished on the flanks of high mountains and flow down the mountain sides and sometimes are also called mountain glacier or Alpine glaciers. Cirque is a common feature. They are formed as snow falls from high slopes in amphitheatre like hollows. As the ice flows down the course of the valley in some places passing over uneven slopes the smooth surface of the ice may break up forming crevices. As the glacier moves down the valley, it may be joined by tributary valley glaciers. Where smaller glacier tributaries fall in a main glacial valley a hanging valley is formed. According to the shape adopted by the valley glaciers, they may be divided into cliff or cornice and cascade glaciers. The cliff glaciers are found on the face of the cliff. Cascade glaciers are those that end on the face of a cliff.

According to expanse and form there are six categories of valley glaciers:

Expanded foot type of glaciers are characterized by overflow of the valley. With overflow a separate ice apron is produced. Toku glacier of Alaska is typical example.

Dendritic type of glaciers are formed by the joining of tributary glaciers in the pattern of a tree. Radiating type of pattern develop where smaller glaciers radiate from the main glacier but converge at a point below so that the whole phenomena appears joining sticks arranged in Lady's Fan.

Horse Shoe type of glaciers are smaller glaciers whose front is curved on the inside. These are remnant of some bigger glaciers and go on sticking to the cirque wall. Inherited basin glaciers are those where glacier glow begin in the valley already formed by the other agency. Spitzbergen glacier is that in which the whole valley is filled with such overflowing ice that nothing appears in the valley except the slightly protruding divides.

II. Piedmont Glaciers

These glaciers are formed at the foothills when two or more glaciers emerge from the valleys and coalesce to form an apron of moving ice on the plain below.

III. Ice Sheet

When a glacier spread over small plain area. These are broad mound like masses of glacial ice that tend to spread radially

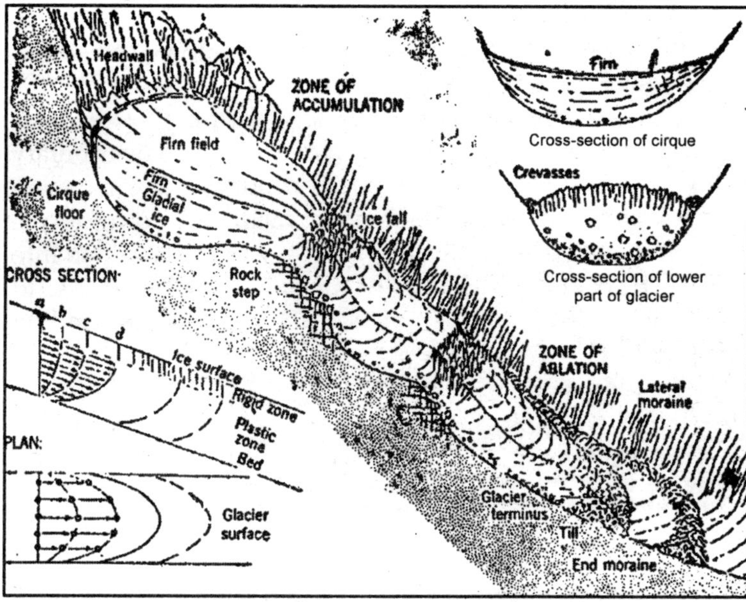

Fig. 9.1: Structure and flowage of a simple alpine glacier.

under their on weight. Vatana glacier of Iceland is a typical example. A localized sheet of sort is called Ice cap.

IV. Continental Glacier

Enormous, thick sheets of glacial ice cover nearly whole of Greenland and the continent of Antarctica. These glaciers are not confined to valleys, but cover all the land surface. About 80% of the area of Greenland is covered with continental glacier. On Greenland ice exceeds 3000 metres in thickness near the centre of the Ice cap.

Glacial Landforms

These are the landforms formed by the erosional processes of the glacier. The erosional effects of glaciers are not limited to the fine polish and striations, however, glaciers on grand scale produce spectacularly sculptured peaks and valleys in the mountainous areas of the world.

Fig. 9.2

Features developed by the erosional work of a glacier are as follows:

(*A*) *Cirques*: Increase in temperature melts some of the snow. The melt-water seeps down to the bedrock and trickles along the margin of the snow bank. Some of the water works its way into cracks in the bedrock, where it freezes again producing pressures that loosen and pry out fragments of rock. These fragments are moved off by solifluction, forming a shallow basin. As the basin gradually grows deeper a Cirque eventually develops. Continued accumulation of snow leads to the formation of firm, if the basin becomes deep enough the firm is transformed into ice. Finally, the ice begins to flow out of the Cirque into a valley below and a small glacier is born (Figs. 9.1—9.3).

The actual mechanism by which a Cirque is enlarged is still a matter of dispute. Cirque is a focal point for the glacier's nourishment. The lake that forms in the bedrock basin of the Cirque is called a Tarn.

(*B*) *Bergschrund*: A firm field is separated from the steep ice walls of surrounding peaks by a gaping crack which can be seen round the head of the basin. ln other words, it is a slender snow bridge.

A. Before glaciation sets in the region has smoothly rounded divides and narrow V-shaped stream valleys

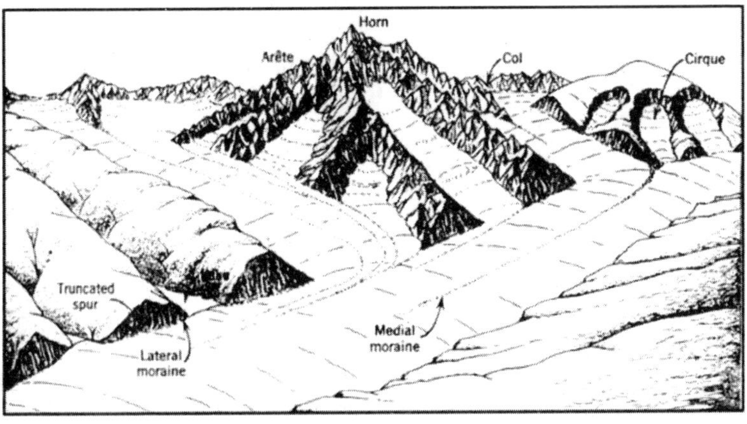

B. After glaciation has been in progress for thousands of years new erosional forms are developed : Landforms produced by Alpine glaciers. (by W. Arrow M. Dews and A.K. Loback)

Fig. 9.3

(*C*) *Horns* : A horn is a spire of rock formed by headward erosion by a number of cirques around a single high mountain. When the glaciers finally disappear, they leave a steep, pyramidal mountain outlined by head walls of the Cirque.

(*D*) *Arêtes* : It is a French word, meaning "Fish borne" ridge or sharp edge. It is formed when a number of cirques grow

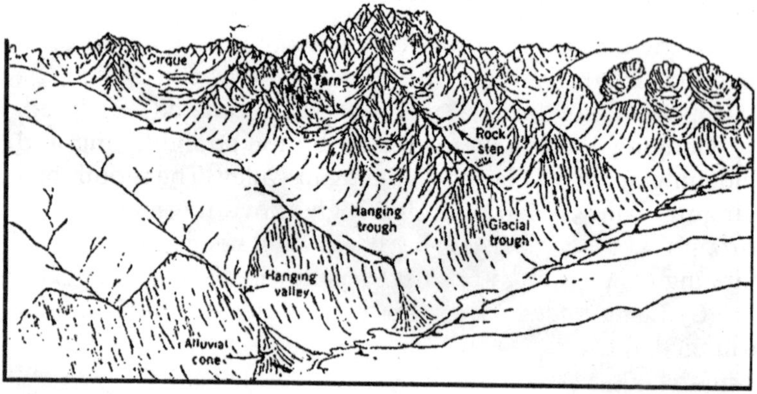

Fig. 9.4

into a ridge from opposite side. The ridge become knife-edged rugged and serrated.

The Swiss Alps appear from the air as a sea of sharp arêtes and tooth like horns. In the foreground is a circue (Swissair Photo). (Fig. 9.4)

(*E*) *Col* (*From the Latin word Collum mean 'Neck'*): A Col or pass is formed when two cirques erode headwards into a ridge from opposite sides. When their headwalls meet they cut a sharp edged gap in ridge (Figs. 9.1—9.3).

(*F*) *Glaciated Valleys*: Instead of fashioning their own valleys, glaciers probably follow the course of the pre-existing valleys, modifying them in a variety of ways. Usually, the resulting valleys have a broad U-shaped profile, because the tongue of an advancing glacier is relatively broad. It tends to broaden and deepen the V-shaped stream valley, transforming them in the broad U-shaped trough. Glacier also straightens the valley formed by the stream.

In this process of straightening the ice units of the masses of the spurs form truncated spurs (Figs. 9.1—9.3).

A glacier also gives a mountain valley a characteristic longitudinal profile from the cirque downwards. The course of a glaciated valley is marked by a series of rock basins, probably formed by plucking in areas where the bedrock was shattered or closely jointed. Between the basins are relatively flat stretches of rock that were more resistant to plucking. As the time passes, the rock basin may fill up with water, producing a chain of lakes that are sometimes referred to as paternoster lakes as resembling a string of beads. With the disappearance of the ice a system of glacial troughs is exposed.

Hanging Valleys were created by the straightening and widening action of glacier on the main valley. The mouth of a hanging valley is left stranded high above the main valley through which a glacier has passed. As a result stream from hanging valley plunges into main valley in a series of falls.

(G) *Comb-ridges*: When the rock arêtes are horizontally arranged, they are known as comb-ridges. These characteristic forms belong to the upper slopes above the level of the glacier's surface. There the erosion does not proceed by plucking and abrasion. Indeed the erosion at these high levels is accomplished by sapping and undermining.

(H) *Roche Mountain*: In many places glacial erosion of bedrock produces small rounded asymmetric hills with gentle, striated and polished slopes on one side and steeper slopes lacking polish and striations on the opposite side. The new gentle slope faced the advancing glacier and was eroded by abrasion. The opposite slope was steepened by the plucking when a glacier flows across a fractured and jointed stretch of bedrock. It may lit up large, blocks of stones and move them off. This process is known as plucking action of ice. The resulting form is known as Roche Mountain resembling a sheep skin.

(I) Crag and Tail: It is formed due to some hard obstructive-mass of rock, the crag, which lies in the path of oncoming ice, This mass protects softer rocks in its Lee from glacial erosion, for the ice comes to have moved over and ground the 'crag' leaving a gently sloping Tail in its Lee.

(J) Ice Ways: These landforms are the result of the erosion of ice sheets. In some areas ice ways in the form of linear hollows were formed by the motion of lobes from a low-land ice sheet.

Depositional Landforms

The debris carried by a glacier is deposited either because the ice that holds it melts or commonly because moving ice plasters the debris across the land surface. The general term 'Drift' is applied to all deposits that are laid down directly by a glacier or as a result of glacial activity.

Drift can be divided into two general categories : unstratified and stratified deposits:

I. Unstratified Deposits

These are the deposits which are laid down directly by a glacial ice. The following are the main unstratified deposits:

(*a*) *Till*: It consists of an unstratified mass of material with a mixture of clay and sand, containing rock fragments. Many of the large pieces are striated, polished and faceted as a result of the wear they underwent during transportation by the glaciers. Some of the material picked up along the way was smeared across the landscape during the glacier progress, but most of it was dropped when the rate of wastage began to exceed, the rate of nourishment and the glacier gradually melts away. On the basis of material Till can be divided into two types, e.g. clay tills are those tills where the clay size particles predominate with only a scattering of larger units. Stony or boulder Till is composed of large rock fragments and boulders. According to its location tills may also be divided into two main types: Lodgement Till, deposited beneath the active ice and ablation Till, left *in situ* as the ice decays and melts.

(*b*) *Erratic*: There are boulders which have been transported from its source by a glacier, sometimes over a considerable distance and has been left stranded when ice is melted.

(*c*) *Boulder Train and Valley Trains*: They consist of series of erratic usually worn from the same bed-rock source, carried forward by moving ice and deposited. Either in fan pattern, with the apex pointing to the origin of material or in more or less straight line.

Some drumlins contain a rock core, around which the till has been plastered. If the core is very thin a feature so formed will be known as rock false drumlin.

II. Stratified Deposits

In stratified drift ice transported sediment that has been washed and sorted by glacial melt waters. Because water is much more selective sorting agent than ice, deposits of stratified drift are laid down in recognizable layers unlike the random arrangements of particles typical of till. The following are the main stratified deposits:

(*a*) *Outwash Deposits*: The sand, gravel and clay carried outward by melt water froth the front of a glacier are called outwash. If material is deposited over an extensive lowland from a broad ice margin, the feature so formed is termed as outwash plain. If the zone of wastage happens to be located in valley, the outwash deposits are confined to the lower valley and form a valley train.

(*b*) *Kettles*: Sometimes, a block of stagnant ice becomes isolated from the receding glacier during wastage and is partially or completely buried in Till or Outwash, before it finally melts. When it disappears, it leaves a Kettle, a pit or depression in drift. These depressions range from a few metres to several kilometres in diameter and from a few metres to over tens of metres in depth.

(*c*) *Kame* : Kames are steep sided mound of glacial deposits. Unlike drumlins, Kames are of random shape and the deposits that composed them are stratified.

A kame terrace is a deposit of stratified sand and gravel that has been laid down between a wasting glacier and an adjacent valley wall. When glacier disappears, the deposit stands as a terrace along the side of the valley.

(*d*) *Eskers*: An esker is an elongated ridge of stratified deposits of sand and gravel, sometimes straight, sometimes sinuous. They range from a fraction of a kilometre to hundreds of kilometres in length, but they are only a few metres wide.

Most of the investigators believe that eskers were formed by the deposits of the streams running through tunnels beneath stagnant ice. When the body of the glacier finally disappeared, the old stream deposits were left standing like a ridge.

(*e*) *Crevasse Filling*: They were formed by the filling of crenice in stagnant ice. Their shape resembles with eskers.

(*f*) *Varves*: A varve is a pair of thin sedimentary beds; one of coarse and one of fine sediments. During the summer water from a melting glacier carries large amounts of clay, fine sand and silt out into lakes. The coarser particles sink fairly rapidly and blanket the lake floor with a thin layer of silt and silty sand. But as long as the lake is unfrozen, the wind creates currents strong enough to keep the clay particles in suspension. When the lake freezes over in the winter these wind generated currents cease, and the fine particles sink through the quite water to the bottom, covering the somewhat coarse layer. A varve is usually a few millimetres to one cm. thick but in rare cases thickness of 30 cm. or more are also known.

Moraines

Moraines are the glacial deposits comprising of the sediments which are carried by the glacier either along the sides or along its bed, between the body of the two glaciers or at the front of the glacier. These materials when deposited by the melting of the glacier take different shapes. The moraines are classified on the basis of their location as under:

(i) Lateral moraines.
(ii) Medial moraines.
(iii) Terminal moraines or end moraines.
(iv) Stadial moraines or recessional moraines.
(v) Bed moraines.

Lateral Moraines

Glacier carries some sediment along the sides grinding between the glacial ice and valley wall. When the glacier melts this material is deposited along the side of the valley in a line of boulders and other material. These are known as lateral moraines.

Medial Moraines

When two glaciers meet their inner lateral moraines also meet with each other. These glaciers move side by side the material between them is also moving. When the glaciers melt the intervening material is deposited in the form of medial moraines.

Terminal Moraines

A terminal or end moraine is a ridge of till that mark utmost limit of glacier advance. These ridges vary in size from ramparts scores of metres high to very low, interrupted walls of debris. These moraines are formed when the glacier reaches the critical point of equilibrium.

Recessional Moraine

Behind the terminal moraine, and at varying distances from it, a series of smaller ridges known as recessional moraines, are build up. These ridges mark the position where the glacier front was stabilized temporarily during the retreat of the glacier. Not all the rock debris carried by a glacier finds its way to the terminal and recessional moraines. These are end moraines at different stages.

Bed Moraines

Some of the sediments are moved along the bottom of the glacier. This material is grinding against the bed of the glacier. When the glacier melts the debris lay in the form of a rolling boulder plain and called bed moraine.

Drumlins

Drumlins are smooth, elongated hills composed largely of till. The ideal drumlin shape has asymmetric profile, with a bluntness pointing in the direction in which the vanished glacier advanced, and with a gentler, long slope pointing in the opposite direction. In most areas drumlins occur in clusters, or drumlin fields.

Causes of Glaciation

For glaciation to happen in the first place, we must have sizable land masses near the poles and those land masses must have high elevations.

The land masses get there via plate tectonics.

Once the land masses are situated correctly for glaciers to form, the average temperature must decrease so they do form. This is where the orbit of the earth and its variations take over.

In 1930, Milutin Milankovitch proposed that variations in three parameters of the earth's orbit caused glacial fluctuations:

1. *Orbital eccentricity*—The orbit of the earth around the sun is not a circle, but is elliptical and also varies. This eccentricity is a minor cause for seasons.
2. *Tilt variations in the axis of rotation (obliquity)*—The tilt of the earth's rotational axis varies with time. A tilted axis is the primary cause of seasons.
3. *Precession*—The earth's axis of rotation wobbles which results in minor fluctuations in the amount of solar radiation we receive.

These three factors account for only some of the reasons why the earth receives varying amounts of solar radiation. Other factors are dependent on solar-system and galaxy scale events.

ORIGIN OF ICE AGE THEORY

The idea that in the past glaciers had been far more extensive was folk knowledge in some alpine regions of Europe. Imbrie and Imbrie (1979) quote a woodcutter by the name of Jean-Pierre Perraudin telling *Jean de Charpentier* of the former extent of the *Grimsel* glacier in the *Swiss Alps*. Macdougall (2004) claims the person was a Swiss engineer named *Ignaz Venetz,* but no single person invented the idea. Between 1825 and 1833, Charpentier assembled evidence in support of the concept. In 1836, Charpentier, Venetz and *Karl Friedrich Schimper* convinced *Louis Agassiz,* and Agassiz published the hypothesis in his book *Étude sur les glaciers* (Study on Glaciers) of 1840. According to Macdougall (2004), Charpentier and Venetz disapproved of the ideas of Agassiz who extended their work claiming that most continents were once covered by ice.

At this early stage of knowledge, what was being studied were the glacial periods within the past few hundred thousand years, during the current ice age. The existence of ancient ice ages was as yet unsuspected.

Evidence of Ice Ages

There are three main types of evidence for ice ages: geological, chemical, and paleontological.

Geological evidence for ice ages comes in various forms, including rock scouring and scratching, *glacial moraines, drumlins,* valley cutting, and the deposition of *till* or tillites and *glacial erratics*. Successive glaciations tend to distort and erase the geological evidence, making it difficult to interpret. It took some time for the current theory to be worked out.

The **chemical** evidence mainly consists of variations in the ratios of *isotopes* in fossils present in sediments and sedimentary rocks, ocean sediment cores, and for the most recent glacial periods, *ice cores*. Because water containing heavier isotopes has a higher *heat of evaporation,* its proportion decreases with colder conditions. This allows a temperature record to be constructed. However, this evidence can be confounded by other factors recorded by isotope ratios; for example, a *mass extinction* increases the proportion of lighter isotopes in sediments and ice because biological processes preferentially use lighter isotopes, so a reduction in land or ocean *biomass* results in a sudden bio-induced displacement of isotope equilibrium towards larger proportions of lighter isotopes available for deposition.

The **paleontological** evidence consists of changes in the geographical distribution of fossils. During a glacial period cold-adapted organisms spread into lower latitudes, and organisms that prefer warmer conditions become extinct or are squeezed into lower latitudes. This evidence is also difficult to interpret because it requires: (1) sequences of sediments covering a long period of time, over a wide range of latitudes and which are easily correlated; (2) ancient organisms which survive for several million years without change and whose temperature preferences are easily diagnosed; and (3) the finding of the relevant fossils, which requires a lot of luck.

Despite the difficulties, analyses of ice core and ocean sediment cores has shown periods of glacials and interglacials over the past few million years. These also confirm the linkage between ice ages and continental crust phenomena such as glacial moraines, drumlins, and glacial erratics. Hence the continental crust phenomena are accepted as good evidence of earlier ice ages when they are found in layers created much earlier than the time range for which ice cores and ocean sediment cores are available.

10

COASTAL LANDSCAPES

Coast is a zone of contact between land and sea and the zone between the low tide mark and the highest wave mark is called shore. The line that separates land and water is called the shoreline. Shorelines are formed in different ways. They can be divided into four principal types:

(i) Shorelines of submergence
(ii) Shorelines of emergence
(iii) Neutral shorelines
(iv) Compound shorelines.

Beaches are deposits of loose material within the reach of waves along the shore. Coastline itself is demarcated either by the cliff line or by the line reached by storm waves. The shore is divided into two zones: The fore shore extends from the lowest water line to the average high water line, while the back shore extends from the high water line to the coast line.

Classification of Coasts

Prof. Francis Shepard classified coasts into two systems, e.g. primary coasts and secondary coasts. Coasts that have been modified only slightly by marine processes are called primary coasts; and the coast that have undergone significant modification by marine erosion, deposition or growth of marine organisms, are called secondary coasts.

Primary coasts are further classified into following types:

(*a*) *Ria Coasts*: Named for the Spanish word for River 'Ria' coasts are characterized by bay or estuaries formed when the ocean rose and 'downed' the valley of the river flowing into the ocean. A good example of this type of coast which frequently has a shoreline resembling the outline of an oak leaf is Chesapeake Bay.

(*b*) *Fjord Coasts*: Named for Norwegian embayment, fjord coasts result from glacial erosion of river valleys. Western coast of Canada is good example of fjord coast.

(*c*) *Volcanic Coasts*: Most of the Islands in the deep ocean are formed of the volcanoes that rise from the ocean floor. The slope of the ocean floor just off volcanic coasts is usually similar to the slope of Lava flows that flow into the ocean at the edge of the island. Most of the coasts of the islands of Japan are volcanic.

(*d*) *Deltaic Coasts*: In some areas sediments carried in large volumes by a river get deposited at the shore resulting in the formation of a delta. A classic example of such a coast is the Deltaic coast of the Mississippi river.

(*e*) *Glacial Deposition Coasts*: In high latitudes continental glaciers once extended beyond the present shoreline up to the sea shore. When the glaciers melted they deposited the sediments near the coastline. Such shoreline is commonly the product of deposits laid down by the glaciers as they melted. Long Island is a glacial moraine, a depositional ridge, laid down by the retreating glacier. It was subsequently surrounded by the ocean which advanced as the glacier retreated. Boston harbour is best example of such a coast.

(*f*) *Landslide Coast*: Along the California and Oregon coast one finds irregular shaped lobes of landslide material that have been carried into ocean after the collapse of coastal rocks. At the shoreline, boulders eroded from the slide material usually extend out into the ocean.

Diastrophic Coasts

Where folding and faulting of coastal rocks is occurring diastrophic coasts will be formed. The Gulf of California is best example of this type of coast.

Secondary coasts are also subdivided into two types:

(*i*) *Wave Erosion Coasts*: When high energy waves erode shoreline created by the submergence of coastal features sea cliffs are created. The shape of the cliff is controlled by the hardness of the rock.

(*ii*) *Marine Deposition Coasts*: These are those coasts which are probably formed by the marine depositions. The dominant features of marine coasts are Barrier Islands. Coast along gulf of Mexico is a good example.

Coast Built Forward by Organism

The modification of coastal areas by the growth of organisms is achieved primarily in intertropical latitudes by the development of mangrove trees and coral reefs.

Erosional Landforms of the Coastal Landscape

The waves operate agents of erosion in several ways. The hydraulic action of a mass of water has direct shattering effect on the rocks. Due to refraction, the wave energy is concentrated on the headlands that protrude from the continent, while the amount of the energy reaching the shore in bays is reduced. As the waves concentrate their energy on the headlands, erosion occurs and the shoreline retreats. During the rare periods when storm waves may batter the shore more erosion may occur across the entire shore in one day than may be achieved by average wave conditions in a year.

The undercutting may be evident in the form of a notch at the base of the cliff that may be characterized as sea cave. Caves may develop into openings running through the headlands called sea arches. Continued erosion and crumbling of arches

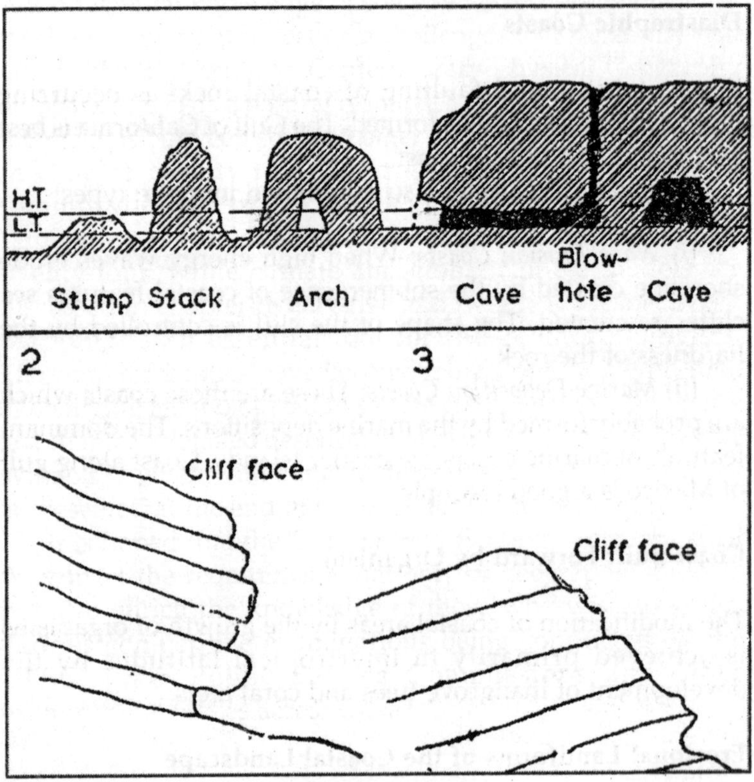

Fig. 10.1: Coastlines,

will produce stacks (Fig. 10.1). Their remnants rise from the relatively smooth wave cut benches cut from the bedrock by wave erosion.

The rate of wave erosion will be determined by a number of factors such as :

(1) *Degree of exposure of the coastal region to the open ocean*: Coasts that are fully exposed receive higher energy wave action and are likely to have rugged cliffs.

(2) *Tidal Range*: Given the same amount of wave energy, a region that possesses a small tidal range will erode much more rapidly than one with a large tidal range that allows the wave energy to be spread over a much broader shore.

(3) *Composition of Coastal bedrock*: Crystalline igneous rocks, such as granite, are relatively resistant and generally produce rugged shoreline topography. Sedimentary rocks, such as sandstone and shale are more easily eroded, and the topography is associated with more extensive beach deposits.

Depositional Landforms of Coastal Landscape

Waves, currents, wind and tidal movements are main agents of transportation of material along the shore. The material found in a beach deposit will depend on the source of the sediments which are transported by the long shore drift. In areas where the sediment is provided by the coastal mountain ridges the beaches will be composed of quartz and the feldspar fragments with smaller amount of other common rock forming minerals. It has relatively coarse texture. If the sediments are provided primarily by the rivers that drain lowland areas, sediments that reach the coastal region will normally be of fine texture. In many cases mud flats develop along the shore because only clay and silt sized particles are emptied in the ocean.

The relationship of particle size to beach slope is shown by Wentworth: Cobble at 24°, Pebble at 17°, Granule at 11°, Very coarse sand at 9°, Coarse sand at 7°, Medium sand at 5°, fine sand at 3° and very fine sand at 1° slope.

Types

(1) *Spit*: A spit is a linear ridge of sediments attached at one end of the land. The other end of the deposit point in the direction of long shore drifts and ends in open water. Spits are simply extension of beaches into deeper water near the mouth of bay. The open water end of spit will normally curve into the bay as a result of the action of the current (Fig. 10.1).

(2) *Tom bolo*: A tom bolo is a sand ridge that connects an island with another island or mainland. Such a feature probably results from the growth of a spit.

(3) *Bay Barrier*: With the action of currents and waves spit may eventually extend across the bay and tie to main land, completely separating the bay from the open oceans. The spit then becomes a bay barrier.

(4) *Barrier Islands*: There are long offshore deposits of sand lying parallel to the coast. The origin of a barrier island is complex, and it may be that there are numerous explanations for their existence. However, it appears that many such structures have developed by raising the sea level during ice age 18000 years ago.

(5) *Deltas*: Some rivers carry more sediment than can be distributed by the long shore current. Such rivers develop delta deposits as the sediment settles out at the river mouth. Mississippi formed largest such delta.

(6) *Offshore Bar*: An offshore bar lies roughly parallel to the shore line. The origin of such a bar is disputed. They are probably formed by the interaction of number of factors. They may be formed due to wave action, so that material drifted along the shore is deposited just inside the line where the incoming waves first break (hence the term break point bar) or they may be formed from the material combed directly down the beach by the backwash. Possibly too the scour of long shore currents helps. This results in the formation of a bar, behind which marshes, mud flats and lagoons may develop.

(7) *Dune Belts*: Extensive sandy beaches are nearly always backed by sand dunes because strong coastal wind can readily move some of the sand which dries out at low tide.

(8) *Mud Flats and Salt Marshes*: Coastal marshes are found either behind shingle bars and sand spit or in sheltered parts of estuaries and embayment. The basis of the formation of coastal mudflats and marshes is the deposition of fine silt by the tides in the sheltered area helped in some cases of alluvium brought down by rivers.

11

AEOLIAN LANDSCAPE

Aeolian landscape stands for landscapes which are developed by the action of wind in desert areas of the world. A third of the land surface of the globe experiences desert or semi-desert conditions not including the polar and sub-polar lands which are sometimes called cold desert. Hot deserts are characterized by the absence of cold season with mean monthly temperature always remaining above 6°C. A desert is characterized by a lack of moisture. The basic fact responsible for the existence of a desert is aridity. Much more extensive topographic deserts are the sub-tropical deserts lying in zones that range 5° to 30° north and south of equator.

At the earth's surface the descending air moves laterally away from the high pressure zones. Pole-ward, the winds become prevailing westerlies in the northern hemisphere and the south-east trades in the southern hemisphere. It is the continuous circulation of the air that creates the great deserts on the either side of the equator. These include the Sahara desert of North Africa, the Arabian desert of Middle East, the Victoria desert of Australia, the Kalahari desert of south-west Africa and Botswana, the Sonora desert of north-western Mexico, southern Arizona and California, the Atacama desert of Peru and Chile and the desert of Afghanistan, Baluchistan and Thar desert of India and Pakistan.

In all these deserts Aeolian topography is created by the erosional and depositional processes of wind. Wind velocity increases rapidly with height above the ground. The general movement of wind is forward, across the surface of the land

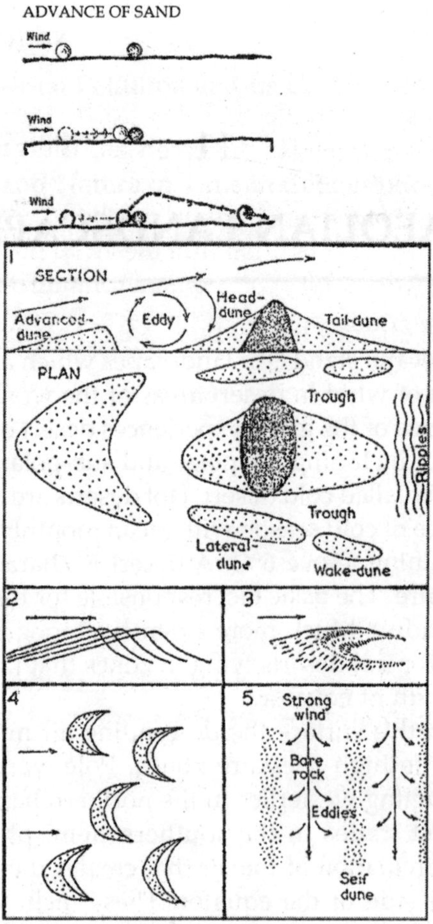

Fig. 11.1: Sanddunes.

but within the general movement the air is moving upward and downward and from side to side.

Material blown by the wind usually falls into two groups based on size of sand and dust. The average diameter of sand grain is 0.15 to 0.30 mm, with some grains as fine as 0.06 mm. All particles smaller than 0.06 mm are classified as dust; Sand grains move forward in a series of jumps, in a process known as saltation. Sand particles are thrown into the air only under the impact of other particles (Fig. 11.1). When the wind reaches a critical velocity, grains of sand begin to collide with each other. The impact may lift either the second particle or first into air. Once in the air the sand grain is subjected to two forces. First, gravity tends to pull it down to earth again, and eventually it succeeds. But even as, the grain falls, the horizontal velocity of the wind drives it forward.

The course of the sand grain is parabolic between the point from where it was first lifted into the air to the point where it finally hits the ground.

When the grain strikes the surface, it may either bounce off a large particle and be driven forward once again by the wind or bury itself in the loose sand perhaps throwing another grain into the air by its impact.

Erosion

Erosion by the wind involves three processes:

- (i) Deflation means to blow away the material by the blowing of wind/movement of wind.
- (ii) Abrasion leading to corrosion and the smoothing and polishing of rock by the particles of sand which are carried by wind.
- (iii) Attrition production of rounded sand by the movement of wind.

Deflation: The word 'deflation' is derived from the Latin for "To blow away". It is an erosive process of wind carrying off unconsolidated sediments.

The process creates several recognizable features in landscape. For example, it often scoops out basins in soft unconsolidated deposits ranging from a few metres to several kilometres in diameter. The basins are known as Blow outs for obvious reasons. Deflation is a sand attack on rock surface detaching. Small fragments especially from sandstone that weather to form loose grains. Deflation removes only the sand and dust particles from the deposit. It leaves behind the larger particles of pebble or cobble size. As the time passes, these

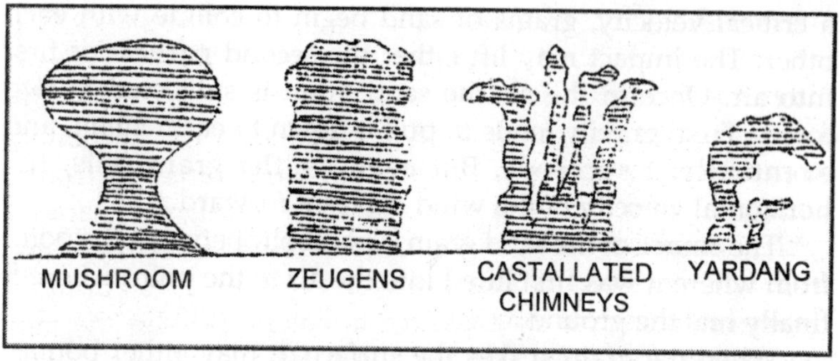

Fig. 11.2: Wind abrasion.

stones form a surfacial cover. When the pebbles become closely fitted with each other desert pavement is formed (fig. 11.1).

Abrasion: The particles carried by the wind are highly effective abrasive agent in eroding rock surface. The sand blast effect of wind carrying a load of hard quartz grains has great potency. It can smooth and polish a surface of 'Homogeneous Rock', where horizontal beds of uniform hardness occur. The most common products of abrasion are certain pebbles, cobbles and even boulders that have been eroded in a particular way. These pieces of rock are called vent facts. A particular type of vent facts with three wind faceted surfaces is called dreikanter.

The surface of the pebbles may be so wind polished as to form a desert pavement or desert mosaic. Wind abrasion produces pits, grooves and hollows in the rock. The sandblast action of wind against exposed rock surfaces is limited to the basal few feet of a cliff, hill or other rock mass rising above a relatively flat plain, because sand grains do not rise high into the air. Where a small rock mass projects above the plain it may cut away at the base to make a pedestal rock delicately balanced upon a thin stem. Most pedestal rocks or mushroom rocks are, however, produced by weathering processes. (Fig. 11.2)

The characteristic mushroom blocks undercut at the base are known as gour (Singular gara) in Sahara desert. Where tabular mass of more resistant rock resting on undercut pillars after removing the softer material are called zeugen. They are very often elongated in the direction of the prevailing wind.

These zeugens sometimes stand up prominently 30 metres (100 ft.) or more above the surrounding level. When rocks of differing resistance occur in bands roughly parallel, to the direction of the prevailing wind a ridge and furrow relief may be produced. Well worn ridges with blunt rounded fronts, facing the wind and with sharp keel like crests are separated by shallow grooves or furrows. In some deserts of Central Asia, the ridges have been borne into rock ribs known as yardang. These yardangs may be 6 metres in height. Besides the rain water tends to sharpen the yardangs so that they begin to look like pointed needles or a series of castellated chimneys. (Fig. 11.2)

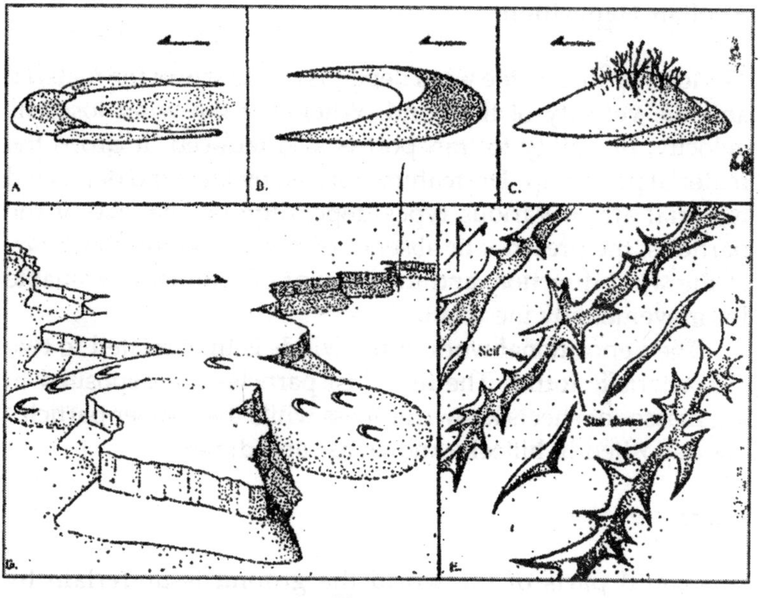

Fig. 11.3

Major dune forms. All major Aeolian landforms have elements of two basic classes: longitudinal dunes and barchans. In Fig. 11.3A, this *parabolic dune* may have formed because the end (*horns*) became anchored by some obstacle or by a rise of the water table; the perennially dry pan or playa basin on the lee side is a common associate of parabolic dunes and often is an expression of the seasonal water-table level. Some parabolic forms may be former barchans that inverted as the horns stabilized and the center "blew through".

11.3B. A typical barchan. 11.3C. A *coppice dune,* or *shrub coppice dune* if vegetated. 11.3D. *Climbing dunes* (left) and *falling dunes* (right). A few barchans as may re-form when sand streams across the uplands, as shown here. 11.3E. A series of longitudinal dunes (seifs) depicted as sinuous, parallel ridges; crossing wind patterns may be in barchanoid and longitudinal elements and form *pyramidal* or *stellate dunes* where wind shifts direction either seasonally or annually. However, the position of the stellate dune remains relatively fixed, or static. Arrows indicate prevailing wind direction.

Attrition: Wind borne material is in a constant state of movement. Impacting both against the rock surface it meets and also one grain against another. This leads to produce characteristic rounded sand.

Aeolian Deposition

The load carried by the wind depends on the size of the material and the velocity of the wind. Whenever the wind loses its velocity, its ability to transport is also reduced. It drops the material picked up. The features formed by the wind deposited material are of various types depending on the size of the particles, the presence or absence of vegetation the constancy of the wind direction, and the amount of sediments available for movement by the wind.

The depositional work of the winds is thus related to two characteristic forms. The fine dust particles are deposited in big masses or sheets known as loess while the coarser particles are deposited in hillocks known as sand dunes.

Loess

In several parts of the world the ground is underlain by deposits of wind transported silt which has settled out from dust storms over thousands of years. The material thus formed is known as loess. To it the name loess has been given, from a village in Alaska where somewhat similar deposits occur loess has a tendency to break or cleave along vertical cliffs wherever it is exposed either by stream or by man. The cleavage is possibly produced by a slight shrinkage of entire mass.

Perhaps the greatest deposits of the loess are in China, where thickness of over 100 ft (30 metres) are common and maximum of 300 ft (90 metres) has been measured. It covers parts of Gobi desert of Central Asia and Mississippi valley in the United States. These deposits are also found in Argentina, Central Europe and New Zealand.

Sand Dunes

A dune is a hillock or accumulation of sand deposited by the wind. Dune may be active or live when there is no vegetation and constantly keeps changing from under wind currents (Figs. 11.3—11.4). They may be inactive or fixed dunes where area is covered by vegetation that has taken root. Vegetation prevents further shifting of the sand dune.

According to shape following are the main types of dunes:

Crescentic Dunes or Barchan Dunes

The name barchan (also spelt barkhan, or barchane) is derived from the deserts of Turkey. They are crescent shaped with the horn pointing in the direction of the wind.

These dunes retain their shape while they continue to move at 75-15 metres per year. They develop best where wind is constant and of moderate velocity. The underlying surface is smooth and the sand supply is adequate but not abundant, possibly patches of pebbles or a sudden wind fluction, may cause the accumulation of a low heap of sand from which the features of the characteristic barchan can develop. The barchan vary in height up to more than 100 ft. (Fig. 11.3)

Climbing and Falling Dunes

In some areas barchan dunes are being blown by deflation from one basin to another across low passes in mountain ranges. Though the barchan form is destroyed during this inter-basin transfer of sand but the barchan reforms as the sand reaches the flatter floor of the receiving basin. They are called climbing dunes where they rise up hill and falling dunes where they cascade down the opposite side of the range.

Transverse Dunes

They are formed in sand abundance areas, where sand covers the entire ground the wave like ridges of sand separated by furrow like depressions are called transverse sand dunes. It is called transverse dune because the crests tend at right angles to the direction of wind. In the Sahara desert some transverse dunes measure over 100 km in length and 1 to 3 km in width. The entire area may be called sand sea. A large expanse of dunes is called 'Erg' in the Sahara desert. Individual ridges have sharp crests and are asymmetrical, the gentle slope being on the windward side and the steep slope on the lee side.

Longitudinal Dunes

Longitudinal dunes are long ridges of sand running in the general direction of wind movement. The smaller types are about 3 metres high and about 60 metres long. In Arabian deserts, however, they commonly reach a height of 100 metres and may extend for 100 km across the country. There they are known as Seif-dunes from the Arabic word, for "Sword". In some areas, the longitudinal dune is produced by extreme development of hairpin dune, such that the parallel side ridges become the dominant form.

(*a*) *Sand Drift*: This belongs to the family of longitudinal dunes. It is longitudinally oriented with respect to wind. It is a long tapering, sharp-crested ridge of sand extending downward from some topographic obstacle, such as hill that might rise above a desert plain.

(*b*) *Sword Dune*: It is a huge tapering sand ridge in the shape of a sword. Its crest line rises and falls in alternate peaks and saddles. Sides of slope are indented by crescentic slip faces. This dune may be a few hundred feet high and tens of miles long (Fig. 11.3).

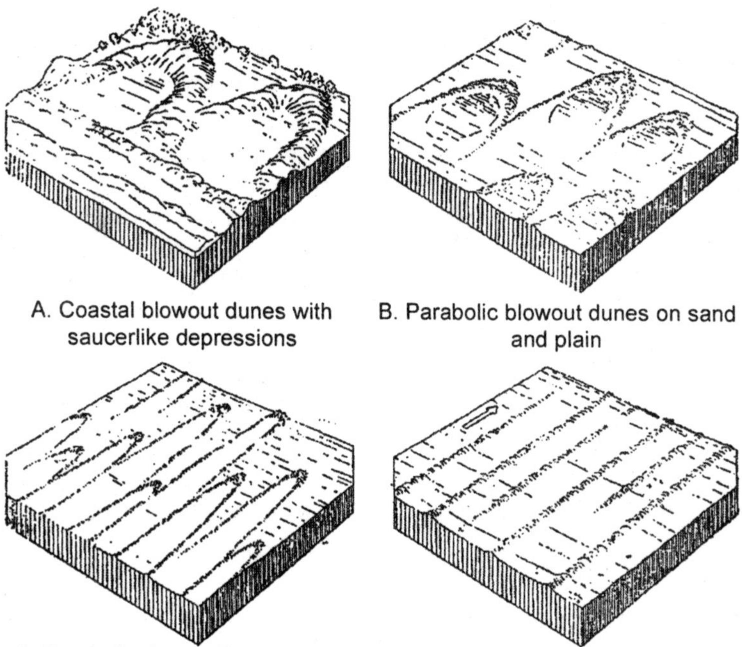

A. Coastal blowout dunes with saucerlike depressions

B. Parabolic blowout dunes on sand and plain

C. Parabolic dunes of hairpin form

D. Longitudinal dune ridges on a desert plain

Fig. 11.4

(*c*) *Star Dune*: Star dune is great hill of sand whose base resembles a many pointed star in plan.

(*d*) *Pyramidal Dune*: This dune resembles with a pyramid so it is called pyramidal dune.

Parabolic Dunes

In this dune the curve of the dune crest is bowed convexly downward. These may be called parabolic dunes.

(*a*) *Coastal Blowout Dunes*: A tentative of this class the coastal blowout dune is formed adjacent to beaches, where large supplies of sand are available and are blown landward by prevailing winds.

(*b*) *Parabolic Blowout Dunes*: These dunes develop to the lee of shallow deflation hallows. Sand is caught by low bushes and accumulates in a broad low ridge. The dunes have no steep slip faces and may remain relatively immobile (Fig. 11.4).

(*c*) *Hairpin Dune*: When the dune ridge migrates downwind, drawing the parabola into long narrow form with parallel side it is known as hairpin dune. Its form resembles with a hairpin. These dunes are stabilised by vegetative growth.

(*d*) *Whale Back Dunes*: Very large longitudinal dunes with flat tops on which barchans or seifs may occur. They are formed by the coalescence of several seifs.

Shrub Coppice Dune

These dunes are very common in desert and semi-arid region where there is a substantial cover of low bush or coarse grass especially along the playa and stream courses.

Work of Water in Desert

Although rainfall is extremely sparse in deserts there is still enough water to act as an important agent of erosion, transportation and deposition. In fact water is probably more effective than even the driving winds in moldings landscape in the desert.

Stream beds in most deserts, however, are dry over long periods of time and carry water only during floods. In some places, however, such broad dry plains slope down from the mountain ranges towards basins called playas, where surface run off collects from time to time. But the playa lakes formed in basins usually dry up in short time or at best exist as shallow, salty lakes.

Sometimes, sudden torrential rain occurs for a short period. This sudden rainfall causes flash floods. These floods carry on immense load of solid material the product of desert weathering. They sweep along so much material that they turn into mud flows and soon come to rest.

A number of dry stream course open into depression. Where a number of alluvial cones coalesce along the edge of the depression, a gently undulating sloping known as Bajada is formed.

In the desert inland drainage is formed where the mountain rims are high enough to receive appreciable precipitation. The landscape so formed in north America is called mountain and bolson desert. Both temporary lakes and hollows they occupy are known as playa, Salins or Salara in USA and Mexico and as 'shotts' in North Africa.

Wadis, ravines, nullahs, washes and arroyos are striking features of rock desert. Wadi is a valley with an intermittent stream, especially as developed in semi-desert areas.

Pediments are broad, gently sloping areas of bedrock, spread as aprons around the base of the mountain. The pediment is formed by running water that builds up an apron of alluvium below as it continues to cut back on erosional platform above. As such, it is an evolutionary product of long continued head-ward erosion of rock (such as sandstone volcanic tuff) that are easily eroded in the desert environment.

Another important feature of desert is an inselberg. The formation of inselberg is still disputed. Most of the scholars agree that inselbergs are the result of wind and water action. There is a marked similarity between the inselbergs and the monadrocks.

12

APPLIED GEOMORPHOLOGY

Applied Aspects of Geomorphology

"Applied Geomorphology has come up as a very important branch of earth sciences particularly with the advancement in its techniques to evaluate and tackle problems associated with man's occupancy of earth space." Of late, it has acquired enormous significance with the application of geomorphic understanding to the analysis and solution of myriad problems concerning land occupancy, resource utilization, environmental management and planning or problems of similar nature. In the beginning, the landforms studies were carried out mostly with aid of qualitative techniques but lately various quantitative techniques have been introduced alongside the use of aerial photographs and satellite imageries for identifying morphological characteristics of landforms and evaluation of terrain which have further enhanced the value of this branch of geography.

It is a common knowledge today that geomorphology tells us about the mountain ranges, degree of slope, the drainage texture, drainage patterns, basin shape and the material exposed in all available areas But it is not wise to limit the role of geomorphology to physical conditions only. The cultural value of geomorphology has been underestimated by most of the people. In fact geomorphology is useful to man in his daily life in a number of ways.

Geomorphology and Environment

The study of geomorphology leads to an increased interest in environment. It has been rightly said that for a better understanding of the environment of which we are a part, we must study the surface features of landforms and intelligently manage them. Thornbury has written that topography obviously plays an important role in determining the most feasible high route, although sometimes cultural features may necessitate routes not in best accord with topography. Soil maps to some degree represent topographical maps. The problem of soil erosion is essentially a problem involving recognition and proper control of such geomorphic processes as sheet erosion, gully erosion, mass wasting, etc. On the other hand, sand and gravel have been used in a number of engineering, commercial and industrial pursuits. For the location of a steel plant due importance is given to the nature of the rock structure underneath, whether it can sustain the load of a steel plant or not. It is a geomorphic study.

The geomorphology is more closely related to geography than to geology insofar as it studies surface forms rather than deposits. The location elevation and nature of landforms on the surface have their effects on settlement sites, water supply, agriculture, transport and communication and almost on all elements of environment. The relationship between the environment and geomorphology is very close. In fact, geomorphology and environment especially physical cannot be separated from each other. Even cultural environment is more often determined by the landforms. The elements of physical environment are the constituent elements of geomorphic changes.

Landforms of different orders which are considered as the cause of geomorphic studies are also the primary constituents of physical environment. The earthquakes and volcanoes have definite bearing on our environment. These sudden forces bring about lot of changes in the environmental conditions. Due to the earthquakes a series of folds, faults, rift valleys, etc., will be formed. Similarly, a volcanic eruption may change the atmospheric temperature of the surrounding areas. In the same way geomorphic agents such as rivers, glaciers, wind, waves, etc., can bring about tremendous changes in the environment.

A little more than 97 per cent of the total water of the earth is found in oceans. Every year 0.03 million km^3 of water evaporates. Most of it returns back to oceans. About 89 per cent returns to oceans by direct precipitation. Only 11 per cent is transferred on the land surface. When it moves in the form of rivers or glacial melt they bring physical changes, i.e. in the mountain by erosion and in the plains by deposition and floods. The interaction among the processes of erosion, sedimentation and compensating uplift can be seen as a part of environmental cycle. Wind has also brought about environmental changes in many areas by moving the desert, example can be given of Loess deposits of north China. An area located in the heart of high mountain ranges is having closed ecosystem. Its environment is different and by nature the people remain orthodox the degree of which increase with remoteness. The people of such areas are narrow-minded and are rigid in thinking.

Geomorphology and Agriculture

Land use is a term used for various functional and non-functional uses to which an accessible piece of land is put by man. It can be put under single or double cropping or it can be left fallow to rest for a few years. Various land uses are dependent on geomorphic conditions of the area in question. For example an area located at very high altitude cannot be used for cultivation twice in a year because of low temperature conditions. Whereas in plain area of the tropics, a piece of land can be cultivated more than once.

Agricultural activity is mostly rich in river valley where the alluvial soils have been deposited by the streams. Secondly, the agricultural activity in the initial stages will flourish only where there is an ample scope of irrigation water.

Knowledge of landforms and morphometry is essential for planning what is to be grown and where it is to be grown. For example, we cannot grow rice where the possibility of irrigation is zero. Secondly, steep slopes cannot be put to any agricultural use effectively as it may lead to soil erosion and landslides.

In the Himalayan region where the slopes are steeper fruit cultivation is preferred because continuous digging is not needed. The grass which grows is not to be disturbed as it protects the soil from erosion. Less sloping areas are put under the cultivation of maize and other dry crops. Wherever irrigation is possible terraces are prepared in gently sloping areas to cultivate rice. In highly mountainous areas where there is little scope of plain surfaces the river terraces, however, smaller in size, are the major tracts of agricultural activity, the agricultural activity diminishes away from these terraces. The degree of application of modern technology in agriculture is also determined by the geomorphic features. Irrigation is a pre-condition for the application of modern technology followed by tracterisation and use of other heavy machinery and chemical fertilizers and pesticides. In an area which is having high degree of ruggedness the degree of application of modern technology will be low.

In short fluvial morphology plays facilitative, permissive, restrictive and prohibitive relationship to land resources. To a large extent it determines the land use and cropping pattern. The large scale alluvial deposition, meandering, flooding etc. cause lot of destruction to cultivable lands, development of marshes and increasing the proportion of current fallow land. A number of examples can be given from Asiatic countries. Beginning at home large scale siltation by Ganga, Brahmaputra and destruction used by the Mekong and Hawang Ho are other examples.

In the fluvial depositional plains the process of alluviation takes place through a number of streams and their tributaries and hence they have various forms of texture. Due to the behaviours of drainage courses of number of streams involved in the process soil texture is always changing in nature as the loam changes slowly into silt loam, silt clay loam and even clay loam. Royal Commission on Agriculture in India has written in its report that in any specific area the variations are usually associated with the depth of soil and lateral changes are not rapid but when subjected to the action of spilt flooding soil texture may vary within the confines of the land.

Fluvial morphology also influences the cropping pattern of the alluvial plains. The riverside levee uplands have different crops than the interfluvial lowlands behind the levees. The River side levee uplands are usually marked with rich and fertile soils the crops grown are such which need well drained soil. Example can be given of north Bihar plain. K.L. Das, writes "influence of alluvial morphology is clearly discernible here with rice bhadai grown on levee uplands." It can be concluded that morphological characteristics greatly influence the agriculture of a region. They influence the field pattern, possibilities of irrigation, size and distribution of holding, continuity of cultivation and above all, the most important aspect of agriculture, the crops, their nature, yield, etc.

Geomorphology and Settlements

R.K. Rai writes "Location of human settlements is an expression of the geographical factors favourable at a particular spot on the landscape."

Rural settlements are directly or indirectly influenced by the nature of terrain, soil, topography, degree of slope, availability of drinking water, accessibility to the region, and climate of the region. The earliest possible settlement have come into existence on a river bank where water for various purposes was easily available. Even in the modern times man's first preference for settlement is water. Secondly, due consideration is given to the flow of water streams with perennial flow of water is preferred over the seasoned streams. Sometimes, the settlements are located on hilltops for security purposes as is the case in most of the settlements in Rajasthan.

Such areas where the topography is highly dissected forming rills, gullies and bad land, have very little proportion of cultivable land available. The soils in such areas are relatively poor due to high degree of sheet and gully erosion. These areas are sparsely settled.

Settlements are also sparse in areas where accessibility is poor. Remote and mountainous areas where one has to cover the distances on foot. The settlements are very sparse. On the other hand, where accessibility is not a problem the settlements tend to be concentrated.

Settlements are never preferred along the steep slopes for fear of landslides or snow slides in humid areas. Another indirect negative factor along the steep slopes is the poor soil. Since incidence of soil erosion is higher in steep slopes the agricultural activity is poor which leads to poor settlements. Studies on the Himalayan regions e.g. "Himalayan Beas Basin" by S.L. Kayastha, "Himalayan Chenab Basin" by R.C. Bhagat and N.S. Raina, "Geomorphology of Rural Settlements in Meghalaya" by R.K. Rai, and many others reveal that the settlements are by and large concentrated in the valley bottom. The hilltops and slopes are either devoid of settlements or have sparse settlements. Secondly, most of the settlements are located along the streams or springs. Thirdly, the settlements are concentrated in and around the agricultural land which is controlled by a number of geomorphic factors already mentioned above.

Geomorphology and Transportation

When the development of transport facilities is planned due consideration is given to the study of geomorphic features of that area. The very first requirement of the surveyor is a topographic map or a contour map. The pilot survey is done where again importance is given to rock strata. Softer rock strata are generally avoided while surveying the transport lines in hilly areas. Many a times, the construction of roads in hilly tracts leads to the initiation of moving slips or landslides.

A tract which involves crossing of large number of streams is costlier as it needs at many places. It has been seen that lying of roads is cheaper and far easier in plain areas. It is a costly affair in plateau regions and costlier and difficult in mountainous areas.

Construction of roads is also avoided on steep slopes, wherever possible. Construction of roads is planned in such a way that a road or a railway line touches or links maximum number of settlements that is why the roads usually spread along the streams.

I agree with Kaul when he says that till plain is ideal for road construction but disagree when he says that a route over karsts plain necessitates cut and fill. In my opinion a karsts plain is to be avoided for laying a transport line whether it is a railway or a road. However, there is no denying the fact that geomorphology of a region imposes serious constraints on the construction of transport lines, e.g. railways and roadways.

Geomorphogy and Minerals

Exploitation of minerals is also determined to a greater extent, by the morphology of the area. A mineral can be exploited if its exposure is adequate, accessibility is better and cheaper and easy. The presence of minerals in very remote areas of a stream is not economical and hence cannot be exploited because their quality is poor and is not accessible easily. The safire mining in Paddar (J&K) has still remained in its crudest form because of its poor accessibility. Petroleum may be present in some of the anticlines of the lesser Himalayas or the greater Himalayas but since the accessibility is zero it is very difficult to assess and exploit.

From the Paddar mines of Doda district in Jammu and Kashmir crude safire used to be brought on ponies to Kistwar and transported to refining centres. This entire passage of difficult approach is a negative factor in the mineral exploitation. On the other hand in most of the faulted valleys of the Deccan Plateau the faulting has made the outcrop visible. Here geomorphology has made the job of exploitation easier.

To conclude the role of geomorphology in the daily life of man is of paramount importance, either directly or indirectly. Directly the geomorphic features of the surface determine the agricultural area, accessibility of a region, irrigation facilities through streams. Indirectly the surface configuration controls humans thinking. In case of closed ecosystem the people are more and more orthodox and are less susceptible to change. In fact geomorphology has its bearing on all the aspects of life.

Military Geomorphology

Military Geomorphology refers to the application of geomorphic concepts, principles, and technologies to military operation. Warfare causes rapid and widespread terrain alteration. Physical landscape modifications, intense vehicular movement, construction of obstacles and fortification and deliberate destruction are common consequences of war.

Munitions and vehicles manoeuvres can change the upper soil profile, destroy vegetation and change natural drainage pattern. Cave, ditches, trenches and bunkers remain on the physical landscape long after their military usefulness in past. Caves and bunkers built or modified for military purposes such as in Tora Bora region in Afghanistan and desert of Iraq remain an intricate part of contemporary landscape morphology. Traditionally, military geomorphology is viewed from a perspective of the power influences terrain morphology has on military operation. The study of Digital terrain models is the part and parcels of military geomorphology.

13

CLIMATOLOGY

Today, the study of climate is a fundamental point of human survival. Human society has developed to such an extent that its activities unintentionally produce weather and climate modification at large scale. To quote J.E. Hobbs: "To these can be added the role of man himself as a climatic factor of growing interest and concern, particularly with regard to his role as a cause of atmospheric pollution, perhaps, leading to climate change."

Weather and Climate

In climatology the terms 'weather' and 'climate' have different connotations. Weather refers to the condition of atmosphere at any given time indicating the short time variation of atmosphere in terms of temperature, pressure, wind, moisture, cloudiness, precipitation and variability. In simple words, weather can be defined as the sum total of its atmospheric condition, changing from hour to hour and day-to-day. Daily weather report is forecasted by the Meteorology department in different TV channels.

Climate, on the other hand, is the sum total of the variety of weather conditions of an area or place. Thus, climate may be defined simply as average weather. The term climate denotes a description of aggregate weather conditions. To define climate one has to consider the situation of all the elements of weather over a sufficiently longer period of time, e.g. 31 years or more.

Some of the definitions of climate given by eminent climatologists:

Critichfield: "The process of exchange of heat and moisture between the earth and atmosphere over a long period of time result in condition which we call climate. Climate is more than a statistical average; it is a aggregate of atmospheric conditions involving heat moisture and air movement. Extremes must always be considered in any climatic description in addition to means, trends, and probabilities."

C.W. Thornthwaite in the Presidential Address to the 57th annual meeting of the Association of American Geographers said that the climatic characteristic of any region of the world depends on the nature of exchange of momentum heat as well as moisture between earth's surface and atmosphere. Thus the climate of any place or locality represents the equilibrium between receipt and the expenditure of the radiant heat as well as moisture. Means it is heat and the moisture budget of specified area that explains its climate.

G.F. Taylor: "Climate is the integration of weather; weather is the differentiation of climate. The distinction between weather and climate is, therefore, mainly one of time."

Koeppe and De Long: "Climate is a summary, a composite of weather condition over a long period of time; truly portrayed, it includes details of variations—extreme frequencies, sequences—of the weather elements which occur from year to year, particularly in temperature and precipitation. Climate is the aggregate of the weather."

Trewartha: "Climate represents a composite of the day-to-day weather conditions of the atmospheric elements within a specified area over a long period of time. It is more than average weather for no adequate concept of climate possible without appreciation of seasonal and diurnal change and of the succession of weather episodes generated by mobile atmosphere disturbances. While in a study in climate emphasis may be given to the averages, still departures, variations, and extremes are also important."

Climatology: Definition

Word climatology is derived from two Greek words, *Klima+ logos; Klima* means **slope** of the earth and *logos* means a discourse or study. The slope of the earth denotes to **latitude.** The latitudinal difference plays a vital role for climatic variation on the surface of the earth when they made use of the term *Klima* for climate. Of all the climatic elements, temperature was considered as the most important by the Greeks. The principal cause of various types of climate found on the study was the inclination of the solar rays. At present climatology is a broad and diverse field; it seeks to describe and explain the nature of climate. In concise climatology may be defined as the scientific study of climate. It is a study of diversities of climates originated on the earth and their distribution over its surface.

According to **Austin Miller,** "Climatology is that branch of science which discusses the average condition of the weather."

Critchfield defined "Climatology broadens the findings of meteorology in space and in time to cover the whole earth and period of time as long as observations and indirect evidence will permit".

Riley and Spolton described "the science of climatology is the study of the weather conditions over a longer period".

According to **Arnold Court** the science of climatology is the compound of parts of three other modern disciplines: meteorology, statistics and geography. Each discipline in his opinion, to a certain extent, is a child of climatology. Climatology must choose from each of these disciplines—the principles, observations, methods and outlook that it needs.

Meteorology and Climatology

Climatology is closely related to Meteorology and is dependent on it for data regarding various elements of weather for effective explanation. The word meteorology is of Greek origin which literally means "discourse on things above". In fact, the word meteorology has been borrowed from Aristotle's

"Meteorologica" dated 350 B.C. What has been mainly discussed in this treatise is the Physics and Chemistry of the atmospheric phenomena. Meteorology is a scientific study of physical processes which are constantly at work in the atmosphere. To understand the processes of the atmosphere proper training in thermodynamics and hydrodynamics is essential. In short, we can say Meteorology is the Physics of atmosphere. The analytical study of the individual air masses with distinctive features of temperature and of humidity, and of the conditions of the upper air have both yielded lot of information which can be utilized by the geographers to assist his climatological descriptions.

Climatology is the branch of atmospheric sciences that deals primarily with the totality of weather and atmospheric behaviour over a number of years in a specific area. It is not concerned with the description of climate but also with analysis of the causes of the climatic differences and changes and their practical consequences.

Subdivision of Climatology

Critchfield has divided climatology into three basic sub-divisions:

1. Physical Climatology
2. Regional (Descriptive) Climatology
3. Applied Climatology

1. Physical Climatology

The physical climatology analyses the statistical data obtained from meteorology on various weather elements such as temperature, pressure, moisture, winds, precipitation etc. and their interrelationships. Physical climatology is a major aspect of meteorology from which most of its basic principles are drawn.

2. *Regional Climatology*

This branch of climatology studies discrete and characteristic weather phenomena of a particular continental or sub-continental region. It is also called Descriptive Climatology, for it is concerned with identification of important climatic characteristics and analyses, the interaction of weather and climatic elements upon the life, health and economics of the people and area. On the size and extent of the climatic regions, the world is divided into following three regions:

A. Macro-climatic regions.
B. Meso-climatic regions
C. Micro-climatic regions.

3. *Applied Climatology*

In applied climatology the main purpose is to find out ways and means to make use of our knowledge of climatic elements for the betterment of human life on the earth. Knowledge of weather and climate is being applied to the solution of such problems as plant and animal production, transportation, communication, industry, structural designs and construction activities, atmospheric pollution and many other economic activities of man.

14

ATMOSPHERE

The atmosphere is a combination of numerous gases. It envelops earth and extends as far as 9600 km. above the earth's surface. This gaseous cover of the earth is held around it by gravitational attraction. The air, sea, and land constitute the major portion of four great material realms or sphere, that comprise the total global environment. These three spheres are shown in Fig. 14.1.

1. Atmosphere
2. Lithosphere
3. Hydrosphere

The Substance

The fourth sphere, the biosphere, encompasses all living organisms of the earth. Human live at the bottom of an ocean of air. They are air-breathers dependent on favourable conditions of pressure, temperature, and chemical composition of the atmosphere that surround them. The surface environment which we may call the life layer, is shallow and complex zone in which atmospheric conditions exert control on the land surface, while on the same time the surface of the land exerts an influence on the properties of the adjacent atmosphere.

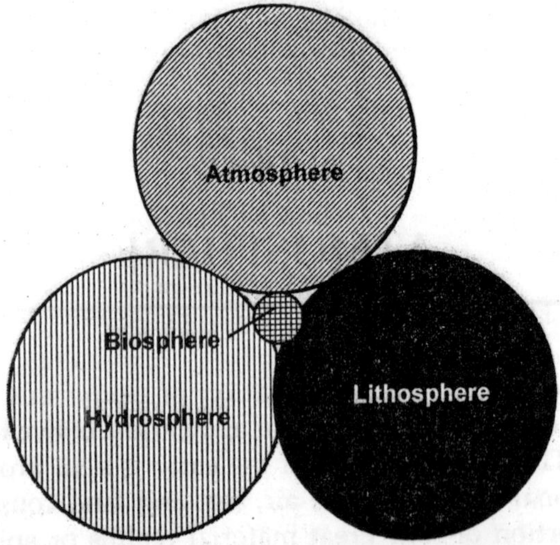

Fig. 14.1: The earth realms shown as intersecting circles. The large outer circles represent the three great inorganic realms; each one overlaps the other two in a small area, suggesting that some of the substance of each realm is held within the other two. The biosphere, or organic realm, draws its substance from its inorganic environment, represented by the circle overlap into the three surrounding realms. The small diameter of the biospheric circle signifies that the total mass of matter held within the biosphere is only a small fraction of that in the other three realms.

Composition of the Atmosphere

The earth's atmosphere consists of a mixture of various gases surrounding the earth to a height of many kms. Held to the earth by gravitational attraction, this envelope of air is densest at sea level and thins rapidly upward. Table 14.1 illustrating the average composition of dry air, shows that three gases—nitrogen, oxygen and argon—account for 99.9 per cent of air by volume. These gases are mixed in remarkably constant proportions up to about 80 kms (50 miles). Of especial importance, despite their relative scarcity, are also called green-house gases, which play vital role in thermodynamics of the atmosphere by trapping long wave terrestrial radiation, producing the greenhouse effect? The concentration of these gases are particularly susceptible to human activities.

Table 14.1: Average composition of the dry atmosphere below 25 km

Component	*Volume*	*Molecular weight*
Nitrogen	78.08	28.02
Oxygen	20.95	32.00
Argon	0.93	39.88
Carbon dioxide	0.035	44.00
Neon	0.0018	20.18
Helium	0.0005	4.00
Ozone	0.00006	48.00
Hydrogen	0.00005	2.02
Krypton	0.0011	—
Xenon	0.00009	—
Methane	0.00017	—

Beside above given gases water vapours comprises up to 4 per cent of the atmosphere by volume (about 3 per cent by weight) near the surface but only 3-6 ppmv (parts per million by volume) above 10-12 km. It is supplied to atmosphere by the evaporation from surface water or by transportation from plants and transferred upwards by atmospheric turbulence. Turbulence is most effective below about 10 km and as the maximum possible water vapour density of cold year is anyway very low.

Layers of the Atmosphere

The earth's atmosphere consists of zones of layers arranged like spherical shells according to altitude above the earth's surface. Each zone has a unique set of characteristics. Since various physical and chemical properties are arranged in altitude zones, the system of layers and nomenclature applied to different layers are based upon the class of properties selected.

According to Pettersen, the atmosphere is divided into the following more significant spheres:

1. Troposphere
2. Stratosphere
3. Mesosphere

4. Thermosphere
5. Exosphere

1. Troposphere

The lowest layer of the atmosphere is called the troposphere. It is the zone where weather phenomena and atmospheric instability are most noticeable, and it contain 75 per cent of the total molecular or gaseous mass of the atmosphere and virtually all the water vapours and aerosols. Throughout this layer there is a general decrease of temperature with height at a mean rate of about 6.5°C/km. The decrease occurs because the air is compressible and its density decreases with height allowing rising air to expand and thereby cool. Additionally, the atmosphere is heated mainly by turbulent heat transfer from the surface not by absorption of radiation. The troposphere is capped in most places by a temperature inversion level (e.g. a layer of relatively warm air above a colder one) in others by a zone that is isothermal with height. The troposphere thus remains to a large extent self-contained, because the inversion acts as a lid that effectively limits convection. This inversion level or weather ceiling is called the *tropopause.* Its height is not constant in either space or time. The height of the tropopause at any point is correlated with sea level temperature and pressure, which are in turn related to the factors of latitude, season and daily changes in surface pressure. There are marked variations in the altitude of the tropopause with latitude as shown in Fig. 14.2.

2. Stratosphere

The stratosphere extends upwards from the tropopause to about 50 km. Although the stratosphere contains much of the total atmospheric ozone (it reaches a peak density at approximately 22 km), the maximum temperature associated with the absorption of sun's ultraviolet radiation by ozone occur at tropopause, where temperature may exceed 0°C.

Ozone: It is concentrated mainly between 15 and 35 km. The upper layers of the atmosphere are irradiated by ultraviolet

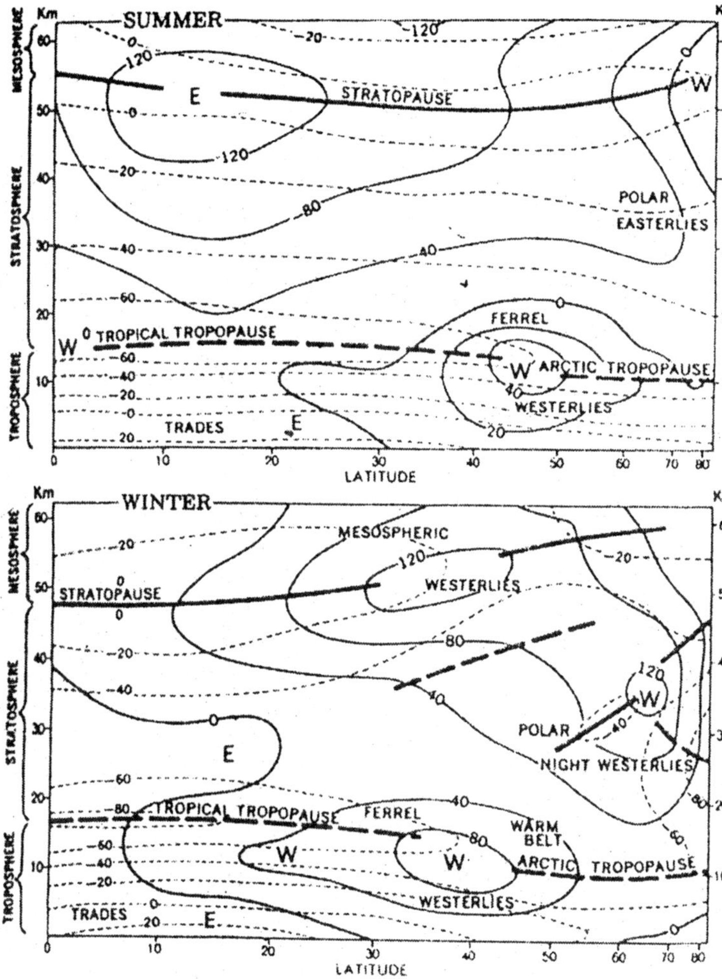

Fig. 14.2: Mean zonal (westerly) winds (solid isolines, in knots; negative values from the east) and temperatures (in °C, dashed isolines), showing the broken tropopause near the mean Ferrel jet stream.

Source: After Boville (from Hare 1962).

Notes: The term 'Ferrel Westerlies' was proposed by F.K. Hare in honour of W. Ferrel (see p. 127). The heavy black lines denote reversals of the vertical temperature gradient of the tropopause and stratopause. Summer and winter refer to the northern hemisphere.

radiation from the sun which causes the breakup of oxygen molecules at altitude of 30 km (e.g. O_2=O+O). These separated atoms (O+O) may then combine individually with other molecules to create ozone as illustrated in Fig. 14.3.

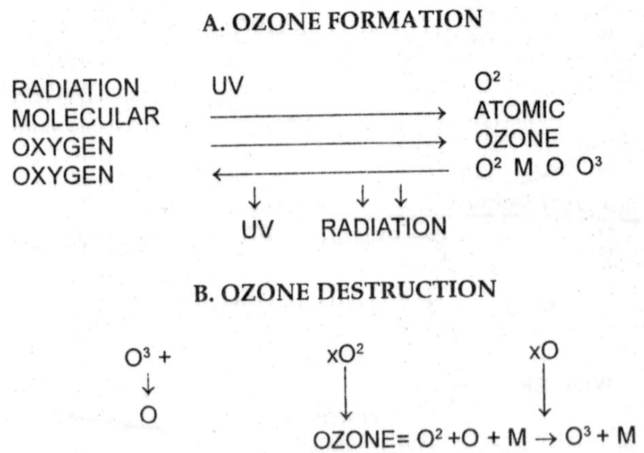

Fig. 14.3

Where M represents the energy and momentum balance provided by collision where a third atom or molecules; this ozone formation cycle was given by Chapman. The normal increase in stratospheric ozone associated with increasing solar radiation in spring apparently fail to develop. Observation in Antarctica shows a decrease in total ozone in September-October from 320 Doson units (10 cubic cm at standard atmospheric temperature and pressure) in 1960 to around 100 in the 1990.

The effect of reduced stratosphere ozone are particularly important for their potential biological damage to living cells and human skin. It is estimated that a 1 per cent reduction in total ozone will increase ultraviolet-B radiation by 2 per cent. The serious threat to the ozone layer is posed by the release in to atmosphere of Freon, synthetic compound containing carbon, fluorine and chlorine atoms, alternative name is chlorofluorocarbons or **CFCs.** In USA many aerosol spray cans used in the household were charged with halocarbons. They are also used as refrigerants, a practice that continues to contribute to the release of halocarbons. In 1985, the Vienna Conference for the protection of the ozone was attended by 43 countries of the world. In 1990, an international agreement was expended and strengthened to a goal of 50% reduction of **CFCs** by 1995 and 100 % by 2000 with the use of substitute gas **HCFCs** (hydro fluorocarbons).

The upper limit of the stratosphere is called stratopause, beyond which lies the Mesosphere.

3. *Mesosphere*

Above the stratopause, average temperature decrease to minimum of about –33°C (140°K) or around 90 km. This layer is termed as mesosphere. The layers between the tropopause and lower thermosphere are now commonly referred to as the middle atmosphere with the upper atmosphere designating the regions above about 100 km altitude. Above 80 km temperatures again begin rising with height and this inversion is referred to as Mesopause. Pressure is very low in the mesosphere, decreasing from about 1 mb at 50 km to 0.01 mb at 90 km.

4. *Thermosphere*

Above the mesopause, atmospheric densities are extremely low. The lower portion of the thermosphere is composed mainly of nitrogen and oxygen in molecular and atomic forms, where above 200 km atomic oxygen predominates over nitrogen. Temperature rise with height owing to absorption of extreme ultraviolet radiation by molecular and atomic oxygen probably approaching 800-1200°K at 350 km, but these temperatures are essentially theoretical.

Above 100 km atmosphere is increasingly affected by cosmic radiation, solar X-rays and ultraviolet radiation which causes *ionization* or electrical charging by separating negatively charged electrons from neutral oxygen atoms and nitrogen molecules leaving the atom or molecule with a net positive charge (an ion). The term ionosphere is commonly applied to

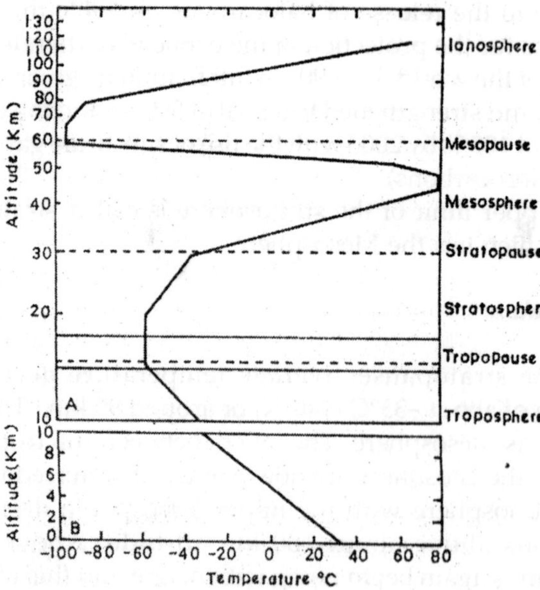

Fig. 14.4: Temperature structure of the atmosphere.

the layers above 80 km. The ionosphere consists of the following ionized layers:

D-layer 60-99 km
E-layer 90-130 km
Sporadic E layer 110 km
E_2 layer 150 km
F_1-layer
F_2-layer 150-380km
G-layer-400 km and above

5. Exosphere and Magnetosphere

The base of the exosphere is between about 500 km and 750 km. Here the atoms of oxygen, hydrogen and helium (about 1 per cent of which are ionized) form the tenuous atmosphere. Ionized particles increase in frequency through the exosphere and beyond about 200 km, in the magnetosphere there are only electrons (negative) and protons (positive) derived from the solar wind-plasma of electrically conducting gas (Fig. 14.4).

REFERENCES

Climate Prediction Center. *Climate Glossary*. Retrieved on 2006-11-23.

Life Stories. *Francis Galton*. Retrieved on 2007-04-19.

Other Forecasting Methods: Climatology, analogue and numerical weather prediction. Retrieved on 2006-02-16.

Kenneth C. Allen. *Pattern Recognition Techniques Applied to the NASA-ACTS Order-Wire Problem*. Retrieved on 2007-02-16.

Weather Associates, Inc. *The Role of Teleconnections and Ensemble Forecasting in Extended to Medium-Range Forecasting*. Retrieved on 2007-02-16.

Thinkquest.org. *Teleconnections: Linking El Niño with Other Places*. Retrieved on 2007-02-16.

California Department of Fish and Game, Marine Region. *El Niño Information*. Retrieved on 2007-06-07.

La Niña.

El Nino forms in Pacific Ocean, CNN.

There Goes El Nino, Here Comes La Nina. The Associatied Press/CBS News (2007-02-28). Retrieved on 2007-03-02.

National Center for Atmosphere Research. *Climate Analysis Section*. Retrieved on 2007-06-07.

15

WINDS

Wind can be defined simply as air in motion. This motion can be in any direction, but in most cases the horizontal component of wind flow greatly exceeds the flow that occurs vertically. The speed of wind varies from absolute calm to speeds as high as 380 kilometres per hour (Mt. Washington, New Hampshire, April 12, 1934). In 1894, strong winds in Nebraska pushed six fully loaded coal cars over 160 kilometres in just over three hours. Over short periods of time surface winds can be quite variable.

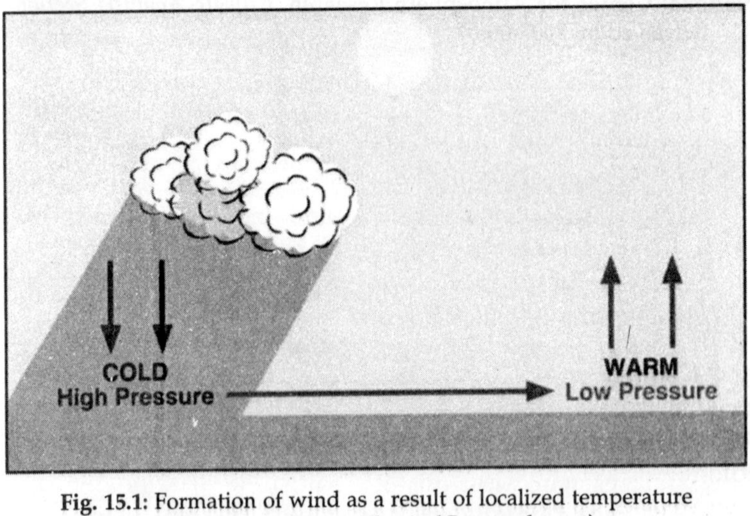

Fig. 15.1: Formation of wind as a result of localized temperature differences. (*Source*: PhysicalGeography.net)

Wind develops as a result of spatial differences in *atmospheric pressure*. Generally, these differences occur because of uneven absorption of *solar radiation* at the Earth's surface (Fig. 15.1). Wind speed tends to be at its greatest during the daytime when the greatest spatial extremes in atmospheric temperature and pressure exist.

Fig. 15.2: Meteorological instruments used to measure wind speed and direction. Wind speed is commonly measured with an anemometer. An anemometer consists of three open cups attached to a rotating spindle. The speed of roation is then converted into a measurement of wind speed. Wind direction is measured with a wind vane. In the photograph above, the wind vane instrument has a bullet-shaped nose attached to a finned tail by a metal bar. (*Source*: PhysicalGeography.net)

Wind is often described by two characteristics: wind speed and wind direction. Wind speed is the velocity attained by a mass of air travelling horizontally through the atmosphere. Wind speed is often measured with an anemometer in kilometres per hour (kmph), miles per hour (mph), knots, or metres per second (mps) (Fig. 15.2). Wind direction is measured as the direction from where a wind comes from. For example, a southerly wind comes from the south and blows to the north. Direction is measured by an instrument called a wind vane (Fig. 15.2). Both of these instruments are positioned in the atmospheric environment at a standard distance of 10 metres above the ground surface.

Wind speed can also be measured without the aid of instruments using the Beaufort wind scale (Table 15.1). This descriptive scale was originally developed by Admiral Beaufort of the British Navy in the first decade of the seventeenth century. The purpose for this system was to allow mariners to determine wind speed from simple observations. The Beaufort system has undergone several modifications to standardize its measurement scale and to allow for its use on land. Users of this scale look for specific effects of the wind on the environment to determine speed.

Table 15.1: Beaufort wind speed scale

Beaufort Code	*Speed Miles per Hour*	*Speed Kilometres per Hour*	*Description*	*Effects on the Environment*
0	< 1	< 1	calm	smoke rises vertically
1	2 - 3	1 - 5	light air	smoke drifts slowly
2	4 - 7	6 - 11	light breeze	leaves rustle, wind can be felt, wind vanes move
3	8 - 12	12 - 19	gentle breeze	leaves and twigs on trees move
4	13 - 18	20 - 29	moderate breeze	small tree branches move, dust is picked up from the ground surface
5	19 - 24	30 - 38	fresh breeze	small trees move
6	25 - 31	39 - 51	strong breeze	large branches move, telephone and power overhead wires whistle
7	32 - 38	51 - 61	near gale	trees move, difficult to walk in the wind
8	39 - 46	62 - 74	gale	twigs break off from trees
9	47 - 54	75 - 86	strong gale	branches break off from trees, shingles blown off roofs
10	55 - 63	87 - 101	whole gale	trees become uprooted, structural damage on buildings
11	64 - 74	102 - 120	storm	widespread damage to buildings and trees
12	> 75	> 120	hurricane	severe damage to buildings and trees

Winds are named according to the compass direction of their source. Thus, a wind from the north blowing towards the south is called a northerly wind. Fig. 15.3 describes the sixteen principal bearings of wind direction. Most meterological observations report wind direction using one of these sixteen bearings.

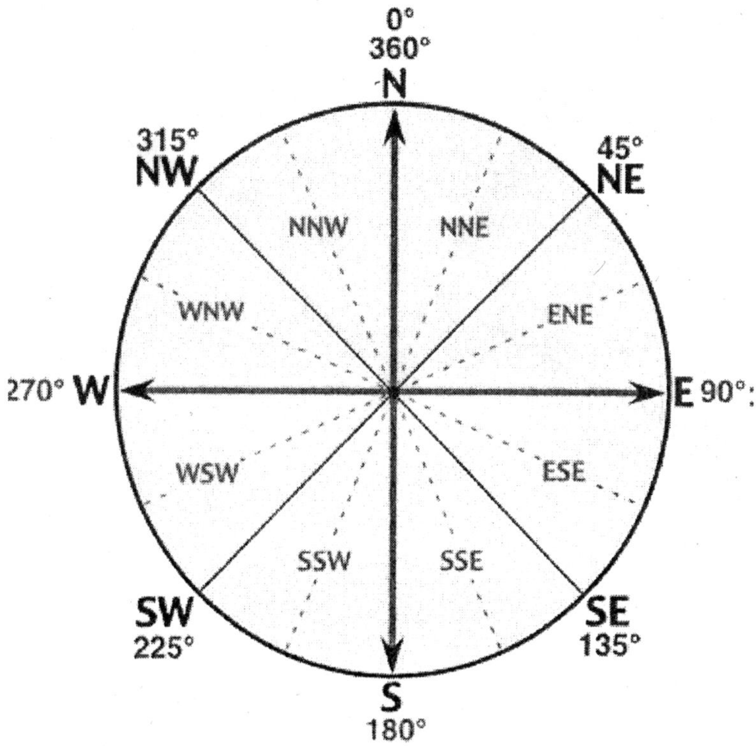

Fig. 15.3: Wind compass describing the sixteen principal bearings used to measure wind direction. This compass is based on the 360 degrees found in a circle. (*Source:* PhysicalGeography.net)

Horizontally, at the Earth's surface wind always blows from areas of high pressure to areas of low pressure (vertically, winds move from areas of low pressure to areas of high pressure), usually at speeds determined by the rate of air pressure change between pressure centres. This situation is comparable to someone skiing down a hill. The skier will of course move from the top of the hill to the bottom of the hill, with the speed of their descent controlled by the gradient or steepness of the slope. Likewise, wind speed is a function of the steepness or gradient of *atmospheric air pressure* found between high and low pressure systems. When expressed scientifically, pressure change over a unit distance is called pressure gradient force, and the greater this force the faster the winds will blow.

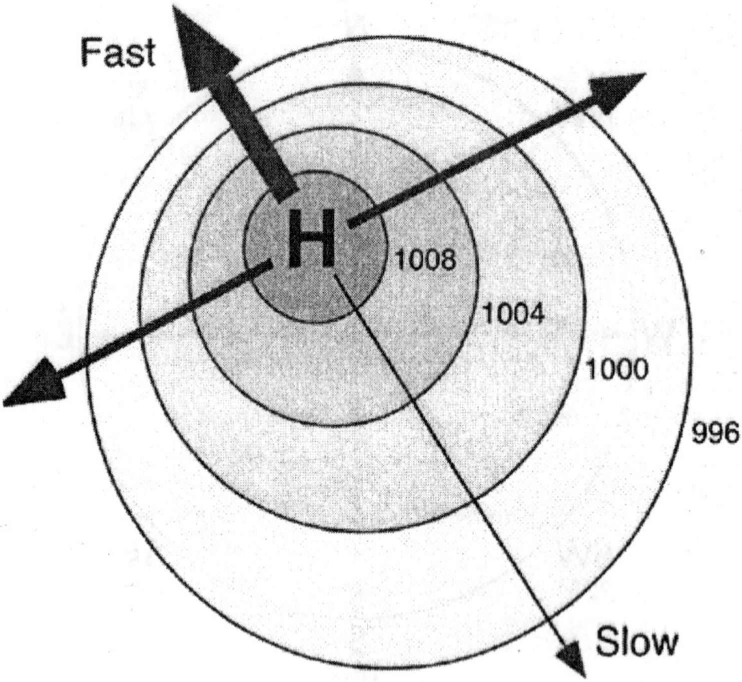

Fig. 15.4: Association between wind speed and distance between isobars. In the illustration above thicker arrows represent relatively faster winds. (*Source*: PhysicalGeography.net)

On weather maps, pressure is indicated by drawing isolines of pressure, called isobars, at regular 4 millibar intervals (e.g., 996 mb, 1000 mb, 1004 mb, etc.). If the isobars are closely spaced, we can expect the pressure gradient force to be great, and wind speed to be high (Fig. 15.4). In areas where the isobars are spaced widely apart, the pressure gradient is low and light winds normally exist. High-speed winds develop in areas where isobars are closer.

Driving Forces

To better understand wind we must recognize that it is the result of a limited number of accelerating and decelerating forces, and that the action of these forces is controlled by specific fundamental natural laws. Sir Isaac Newton formulated these laws as several laws of motion. The first law suggests that an object that is stationary will remain stationary, and an object in motion will stay in motion as long as no opposing force is put on the object. As a result of this law, a puck sent in flight from a blade of a hockey stick will remain in motion until friction slows it down or the goalie makes a save. This law also suggests that once in motion an object's path should be straight.

Newton's second law of motion suggests that the force put on an object equals its mass multiplied by the acceleration produced. The term "force" in this law refers to the total or net effect of all the forces acting on an object. Mathematically, this law is written as:

$$\text{Force} = \text{Mass} \times \text{Acceleration or Acceleration} = \text{Force}/\text{Mass}$$

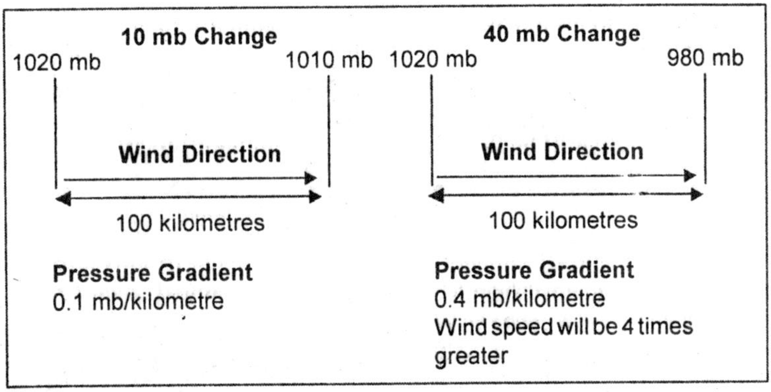

Fig. 15.5: Effect of pressure gradient on wind speed (*Source*: PhysicalGeography.net)

From this natural law of motion we can see that the acceleration of an object is directly proportional to the net force pushing or pulling that body and inversely proportional to the mass of the body. Thus, the greater the force created by the movement of a hockey player's stick the faster the puck will travel. This law also suggests that if the player used a larger (more massive) puck, more force would have to be applied to it to get it to travel as fast as a puck with less mass.

Pressure gradient force is the primary force influencing the formation of wind from local to global scales. This force is determined by the spatial pattern of *atmospheric pressure* at any given moment in time. Fig. 15.5 illustrates two different pressure gradient scenarios and their relative effect on wind speed.

The two diagrams display the relative relationship between pressure gradient and wind speed. This relationship is linear and positive. As a result, quadrupling the pressure gradient increases wind speed by a factor of four. This is what we would expect according to Newton's second law of motion, assuming the mass of the wind is unchanged.

We can also describe pressure gradient acceleration mathematically with the following equation:

$$F(m/s^2) = \left| \frac{1}{D} \bullet \left(\frac{P_1 - P_2}{n} \right) \right|$$

where:

* D = density of air (average density of surface air is 1.29 kilograms per cubic metre)
* P_2 = pressure at point 2 in Newtons/m^2 (N m^{-2})
* P_1 = pressure at point 1 in Newtons/m^2 (N m^{-2})
* n = distance between the two points in metres

From this equation we can determine wind acceleration between two points in metres per second squared by knowing three variables: the density of the moving air; the change in pressure between the points of interest in newtons; and the distance between the two points in metres. For example, to determine the wind speed between two points for moving air with a density of 1.29 kilograms per cubic metre, a pressure difference of 400 Newtons/m^2, and a distance of 300,000 metres, the following calculations would be performed:

$$F(m/s^2) = \left| \frac{1}{1.29} \bullet \left(\frac{400}{300{,}000} \right) \right| = 0.00103 m/s^2$$

From the calculated value of acceleration we can determine wind speed, V, from the formula:

$$V = V_o + F_t$$

Where V_o is the initial velocity of the wind and t is the time during which F is applied.

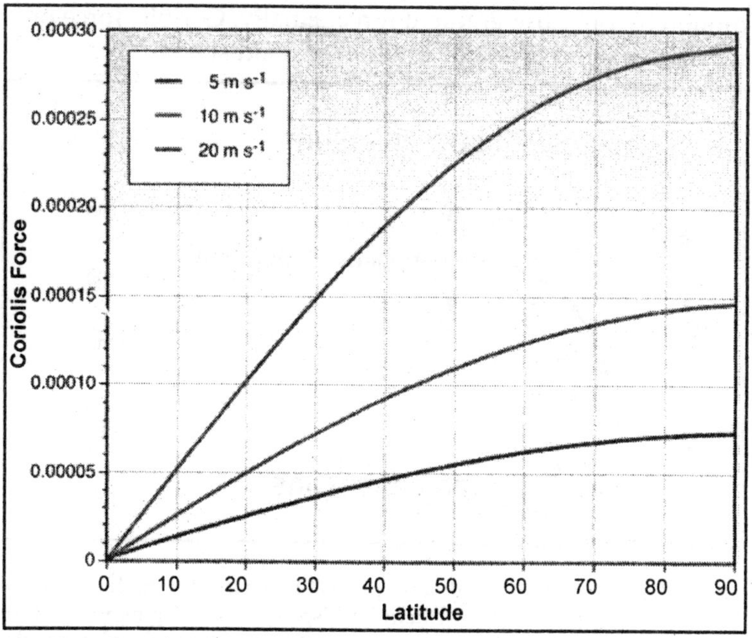

Fig. 15.6: The strength of Coriolis force is influenced by latitude and the speed of the moving object (*Source*: PhysicalGeography.net)

The rotation of the Earth creates another force, termed Coriolis force, which acts upon wind and other objects in motion in very predictable ways. According to Newton's first law of motion, air will remain moving in a straight line unless it is influenced by an unbalancing force. The consequence of Coriolis force opposing pressure gradient acceleration is that the moving air changes direction. Instead of wind blowing directly from high to low pressure, the rotation of the Earth causes wind to be deflected off course. In the Northern Hemisphere, wind is deflected to the right of its path, while in the Southern Hemisphere it is deflected to the left. The magnitude of the Coriolis force varies with the velocity and the latitude of the object (Fig. 15.6). Coriolis force is absent at the equator, and its strength increases as one approaches either pole. Furthermore, an increase in wind speed also results in a stronger Coriolis force, and thus in greater deflection of the wind. Coriolis force only acts on air when it has been sent into motion by pressure gradient force. Finally, Coriolis force only influences wind direction and never wind speed.

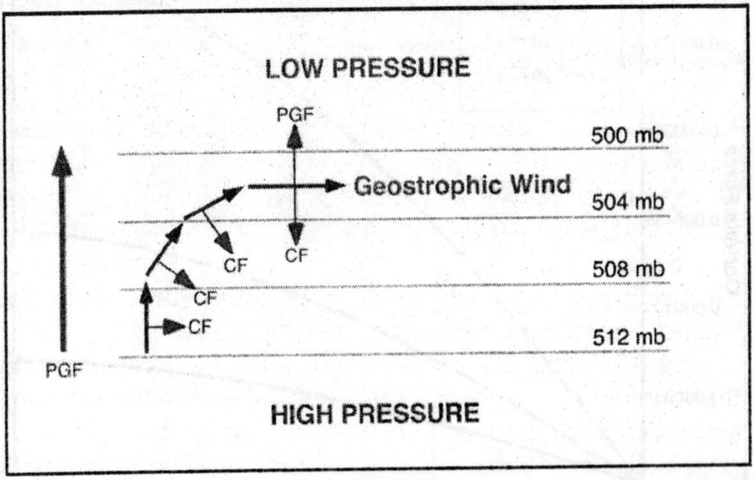

Fig. 15.7: A geostrophic wind flows parallel to the isobars. In this model of wind flow in the Northern Hemisphere, wind begins as a flow of air perpendicular to the isobars (measured in millibars) under the primary influence of the pressure gradient force (PGF). As the movement begins, the Coriolis Force (CF) begins to influence the moving air causing it to deflect to the right of its path. This deflection continues until the pressure gradient force and Coriolis Force are opposite and in balance with each other. (*Source*: Physical Geography.net)

Centripetal acceleration is the third force that can act on moving air. It acts only on air that is flowing around centres of circulation. Centripetal acceleration is also another force that can influence the direction of wind. Centripetal acceleration creates a force directed at right angles to the flow of the wind and inwards towards the centres of rotation (e.g., low and high pressure centres). This force produces a circular pattern of flow around centres of high and low pressure. Centripetal acceleration is much more important for circulations smaller than the mid-latitude cyclone.

The last force that can influence moving air is frictional deceleration. Friction can exert an influence on wind only after the air is in motion. Frictional drag acts in a direction opposite to the path of motion causing the moving air to decelerate (see Newton's first and second laws of motion). Frictional effects are limited to the lower one kilometre above the Earth's surface.

Geostrophic Wind

Air under the influence of both the pressure gradient force and Coriolis force tends to move parallel to isobars in conditions where friction is low (1000 metres above the surface of the Earth) and isobars are straight. Winds of this type are usually called geostrophic winds. Geostrophic winds come about because pressure gradient force and Coriolis force come into balance after the air begins to move (Fig. 15.7).

Fig. 15.7 models air flow in the Northern Hemisphere. In the Southern Hemisphere, Coriolis acceleration acts on moving air by deflecting it to the left instead of the right.

Finally, Buys Ballot's Law states that when you stand with your back to a geostrophic wind in the Northern Hemisphere, the centre of low pressure will be to your left and the high pressure to your right. The opposite is true for the Southern Hemisphere.

Gradient Wind

Wind above the Earth's surface does not always travel in straight lines. In many cases winds flow around the curved isobars of a high (anticyclone) or low (cyclone) pressure centre. A wind that blows around curved isobars above the level of friction is called a gradient wind. Gradient winds are slightly more complex than geostrophic winds because they include the action of yet another physical force. This force is known as centripetal force and it is always directed towards the centre of rotation. Fig. 15.8 describes the forces that produce gradient winds around high and low pressure centres. Around a low, the gradient wind consists of the pressure gradient force and centripetal force acting towards the centre of rotation, while Coriolis force acts away from the centre of the low. In a high pressure centre, the Coriolis and centripetal forces are directed towards the centre of the high, while the pressure gradient force is directed outward.

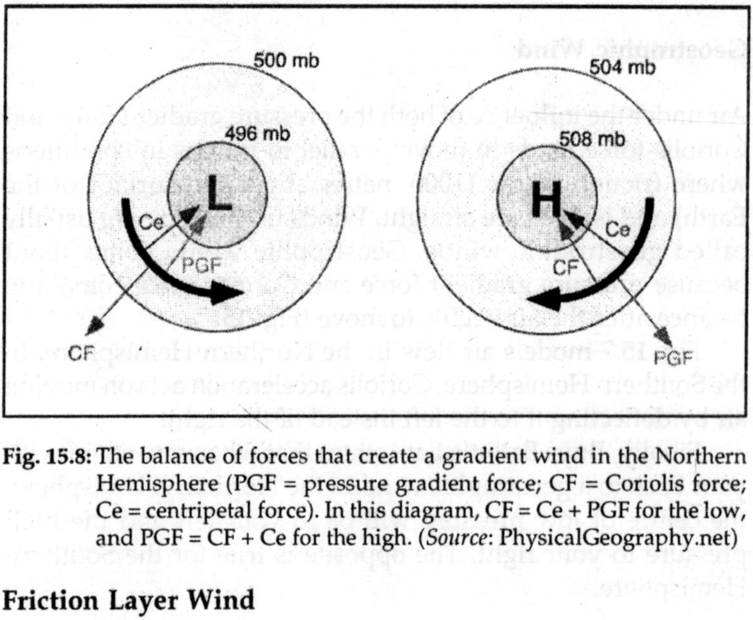

Fig. 15.8: The balance of forces that create a gradient wind in the Northern Hemisphere (PGF = pressure gradient force; CF = Coriolis force; Ce = centripetal force). In this diagram, CF = Ce + PGF for the low, and PGF = CF + Ce for the high. (*Source*: PhysicalGeography.net)

Friction Layer Wind

Surface winds on a weather map do not blow exactly parallel to the isobars as in geostrophic and gradient winds. Instead, surface winds tend to cross the isobars at an angle varying from 10 to 45°. Close to the Earth's surface, friction reduces the wind speed, which in turn reduces the Coriolis force. As a result, the reduced Coriolis force no longer balances the pressure gradient force, and the wind blows across the isobars

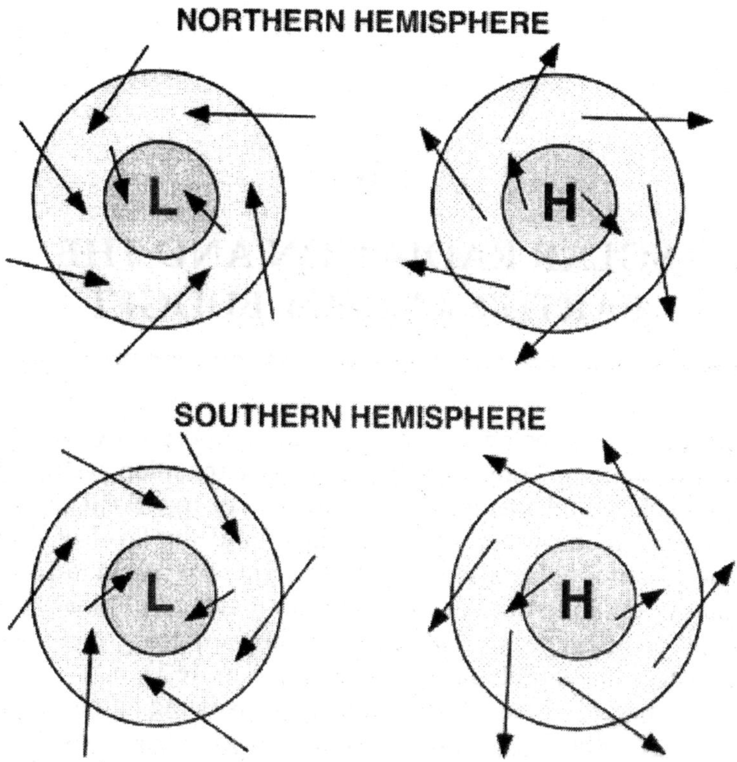

Fig. 15.9: Circulation patterns of high and low pressure systems in the North and South Hemisphere. (*Source*: PhysicalGeography.net)

towards or away from the pressure centre. The pressure gradient force is now balanced by the sum of the frictional force and the Coriolis force. Thus, we find surface winds blowing counter-clockwise and inward into a surface low, and clockwise and out of a surface high in the Northern Hemisphere. In the Southern Hemisphere, the Coriolis force acts to the left rather than the right. This causes the winds of the Southern Hemisphere to blow clockwise and inward around surface lows, and counter clockwise and outward around surface highs (Fig. 15.9).

16

SOLAR RADIATION AND THE EARTH'S ENERGY BUDGET

All life processes on Planet Earth are supported by radiant energy coming from sun. The planetary circulation system of atmosphere and oceans are driven by solar energy. Solar energy is intercepted by our spherical planet and the level of heat energy tends to be raised. At the same time our planet radiates energy into outer space a process that tends to diminish the level of heat energy. The incoming and outgoing radiation processes are simultaneously in action (Fig. 16.1).

In one place and time more energy is being gained than lost; in another place and time more energy is being lost than gain. The prime source of the energy injected into our atmosphere is the sun which is continually shedding part of its mass by radiating waves of Electromagnetic energy and high energy particles into space. The amount of energy received by the earth, assuming for the moment that there is no interference from the atmosphere is effected by four factors:

1. Solar output
2. The sun earth distance
3. The altitude of the sun
4. Day length

1. Solar Output

Solar energy originates from nuclear reactions within sun's hot core (16×10^6K) is transmitted to the sun's surface by radiation and hydrogen convection. Visible solar radiation

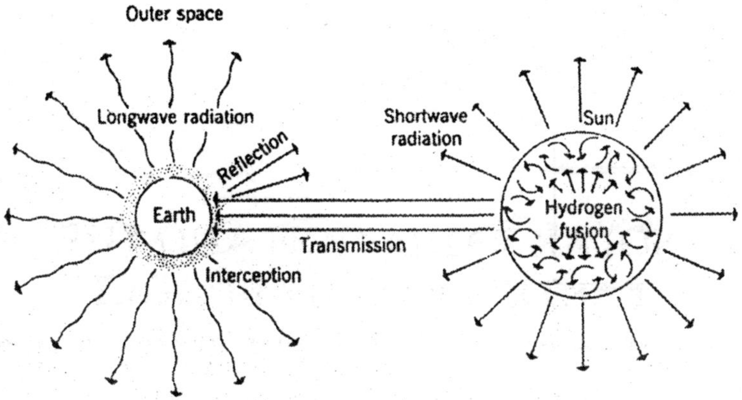

Fig. 16.1: Simplified diagram of the sun-earth-space radiation system. (From A.N. Strahler, "The Life Layer", *Journal Geography*, vol. 69, p. 72. © 1970 by *The Journal of Geography;* reproduced by permission).

(light) comes from a cool (6000 K) K stands for absolute temperature in degree Kelvin. The degree of same unit value as in the Celsius scale, but the zero point of the Kelvin scale is at –273°C or absolute zero) outer surface layer called the Photosphere. Temperatures rise again in the outer chromosphere (10,000 K) and corona (10^6K) which is continually expanding into space. The outflowing hot gases (Plasma) from the sun referred to as Solar Wind (with the speed of 1.5 x 10^6 km hr^{-1}) interact with the earth's magnetic field and upper atmosphere.

The sun behaves virtually as a Black body meaning that it both absorbs all energy received and in turn radiates energy at the maximum rate possible for a given temperature. The sun has a surface temperature of about 6000°C. The highly heated Incandescent gas that comprises the sun's surface emits electromagnetic radiation moving in straight lines radially outward from the sun and requiring about 8-1/3 minutes to travel the 150 million km (93 million miles) from sun to earth. The electromagnetic waves differ in dimensions throughout their entire range or spectrum. Waves are described in terms of either wavelength or frequency (Fig. 16.2). All the waves of EMR spectrum travel at the same speed. The number of waves passing a fixed point in a unit time (1 second) is known as

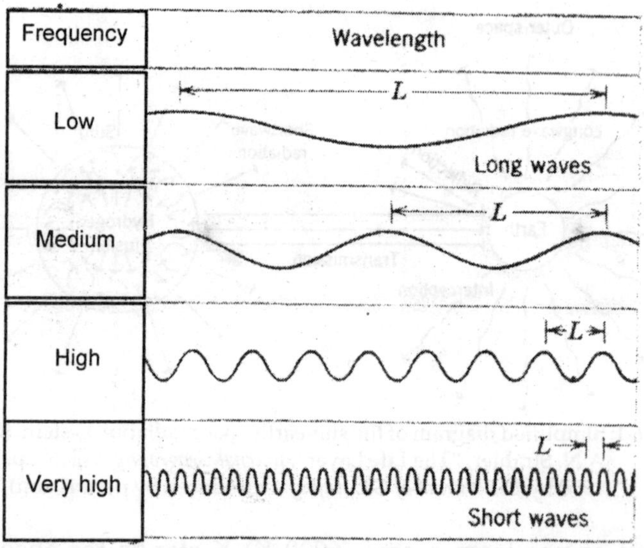

Fig. 16.2: Relationship of wavelength to frequency. (From A.N. Strahler, *The Earth Sciences*, 2nd ed., Harper & Row, Publishers. Copyright © by Arthur N. Strahler).

wave frequency. Frequency depends on the wavelength. Frequency is stated in terms of cycles per second; the unit is measured in the hertz. One hertz is a frequency of one cycle per second; 1 megahertz is a frequency of 1 million cycles per second. Equivalents between wave's length and frequency can be read directly on the scales given in Fig. 16.3 which shows the division of the EMR Spectrum-ranging from the shortest waves at the left and longest waves at the right. The scale is a logarithmic scale (constant ratio scale). Frequency in megahertz is given in powers of 10 ; each scale unit represents a tenfold increase from right to left.

At the end of the spectrum are gamma rays of extremely high frequency and short wavelength are given in unit of Angstroms. One angstrom is equal to 0.00000001 cm (10^{-8} cm) Gamma rays are shorter than 0.03 angstroms. The shorter X-rays described as Hard fall in the range 0.03 to 0.06 angstroms. The longer X-rays described as Soft range from about 0.6 to about 100 angstroms ; they grades into ultraviolet rays which extend to about 4000 angstroms.

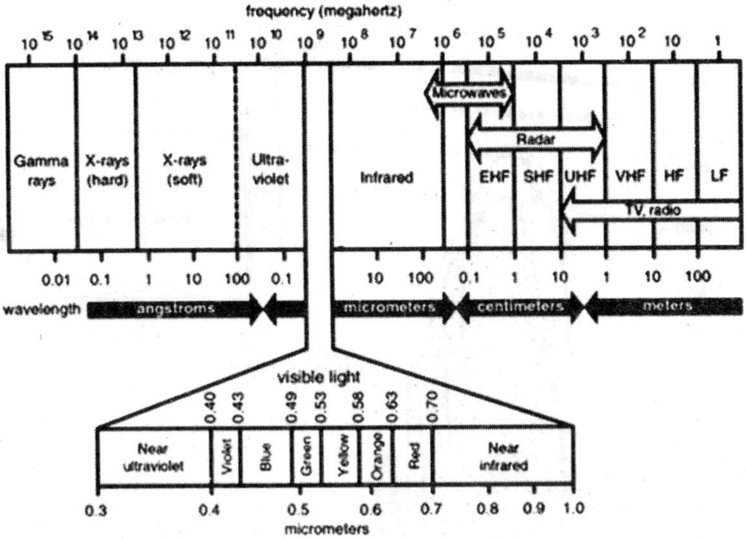

Fig. 16.3: The electromagnetic radiation spectrum. (From A.N. Strahler, *The Earth Sciences*, 2nd ed., Harper and Row, Publishers. Copyright © by Arthur N. Strahler.

Visible spectrum is described in Micron. One micron equals 0.0001cm. One micron thus equals 10,000 angstroms. The visible light portion of spectrum begins with violet at 0.4 micron. The infrared rays consist of wavelengths from 0.7 to 100 microns.

Fig. 16.4 shows that wavelength of range of solar (shortwaves) radiation and longwave length (infrared) radiation emitted by the earth and atmosphere. For solar radiation 8 % is ultraviolet and shorter wavelength emission 39 % visible light (0.4-0.7 micrometre and 53 % near infrared (>0.7 micrometre) the figure illustrates the black body radiation curve for 6000 K at the top of the atmosphere (which slightly exceeds the observed extraterrestrial radiation) for 300 K and for 263 K. The mean temperature of the earth's surface is about 288 K(15°C) and of the atmosphere about 250 K (-23°C). Gases do not behave as black bodies.

The energy received at the top of the atmosphere on a surface perpendicular to the solar beam (ray) for mean solar distance is termed the **solar constant.** This radiation intensity has an approximate value of 1400 watt per sq. metre. The watt

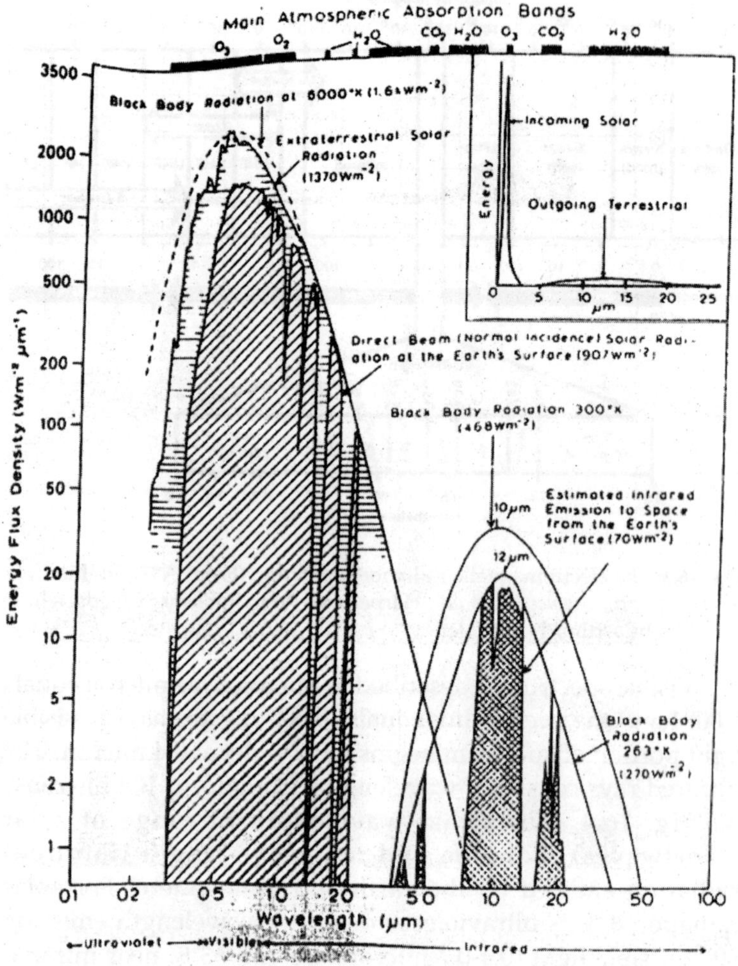

Fig. 16.4: Spectral distribution of solar and terrestrial radiation, plotted logarithmically, together with the main atmospheric absorption bands. The cross-hatched areas in the infrared spectrum indicate the 'atmospheric windows' where radiation escapes to space. The black-body radiation at 6,000 K is that proportion of the flux which would be incident on the top of the atmosphere. The inset shows the same curves for incoming and outgoing radiation with the wavelength plotted arithmetically on an arbitrary vertical scale.

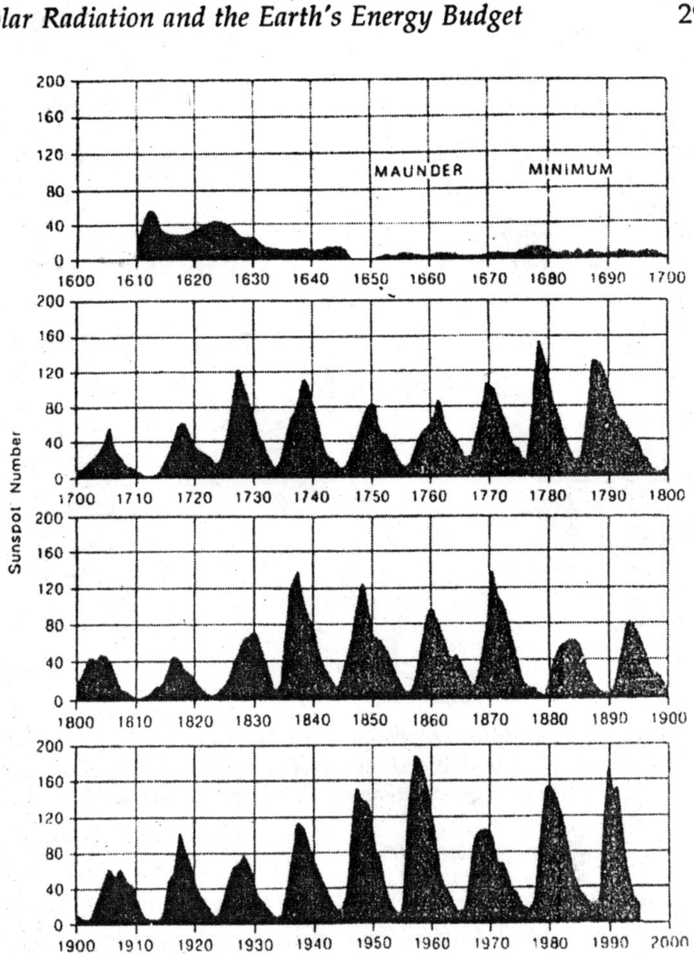

Fig. 16.5: Yearly sunspot numbers for the sun's visible surface for the period 1700-1995.

Sources: Reproduced by courtesy of the National Geophysical Data Center, NOAA, Boulder, Colorado. Data before AD 1700 courtesy of Foukal (1990) and *Scientific American*.

is defined as 1 joule per second (1W=1J/sec). Satellite data show that the solar constant undergoes small periodic variations of about 0.1 per cent related to sunspot activity. **Sunspots** are dark sunspots (i.e. cooler) area visible on the sun's surface. Their number and positions change in a regular manner, known as

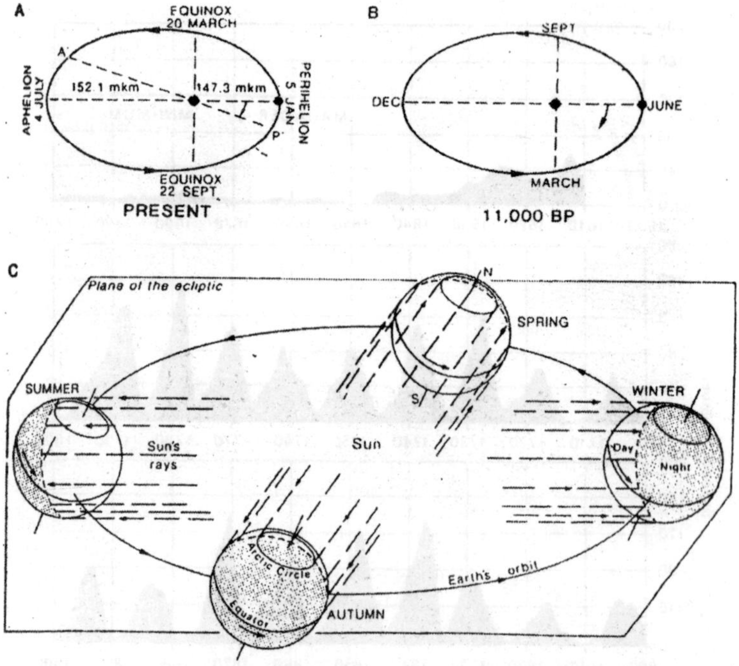

Fig. 16.6: Perihelion shifts. (A) The present timing of perihelion; (B) The direction of its shift and the situation at 11,000 years BP; (C) The geometry of the present seasons (northern hemisphere).

Source: Partly after Strahier, 1995.

the sunspot cycles. Figure 16.5 shows the variation of sunspot activity since 1610. Given data indicates that output within the ultraviolet parts of the spectrum shows considerably variability, with up to twenty times more ultraviolet radiation emitted at certain wavelength during a sunspot maximum than during a sunspot minimum. The relation between sunspot's activity and terrestrial temperature is a matter of some dispute.

2. Distance from the Sun: The annually changing distance of the earth from the sun produces seasonal variation in our receipt of solar energy. Owing to the eccentricity of the earth's orbit around the sun, the receipt of solar energy on a surface normal to the beam is 7 per cent more on 3 January at the **Perihelion** than on 4 July at the **Aphelion** (Fig. 16.6). In theory (that is discounting the interposition of the atmosphere and

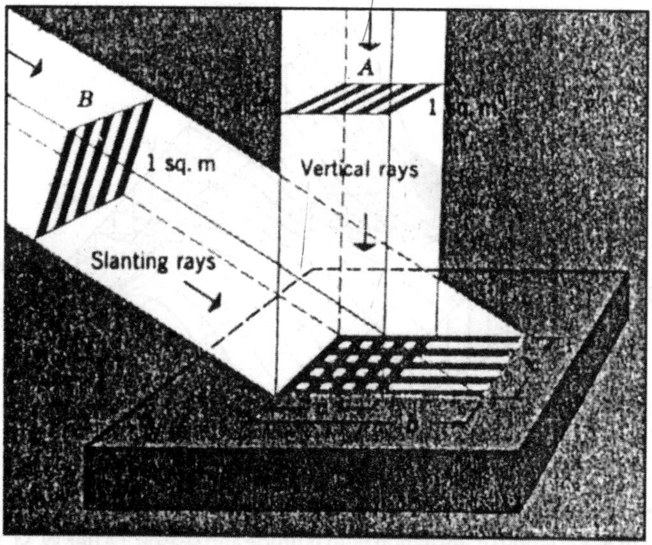

Fig. 16.7: The angle of the sun's rays determines the intensity of the insolation on the ground. The energy of vertical rays *A* is concentrated in square *a*, but the same energy in the slanting rays *B* is spread over a rectangle *b*.

the difference in degree of conductivity between large land and sea masses), this difference should produce an increase in the effective January world surface temperature of about 4°C over those of July. It should also make northern winters warmer than those in the southern hemisphere, and southern summers warmer than those in the northern hemisphere. In practice, the atmospheric heat circulation and the effect of continentality substantial mask this global tendency, and the actual seasonal contrast between hemispheres reversed. Moreover, the northern hemisphere half-year (21 March—22 September) is five days longer than the southern hemisphere summer (22 September—21 March). This difference slowly changes about 10,000 years ago the aphelion occurred in the northern hemisphere winter and northern summers received 3-4 per cent more radiation than today. This same pattern will return about 10,000 years from now.

3. Altitude of the Sun: The angle of incidence (i.e., the angle between its rays and tangent to the earth's surface at the point

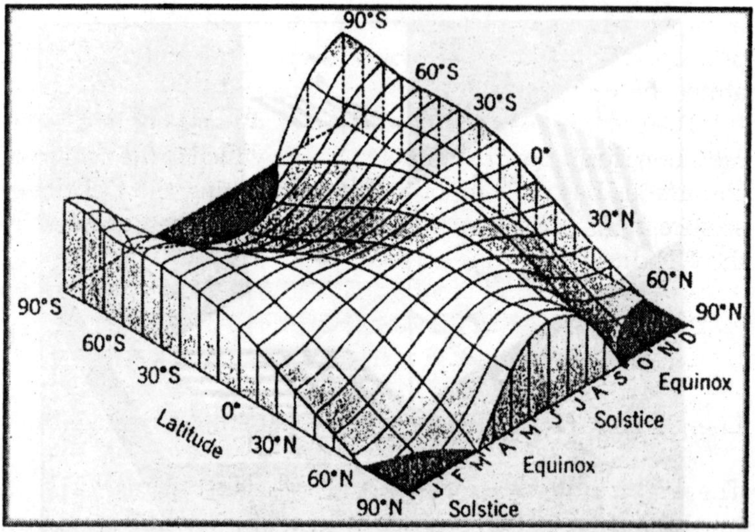

Fig. 16.8: Variation of insolation with latitude and season. (After W.M. Davis)

of observation) also affects the amount of solar radiation received at the surface of the earth. The greater the sun's altitude, the more concentration is the radiation intensity per unit area at the earth's surface and the longer is the path length of the beam through the atmosphere which increases the atmospheric absorption (Fig. 16.7). There are in addition important variations with solar altitude of the proportion of radiation reflected by the surface, particularly in the case of water surface. The principal factors that determine the sun's altitude are, of course, the altitude of the site, the time of the day and season (Fig. 16.8). At the June solstice the sun's altitude is a constant $23^{1/2\,0}$ throughout the day at the northern pole and the sun directly over head at noon at the tropic of cancer $23^{1/2\,0}$ N).

4. Length of the Day: The length of the day light also affects the amount of the radiation received. Obviously the longer the time that the sun shines the greater is the quantity of radiation that a given portion of the earth will receive. At the equator for example, the day length is close to 12 hours in all months, where at the poles it varies between 0 and 24 hours from winter to summer. The combination of all these factors produces the pattern of receipt of solar energy at the top of the atmosphere shown in Figure 16.8.

The polar region receives their maximum amount of solar radiation during their summer solstices which is the period of continuous day. The amount received during the December solstice in the SH is theoretically greater than that received by the NH during the June solstice.

SURFACE RECEIPT OF SOLAR RADIATION AND ITS EFFECTS

Energy Transfer within the Earth-Atmosphere System

The temperature of any particular place depends primarily on the amount of insulations received there. Thus it is evident that latitude is undoubtedly the most important factor that control the distribution of temperature on earth's surface. Heat is a form of energy ; temperature shows how hot a thing is. In other words heat refer to the quantity of energy present in a substance; temperature refers to its intensity or the degree of its hotness. Heat energy can be transferred by the three following mechanisms:

1. *Radiation*: Electromagnetic waves transfer energy (both heat and energy) between two bodies, without the necessary aid of an intervening material medium at speed of 300×10^6 m s^{-1} (i.e. the speed of the light). This is so with solar energy through space, whereas earth's atmosphere allows the passage of radiation only at certain wavelength and restricts at others. Radiation entering the atmosphere may be absorbed by the atmospheric gases in certain wavelengths but most shortwave radiation is transmitted without absorption. Scattering occurs if the direction of the photon of radiation is changed by interaction with atmospheric gases and aerosols. Two types of scattering are distinguished. For gases molecules smaller than the radiation wavelength, **Rayleigh** scattering occurs in all directions. As a result of scattering of blue light (0.4 micrometre) is an order of magnitude (i.e. $\times$ 10) greater than that of red light (0.7 micrometre), thus creating the day time blue sky.

However, when water droplets or aerosol particles with similar size (0.1-0.5 micrometer radius) to the radiation wavelength are present, the most of the light scattered forward is called **Mie** scattering and gives the grayish appearance of polluted atmosphere.

2. *Conduction*: By this mechanism the heat passes through a substance from point to point by means of the transfer of adjacent molecular motions. Since air is poor conductor, this type of heat transfer can be virtually neglected in the atmosphere, but it is important in the ground.

3. *Convection*: This occurs in fluids and gases which are able to circulate internally and distribute heated parts of the mass. The low thickness of the air and its consequent ease of motion make this the chief method of atmospheric heat transfer. It is noticed that forced convection (mechanical turbulence) occurs due to the development of eddies as air flows over uneven surface, even when there is no surface heating to set up free (thermal) convection.

Convection transfers energy in two forms. The first is the ***sensible heat*** content of the air (called enthalpy by Physicists) which is transferred direct by the rising and mixing of warmed air. It is the specific heat at constant pressure (the heat absorbed by unit mass for unit temperature increase). Sensible heat is also transferred by conduction. The second form of energy transferred by convection is indirect involving ***latent heat.*** Here, there is a phase change but no temperature change. Whenever water is converted into vapour by evaporation (or boiling) heat is required. This is referred to as the latent heat of vapourization. When water condenses in the atmosphere the same amount of latent heat is given off as is used for evaporation at the same temperature. In all of these phase changes of water there is an energy transfer.

Effect of the Atmosphere

Solar radiation is virtually all in the short-wave length range less than 4 micrometre. About 18 per cent of the incoming energy is absorbed directly by ozone and water vapour. Ozone absorption is concentrated in three solar spectral bands

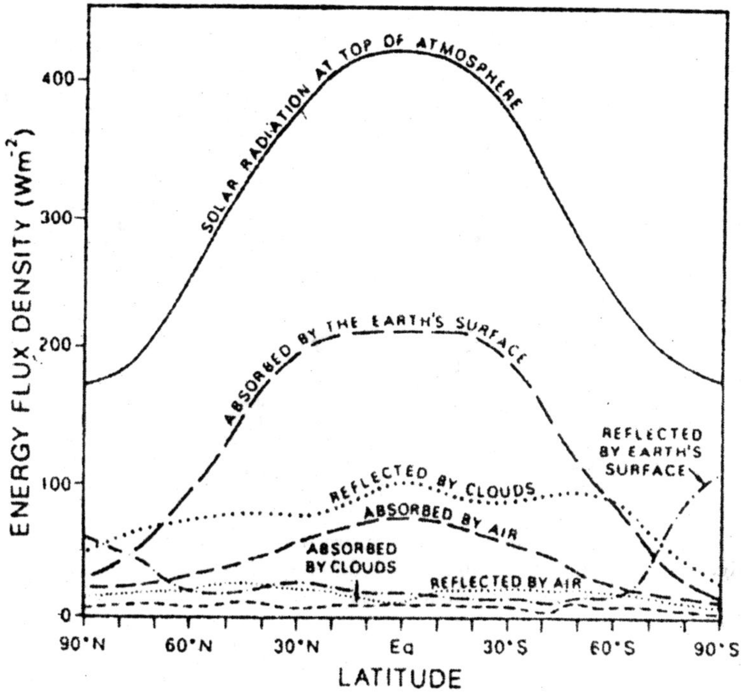

Fig. 16.9: The average annual latitudinal disposition of solar radiation in W m^{-2}. Of 100 per cent radiation entering the top of the atmosphere, about 20 per cent is reflected back to space by clouds, 3 per cent by air (plus dust and water vapour), and 8 per cent by the earth's surface. Three per cent is absorbed by clouds, 18 per cent by the air, and 48 per cent by the earth.

Source: After Sellers, 1965.

(0.20-0.31, 0.31-0.35 and 0.45-0.85 micrometre) while water vapour absorbs to a lesser degree in several bands between 0.9 and 2.1 micrometre. Solar wavelengths shorter than 0.285 micrometer scarcely penetrate below 20 km altitude, where those > 0.295 micrometre reach the surface. Thus, the 3 mm (equivalent) column of stratospheric ozone attenuates ultraviolet radiation almost entirely, except for a partial window around 0.20 micrometres, where radiation reaches the lower stratosphere. About 30 per cent is immediately reflected back into space from the atmosphere, clouds and the earth's surface, leaving approximately 70 per cent to heat the earth

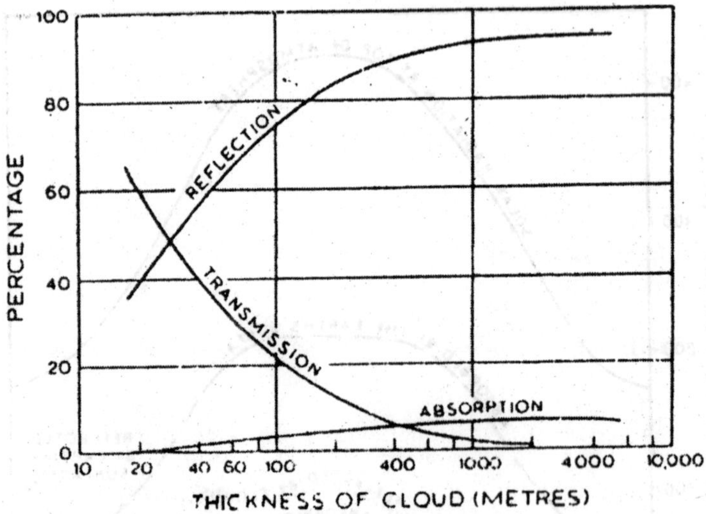

Fig. 16.10: Percentage of reflection, absorption and transmission of solar radiation by cloud layers of different thickness.
Source: From Hewson and Longley, 1944.

and its atmosphere. Of this, greater part eventually heats the atmosphere, but much of this heat is received second hand by the atmosphere via the earth's surface. The ultimate retention of this energy by the atmosphere is of prime importance, because if it did not occur the average temperature of the earth's surface would fall by some 40°C obviously making most life impossible. The surface absorbs almost half of the incoming energy available at the top of the atmosphere reradiates it outwards as long (infrared) wave of greater than 3 micrometre. Much of this reradiated long-wave energy can be absorbed by the water vapour, carbon dioxide and ozone in the atmosphere, the rest escaping through atmospheric windows back into outer space, principally between 8 and 13 micrometres. Figure 16.9 illustrates the relative roles of atmosphere, clouds and radiation at different latitudes.

Effect of Cloud Cover

Cloud cover can, if it is thick and complete enough, form a significant barrier to the penetration of radiation. The drop in surface temperature experienced on a sunny day when a temporarily cuts of the solar radiation illustrates our reliance upon the sun's radiation energy. How much radiation is reflected depends on the amount of the cloud cover and its thickness (Fig. 16.10).

The proportion of incident radiation that is reflected is termed the *albedo* or reflection coefficient (expressed as fraction or percentage). Average albedos as determined by satellites, aircraft and surface measurements are given in Table 16.1.

Table 16.1: The average (integrated) albedo of various surfaces (0.3-0.4 micrometre)

Planet earth	0.31
Global surface	0.14-0.16
Global cloud	0.23
Cumulonimbus	0.9
Stratocumulus	0.6
Cirrus	0.4-0.5
Fresh snow	0.8-0.9
Melting snow	0.4-0.6
Sand	0.30-0.35
Grass, cereal	0.18-0.25
Deciduous forest	0.15-0.18
Tropical rainforest	0.09-0.15
Water bodies	0.06-0.10

Note: Increase sharply at low solar angle in case of water bodies.

Effect of Latitude

Latitude is very important control because this will determine both the duration of day light and the distance travelled through the atmosphere by the oblique rays from the sun. However, actual calculation shows the effect of the latter to be negligible in Arctic apparently due to low vapour content of the air limiting tropospheric absorption.

A special feature of the latitudinal receipt of radiation is that the maximum temperatures experienced at the earth's surface do not occur at the equators, as one might expect, but at the tropic. A number of factors need to be taken into account.

The apparent migration of the vertical sun is relatively rapid during its passage over the equator, but its rate slows down as it reaches the tropics. Between 6°N and 6°S sun's remain almost vertically over head for only 30 days during each of the spring and autumn equinoxes, allowing little time for any large build-up of surface heat and high temperature. On the other hand, 17.5° and 23.5° latitude the sun's rays shine down almost vertically for 86 consecutive days during the period of solstice. This longer sustained period, combined with the fact that the tropic experience longer days than at the equator, makes the maximum zone of heating occur nearer the tropics than the equator. Over the continent, the highest values occur at about 23°N and 10-15°S. In consequence the mean annual **thermal equator** (i.e. the zone of maximum temperature) is located at about 5°N. Nevertheless, the mean air temperatures reduced to mean sea level, are very broadly related to latitude.

Effect of Land and Sea

Water has a tendency to store the heat it receives, land contrast, quickly reruns it to the atmosphere. There are several reasons for this. A large proportion of incoming radiation is reflected back into the atmosphere without heating the earth's surface at all. The proportion depends upon the types of surface (as mentioned in Table 16.1). A sea surface reflects a very little unless the angle of incidence of the sun's rays is large. The albedo for the calm water surface is only 2 to 3 per cent for a solar elevation angle exceeding 60° but is more than 50 per cent for land surfaces, the elbedo is generally between 8 to 40 per cent of the incoming radiation. The figure for forest is about 9 to 18 per cent. Fresh snow may reflect as much as 90 per cent of solar radiation. A measurement of the difference between the sub-surtaces of land and sea is given in Figure 16.11.

Air is poor conductor and for this reason a loose sandy soil surface heats up rapidly by day as the heat is not conducted always. Increased soil moisture tends to raise conductivity by filling the soil pores, but too much moisture increases the soils' heating capacity thereby reduces the temperature response.

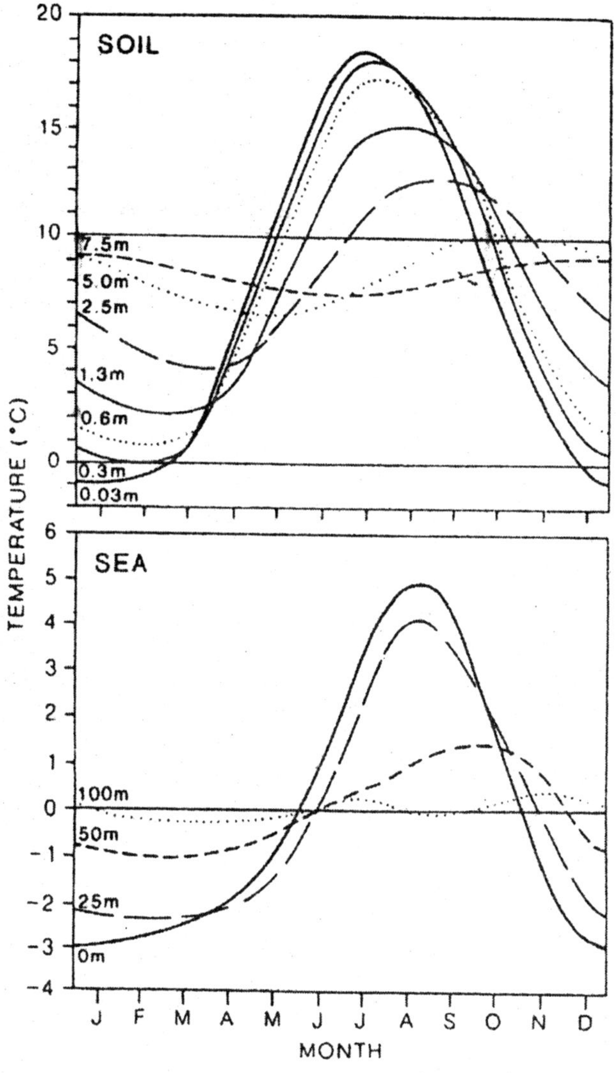

Fig. 16.11: Annual variation of temperature at different depths in soil at Kaliningrad, European Russia (above) and in the water of the Bay of Biscay (at approximately 47°N, 12°W) (below), illustrating the relatively deep penetration of solar energy into the oceans as distinct from that into land surfaces. The bottom figure shows the temperature deviations from the annual mean for each depth.
Source: Geiger 1965 and Sverdrup 1945.

The thermal role of the ocean is important and complex one. The ocean has three thermal layers:

1. The seasonal boundary, or upper mixed layer lying above the thermocline. This is less than 100 m deep in the tropics but is hundreds of metres deep in sub-polar seas. It is subject to annual thermal mixing from the surface.
2. The warm water sphere of lower mixed layer. This underlies '1' and slowly exchange heat with it down to many hundreds of metres.
3. The deep ocean. This contains some 80 per cent of the total oceanic water volume and exchanges heat with '1' in polar seas.

This vertical thermal circulation allows global heat to be conserved in the oceans, thus dumping down the global effects of climatic change produced by thermal forcing. The time for heat energy diffuse within the upper mixed layer is two-seven months, within the lower mixed layer seven years, and within the deep ocean upwards of 300 years. The comparative figure for the outer thermal layers of the solid earth is only 11 years. These differences between land and sea help to produce what is termed **continentality.** Continentality implies, first that a land surface heats and cools much quicker than that of an ocean. Over the land, the lag between maximum and minimum periods of radiation and the maximum surface temperatures is only one month, but over the ocean and at coastal stations the lag is as much as two months. Second, the annual and diurnal ranges of temperature are greater in continental than in coastal location. The third effect of continentality results from the global distribution of the land masses. The smaller area of the northern hemisphere causes the northern hemisphere summer to be warmer, but its winters colder on the average than those of the southern hemisphere (summer 22.4°C *versus* 17.1°C, winter 8.1°C *versus* 9.7°C). Heat storage in the oceans causes them to be warmer in winter and cooler in summer than land in the same latitude, although ocean currents give rise to

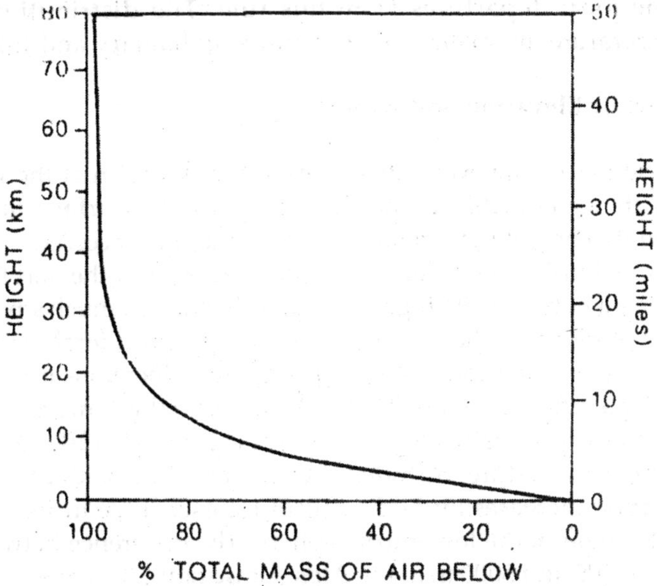

Fig. 16.12(A): The percentage of the total mass of the atmosphere lying below elevations up to 80 km (50 miles). This illustrates the shallow character of the earth's atmosphere.

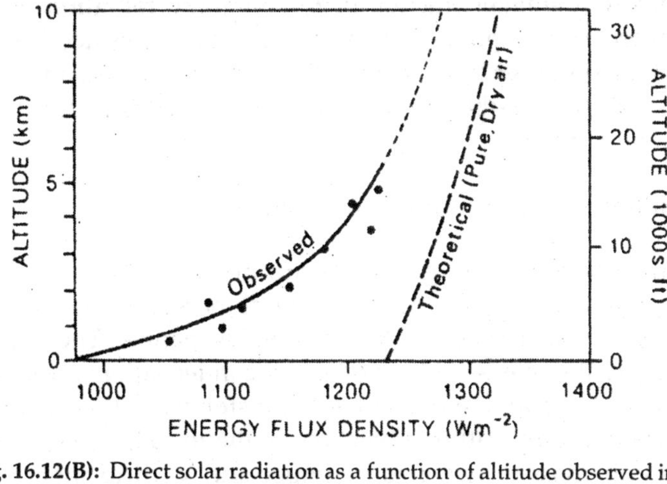

Fig. 16.12(B): Direct solar radiation as a function of altitude observed in the European Alps. The absorbing effects of water vapour and dust, particularly below about 3,000 m, are shown by comparison with a theoretical curve for an ideal atmosphere.

(*Source*: After Albettic, Kastrov, Kimbal and Pope, from Barry, 1992).

some local departures from this rule. The distribution of temperature anomalies for the latitude in January and July.

Effect of Elevation and Aspect

At the local scale even differences in the elevation of the land and its aspect (that is the direction that the surface faces) strongly control the amount of solar radiation received.

Obviously some slopes are more exposed to the sun than others, whereas really high elevations that have a much smaller mass of air above them (Fig. 16.12A) receive considerably more direct solar radiation under clear skies than locations near sea level, particularly below 2000-3000 m due to the concentration of water vapour in the lower troposphere (Fig. 16.12B). On the average in middle latitudes the intensity of incident solar radiation increases by 5-15 per cent for each 1000 m increase in elevation in the lower troposphere. The difference between sites at 200 and 3000 m in the Alps, for instance, can amount to 70 W $^{m-2}$ on cloudless summer days. However, there is also correspondingly greater net loss of terrestrial radiation at higher elevations because the low density of the overlying air results in a smaller fraction of the outgoing radiation being absorbed. The overall effect is invariably complicated by the greater cloudiness associated with mountain ranges, and it is, therefore, impossible to generalize from the limited date at present available.

Variation of Temperature with Height

Vertical temperature gradients are determined in part by energy transfers and in part by vertical motion of the air. The various factors interact in a highly complex manner. The energy terms are the release of latent heat by condensation, radiation cooling of air and sensible heat transfer from the ground. Horizontal temperature advection by the motion of cold and warm air masses may also be important. Vertical motion is depended on the type of pressure system. High pressure area are associated with decent and warming of deep layers if air, hence the decreasing the temperature gradient and frequently causing temperature inversion in lower layer of troposphere. In contrast the low pressure systems are associated with rising air, which cool upon expansion and increase the temperature gradient. This is only part of the story. The vertical decrease of temperature or **Lapse Rate** in the troposphere is as has been stated about 6.5°C/km. However, this is no means constant with height, season and location.

Another factor which causes the lowering of temperature with elevation is the fact that the lower layer of air at or near the surface are dense and contain more water vapour, more water particles and more dust particles than the upper air. Since these substances absorb a large amount of terrestrial radiation, the lower part of the atmosphere has higher temperature than the less dense upper part. The permanent snow caps in higher mountains, even in low latitude indicate the decrease of temperature with altitude.

Heat Budget of the Earth

The net effect of transfer of energy in the earth-atmosphere system averaged over the globe and over an annual period. For convenience we shall regard it as 100 units (Refer Fig. 16.13) incoming radiation is absorbed in the stratosphere (3 units) by ozone mainly 18 units are absorbed in the troposphere by carbon dioxide (1), water vapour (12), dust (2) and water droplets in clouds (3). Twenty units are reflected back to space from clouds, which cover about 62 per cent of the earth's surface, on average. A further 8 units are reflected by from the surface and 3 units are returned by atmospheric scattering. The total radiation reflection radiation is the planetary **albedo** (31 % or 0.31) the remaining 48 units reach the earth either directly (27) or diffuse radiation (21) transmitted via clouds or by downward scattering.

The pattern of outgoing terrestrial radiation is quite different. The black body radiation assuming a mean surface temperature of 288 K, is equivalent to 114 units of infrared (long wave) radiation. This is possible because most of the outgoing radiation is absorbed by the atmosphere; the net loss of the infrared radiation is only 18 units. These exchanges

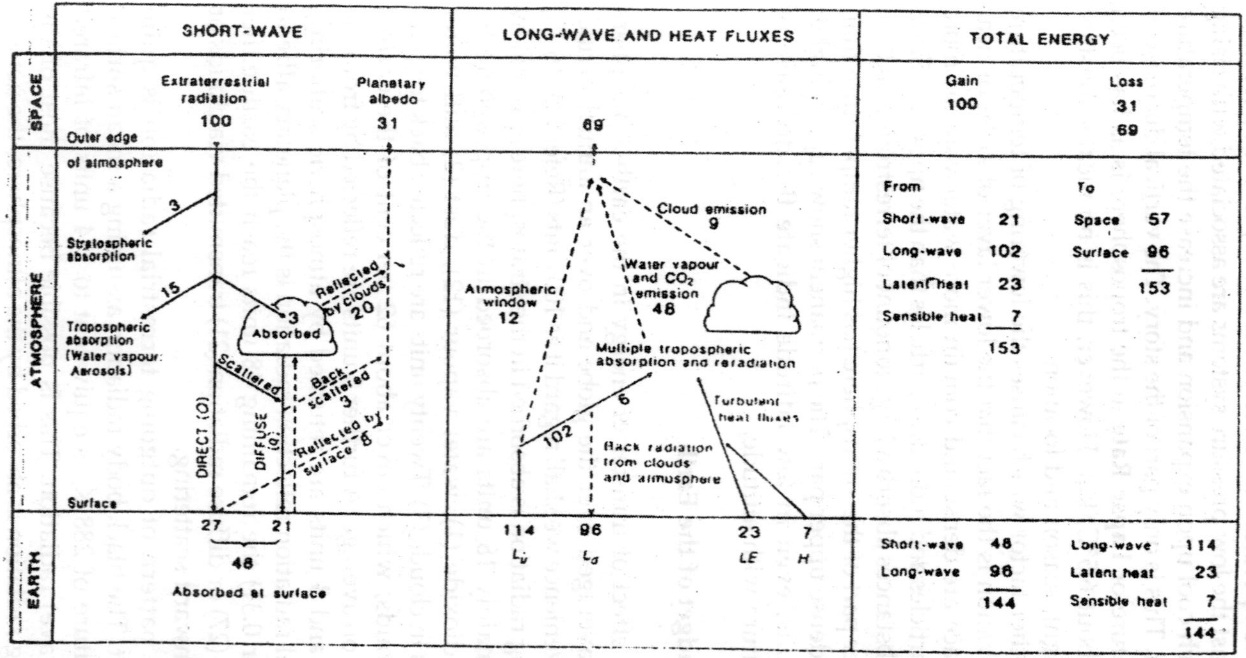

Fig. 16.13: The balance of the atmospheric energy budget. The transfers are explained in the text. Solid lines indicate energy gains by the atmosphere and surface in the left-hand diagram and the troposphere in the right-hand diagram. The exchanges are referred to 100 units of incoming solar radiation at the top of the atmosphere (equal to 342 W m^{-2}).

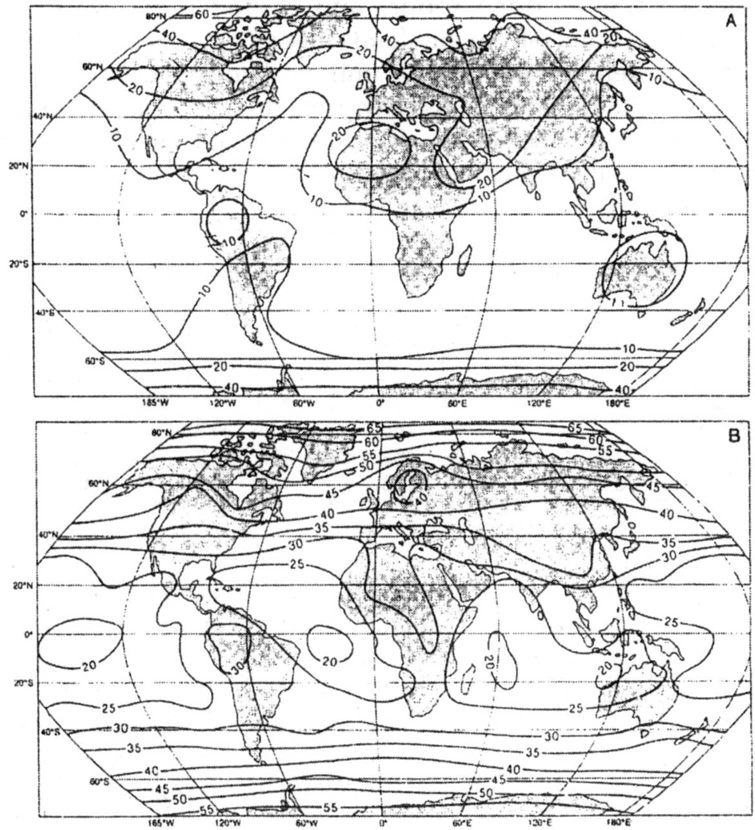

Fig. 16.14: Mean annual albedos (per cent): (A) At the earth's surface; (B) On a horizontal surface at the top of the atmosphere.

Source: After Hummel and Reck; from Henderson-Sellers and Wilson 1983, and Stephens *et al.* 1981.

represent a time averaged state for the whole globe. Recall that solar radiation effects only the sunlit hemisphere, where the incoming radiation exceeds 342 W m^{-2}. Conversely on the night time hemisphere no solar radiation is received. Infrared exchanges continue, however due to the accumulated heat in the ground. Only about 12 units escapes through the atmospheric window directly from the surface, but the atmosphere radiation radiates 57 units to space (48 from the emission by water vapours and carbon dioxide in the atmosphere and 9 from the cloud emission) giving a total of 69 units as well as reradiating 96 units back to the surface.

Satellite measurements now provide global views energy balance at the top of the atmosphere. The incident solar radiation is almost symmetrical about the equator in the annual means as mentioned in Table 16.2.

Table 16.2: Daily solar radiation on a horizontal surface outside the atmosphere: Wm^{-2}

Date	*90°N*	*70*	*50*	*30*	*0*	*30*	*50*	*70*	*90°S*
December 21	0	0	86	227	410	507	514	526	559
March 21	0	149	280	378	436	378	280	149	0
June 22	524	492	482	474	384	213	80	0	0
September 23	0	147	276	373	430	372	276	147	0

Source: After Berger 1996.

The mean annual total on a horizontal surface at the top of the atmosphere are approximately 420W m^{-2} at the equator and 180 W m^{-2} at the poles. The distribution of planetary albedo (Fig. 16.14) shows the lowest values over the low latitude oceans compared with the more persistent areas of the cloud cover continents—the highest values over the polar caps. The resulting the planetary short-wave radiation ranges from 340 W m^{-2} at equator to 80 W m^{-2} at the poles.

The net long-wave (outgoing) radiation, (Fig. 16.15) shows the smallest losses where the temperature are lowest and largest losses over the largely clear skies of the Saharan desert surface and over low latitude oceans.

ATMOSPHERIC ENERGY AND HORIZONTAL HEAT TRANSPORT

There are two other forms of energy which are important: **Geopotential energy** due to gravity and height above the surface and **Kinetic energy** associated with air motion.

Geopotential and internal are interrelated since the addition of heat to an air column not only increases its internal

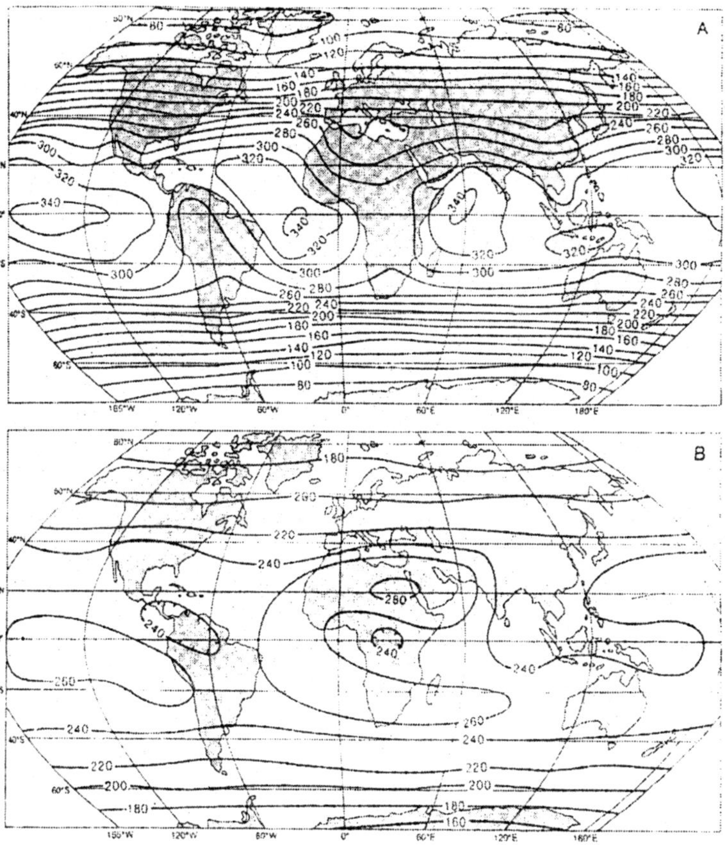

Fig. 16.15: Planetary short-wave and long-wave radiation (W m^{-2}): (A) Mean annual absorbed short-wave radiation for the period April 1979 to March 1987. (B) Mean annual net planetary long-wave radiation (L_n) on a horizontal surface at the top of the atmosphere.

Source: (A) Ardanuy *et al.* 1992 and Kyle *et al.* 1993. From *Bulletin of the American Meteorological Society*, by permission of the American Meteorological Society. (B) Stephens *et al.* 1981.

energy but also adds to its geopotential as a result of the vertical expansion of air column. In a column extending to the top of the atmosphere, geopotential is approximately 40 per cent of the internal energy. These two are therefore usually considered together and termed the total potential energy (PE) for whole atmosphere.

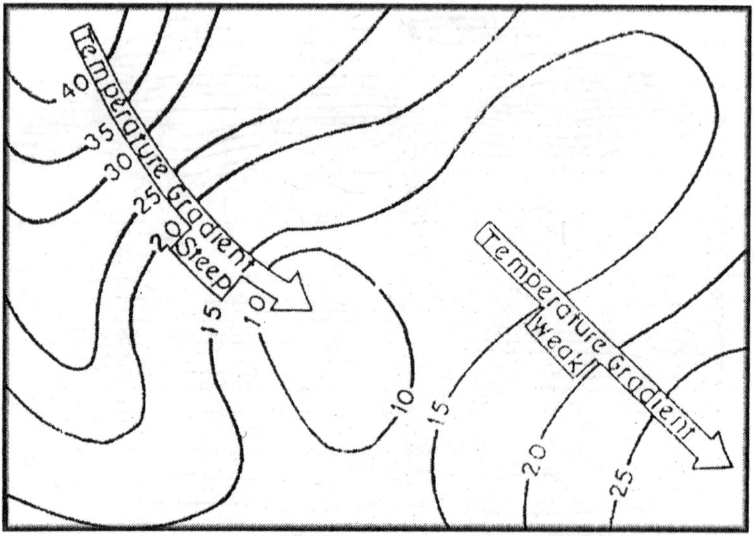

Fig. 16.16: Isotherms and temperature gradient.

Potential energy $=10^{24}$ J
Kinetic energy $=10^{10}$ J

It is evident that the receipt of heat energy is very unequal geographically and that this must lead to great lateral transfers of energy across the surface of the earth.

The Horizontal Transport of Heat

The horizontal distribution of temperature is controlled by number of factors. The effectiveness of insolation in heating the earth's surface depends on latitude. There is general decrease in temperature from equator towards the poles. However, if the effect of the latitudinal variation in the receipt of solar insolation were the only control of temperature distribution, the isotherm on a world temperature map would have been exactly parallel to the parallel of latitude. But in spite of the fact, the isotherms generally trend east-west, they are highly irregular. The isotherm over certain parts of the globe closely spaced, while elsewhere they are widely spaced. This is a clear proof of the wide variation in temperature gradient

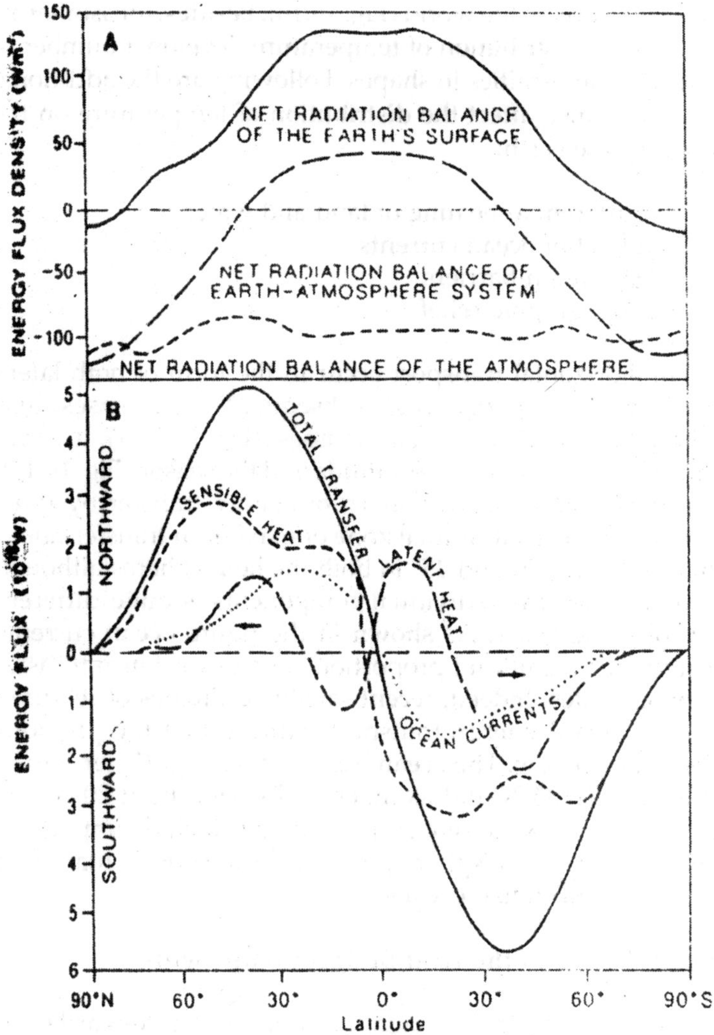

Fig. 16.17: (A) Net radiation balance for the earth's surface of 101 W m^{-2} (incoming solar radiation of 156 W m^{-2}, minus outgoing long-wave energy to the atmosphere of 55 W m^{-2}); for the atmosphere of – 101 W m^{-2} (incoming solar radiation of 84 W m^{-2}, minus outgoing long-wave energy to space of 185 W m^{-2}); and for the whole earth-atmosphere system of zero. (B) The average annual latitudinal distribution of the components of the pole-ward energy transfer (in 10^{15} W) in the earth-atmosphere system.

Source: From Sellers 1965.

in various parts of the world (Fig. 16.16). Besides, the isotherms showing the distribution of temperature develop a number of bends and deformities in shapes. Following are the additional factors that also affect the distribution of temperature on the surface of the earth:

1. Differential heating of land and water
2. Effect of ocean currents
3. Mountain barriers
4. Topographic relief

The horizontal transport occur in the form of both latent (that is water vapour which subsequently condenses) and sensible heat (that is warm air masses) (Fig. 16.17A). It varies in intensity according to the latitude and the season. Fig. 16.17B shows the mean annual pattern of energy transfer by three mechanisms. The latitudinal zone of maximum transfer rate is found between 35° and 45° in both the hemispheres although the pattern for the individual components is quite different from one and other. As shown in the figure ocean currents account for a significant proportion of pole-ward heat transfer in low latitudes. Indeed, recent satellite estimates of required total pole-ward energy transport indicate that the previous figures are too low. The ocean transport may be 47 per cent of the total at 30-35°N and as much as 74 per cent at 20°N; the Gulf stream and Kuro shio currents are particularly important. In the southern hemisphere pole-ward transport is mainly in the Pacific and Indian oceans.

Spatial Pattern of the Heat Budget Components

The annual net radiation magnitude decreases pole-ward from about 25° latitude although as a result of the considerably absorption of solar radiation by the sea, the net radiation is greater over the oceans exceeding 160 Wm^{-2} in latitudes 15-20°—than over land areas, where it is about 80-150 Wm^{-2} in the same latitudes. Net radiation is rather lower in arid continental area than in humid ones, because in spite of the increased insolation receipts under clear skies there is at the same time greater net loss of terrestrial radiation.

TEMPERATURE

One of the most important elements of climate is Air temperature. Human beings are most sensitive to it. Temperature is a relative term which implies a degree of heat. If the heat from one body flows to the other we say that the former has higher temperature. Usually two types of scales are used to measure this temperature known as Fahrenheit and Centigrade scales. In Fahrenheit the boiling point is fixed at 212°, whereas in centigrade it is at 100°. The freezing point in the respective scales is at 32° in Fahrenheit and 0° in centigrade. A third scale known as Reumar scale is used to note the body temperature.

The conversion equation is thus:

$$\frac{C}{100} = \frac{F-32}{180} = \frac{R}{80}$$

F = 180 + 32 and C = 100.

To depict daily temperature of a place, daily mean temperature value has been widely used. It is the daily mean value from which the mean monthly and average annual temperatures are derived. To find out the daily mean temperature we have to add the highest and the lowest values in 24 hours and divide the sum with 2. The daily mean temperature values are added and the sum is divided by the number of days to derive the mean monthly temperature. The mean monthly temperature for the year as a whole is known as Annual Mean of the temperature.

Diurnal Range of Temperature

The difference of temperature between the highest and the lowest of the 24 hours is known as the Diurnal range of temperature or Daily range of temperature. Diurnal range of temperature is higher in inland areas and low in the coastal areas. In areas of hot dry climate, the diurnal range of temperature is higher during calm and clear sky at night.

The difference between the mean temperature of the warmest month and mean temperature of the coldest month

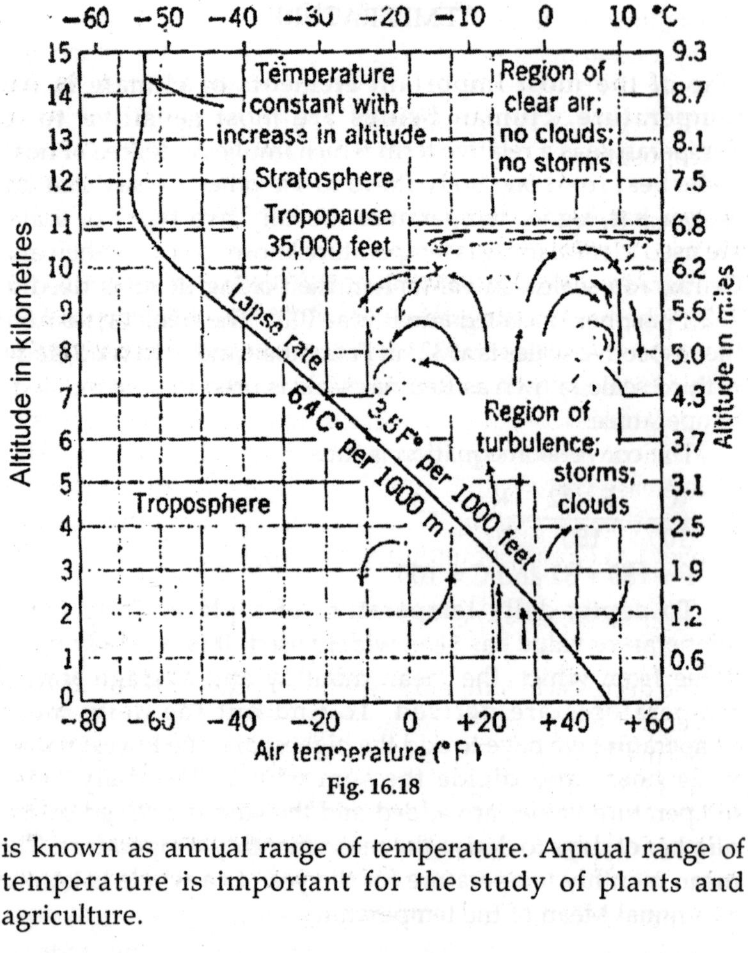

Fig. 16.18

is known as annual range of temperature. Annual range of temperature is important for the study of plants and agriculture.

Vertical Distribution of Temperature

It is an established fact that under normal conditions the atmospheric temperature decreases with increasing elevation. However, the rate of decrease does not remain uniform throughout. It varies with time, space and season. The average decrease in temperature is 3.3°F for 1000 feet of altitude. The fact that lower layer is warmer than the upper layer shows that the layer of atmosphere nearer the earth gets heated more by the heat escape from the earth than from direct radiation.

Another cause of warmth in lower layer is its density. This layer contains water vapours and dust particles which absorb more heat, whereas the upper layers are thinner. Normally, the decrease in temperature continues up to the limits of troposphere.

Lapse Rate

The rate of decrease in temperature with altitude, in the troposphere is called lapse rate. If the rate of decline is 3.3° per 1000 feet of altitude it is mentioned as normal lapse rate but it is an average of many values of lapse rate.

Actual lapse rate is a single observation over a particular location. In some cases the actual observation may not show a decrease with altitude at lower elevations such a lapse rate is called isothermal lapse rate. Such a situation does not occur over large range of altitude or a longer period.

Environmental Lapse Rate

The decrease in temperature with altitude air is still also called vertical temperature gradient or environmental lapse rate or normal lapse rate. The average gradient is 0.6°C per 100 metres or 3.3°F per 1000 feet of altitude.

Dry Adiabatic Lapse Rate

When an unsaturated warm air mass rises upwards it expands and cools adiabatically. The lapse rate occurring under such conditions is known as dry adiabatic lapse rate which is usually 1° for 100 metres or 5.44°F for 1000 feet of altitude.

Wet Adiabatic Lapse Rate

When a saturated warm air mass is uplifted, it expands and cools adiabatically. But the rate of cooling is low due to the release of latent heat from the water particles. This is saturated or wet adiabatic lapse rate which usually ranges from 0-3°C to 0.9°C per 100 metres.

The temperature varies with altitude and the variation have different nature in different layers of the atmosphere. In troposphere, the temperature generally decreases with altitude except for the inversion. In the stratosphere, it starts increasing after 30 kms.

Inversion of Temperature

In some cases, there is an increase in the temperature with altitude in the lower part of the troposphere. In this case we can say that the lapse rate has been inverted. Such a condition is known as inversion of temperature. Temperature inversion may occur either near the surface or above the troposphere. The inversion near the earth surface is caused by:

(i) Radiation of heat from the earth on a clear night.
(ii) Frosts on the surface of the earth cooling the lower layer of the air, whereas the air above is still warm.
(iii) When two air masses of different nature collide with each other the warm air mass overrides the cold air mass.
(iv) Advection of warm air over cold surface.

At higher level, the inversion may occur due to the subsidence of a mass of air during which the air is dynamically heated to cause inversion.

Horizontal Distribution of Temperature

The temperature on the surface of the earth is controlled by a number of factors such as duration of sunshine, latitude, altitude, wind and surface configuration. All these conditions vary from place to place and so is the case with temperature. The places which are nearer the equator have higher temperature than the places located at higher altitude which is colder than the places at lower altitude. Movement of wind regulates the temperature of places. If a warm air mass moves over cold region the severity of cold will be reduced and *vice versa*. Similarly, the ocean currents also have adequate effect on regulating the temperature of places located near the coastal areas visited by them. Gulf Stream is a warm water current which regulates the temperature in North Atlantic Ocean and Scandinavian coast, the kurosivo current has warming effect on Japanese coastal area.

Temperature variations from place to place are as important as the differences from time to time. The increasing number of recording station today has made it possible to workout a regional scheme of temperature for the world as a whole. The temperatures of various places are marked, and places having similar temperature are joined with a line known as Isotherms. Isotherms are then drawn for the entire globe to have a horizontal distribution of temperature on the globe. Generally, the isotherms show a pattern of more or less parallel lines in the direction of the latitudes. However, as the land and water are irregularly distributed on the surface of the earth the latitudinal pattern of isotherms is broken. Land areas get heated and cooled faster than the water bodies and temperature ranges are higher on the land surfaces. This contrast in the temperature of land and water surfaces is because:

(i) Water is always mobile. The heat energy is distributed by vertical as well as horizontal movement of water. In case of land the heat energy is sent downwards by conduction very slowly.
(ii) Water is translucent and the radiant energy penetrates directly deep into the water. Earth is an opaque body and the radiant energy is absorbed by the surface only.
(iii) The specific heat of water is higher than that of land. That means a given mass of water body requires more energy to raise its temperature by 1° than an equal mass of land.
(iv) Reflection of heat from water also leads to slow the rate of heating. However, generally the reflectivity is similar in both the cases.

The general principle is that towards the equator the sun at mid-day is always over head, conditions of summer prevail and towards the poles the sun is always low, winter conditions prevail. The isotherms bend pole-wards over the land and equator-wards over the sea in the vicinity of the equator. Near the polar-areas the bends in isotherms are in opposite direction. Near some middle latitudes the temperature may be about same on land and sea, in that case the isotherms will be about straight lines. Let us trace the 30° isotherm in northern hemisphere from west to east, we find that it bends pole-wards over the Pacific, equator-wards over north America, pole-wards again over the Atlantic Ocean and again equator-wards over the Eurasian Continent. In the southern hemisphere the bends are much less marked, because the land masses are narrow (Fig. 16.18). On the other hand, the isotherm of 80° bends towards the equator over the Atlantic Ocean and pole-wards over Africa and America in the northern hemisphere. The isotherms drawn for the globe possess the following characteristics:

(i) During the summer season isotherms bend towards the equator while passing from land to sea and turn towards the poles during winter.

(ii) On the land where variations in surface features are absent isotherms are parallel to the latitudes, more or less.

However, on the base, temperature conditions divide the globe into 3 different zones:

(i) Torrid Zone extending along the equator tropic of Cancer and tropic of Capricorn,

(ii) Temperate zone lying between 23°N and S latitude to 66°N and south latitude, and

(iii) Frigid Zone lying above 66° N and S latitude.

17

AIR PRESSURE

Air has weight which under normal conditions is about one and a quarter ounce per cubic foot of air. If it has weight it must exert some pressure on the surface below. The pressure of the air depends upon the amount of air above a point. It is a common sense that it will be lesser at the top and greater at the bottom. At sea level under normal conditions the average air pressure will be 14.7 pounds per square inch and it is exerted in all directions.

The air pressure varies from time to time and from season to season at a given place. It also varies from place to place as well as decreases with altitude. Pressure depends upon the temperature of a place. As temperature increases the air gets heated and expands as a result of which the pressure is reduced. The differences in pressure from place to place result in the movement of winds.

Measurement

Measurement of pressure variations is important to understand the flow of winds, storms and related atmospheric phenomena. The air pressure is measured with the help of a barometer. Mercurial barometer is the most accurate instrument to measure the air pressure. The principle is the balancing of the column of air against a column of mercury in a sealed glass tube. It has been found that under normal conditions at sea level the atmosphere balances a column of mercury 29.92 inches high. This means that normally at sea level air pressure is 29.92 inches or 14.7 pounds or 1073.2 mbs. In metric units, the above figures equal to 76 centimetres or 760 millimetres of mercury. Meteorologist commonly uses concept of force, e.g. millibars all-over the world. Use of mbs. was started in 1440 in U.S.A. One millibar is the force exerted by 1000 dynes on a square centimetre. (A dyne is defined as the force which, acting on one gram mass for one second, impart to it a velocity of one centimetre per second.) The physicists are using atmospheres in laboratory work, one atmosphere being the equivalent of 14.7 pounds per square inch or 29.92 inches of mercury or 1013.2 mbs.

Vertical Distribution of Pressure

Just like other gases the air is compressible, hence its density is greater at lower level where it is compressed under the mass of the air above. When a barometer is taken to the top of a mountain we will find that the pressure has reduced simply because the amount of air mass is less at the top. Air pressure always decreases with altitude. Up to an altitude of 5000 feet the average decline in pressure is about one inch per 1000 feet of altitude; from 5,000 to 20,000 feet it varies between 0.61 to 0.97 inches and from 20,000 to 50,000 feet it varies from 1.18 to 2.65 inches (Fig. 17.2).

Horizontal Distribution of Pressure

If we compare the individual station readings of pressure we find that small differences do exist even at sea level from place to place. These small differences in pressure are of great importance for analyzing the weather situation and forecasting of weather. On the globe the pressure at sea level may vary from 29.00 to 30.50 inches of mercury or about 982 mbs. to 1033 mbs. The world distribution of pressure is represented on a world map with isobars. Isobars are the lines connecting all those places where the pressure reading is uniform or similar. Usually sea level isobars are used to depict pressure on the surface. Generally, the isobars may be linked to contour lines on a relief map. The causes of variation in horizontal distribution of pressure are temperature latitude and land and water relationship (Fig. 17.1).

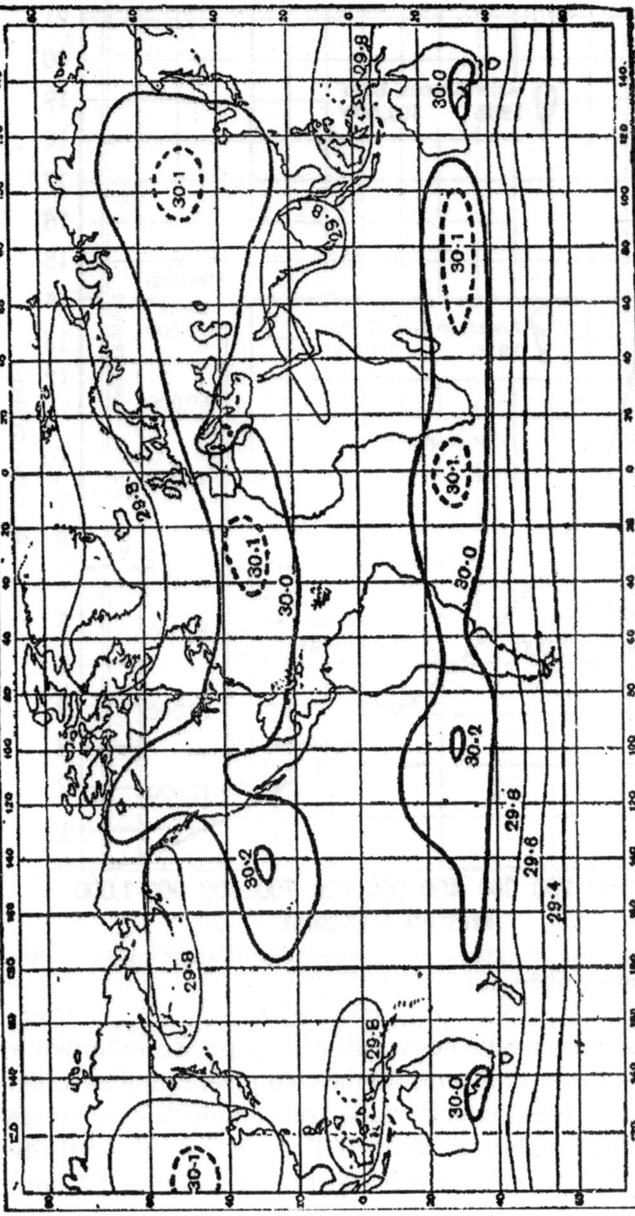

Fig. 17.1: Mean annual isobars (After Buchan and others) (Isobars shown are approximately 995, 1002, 1009, 1016, 1019, 1023 millibars).

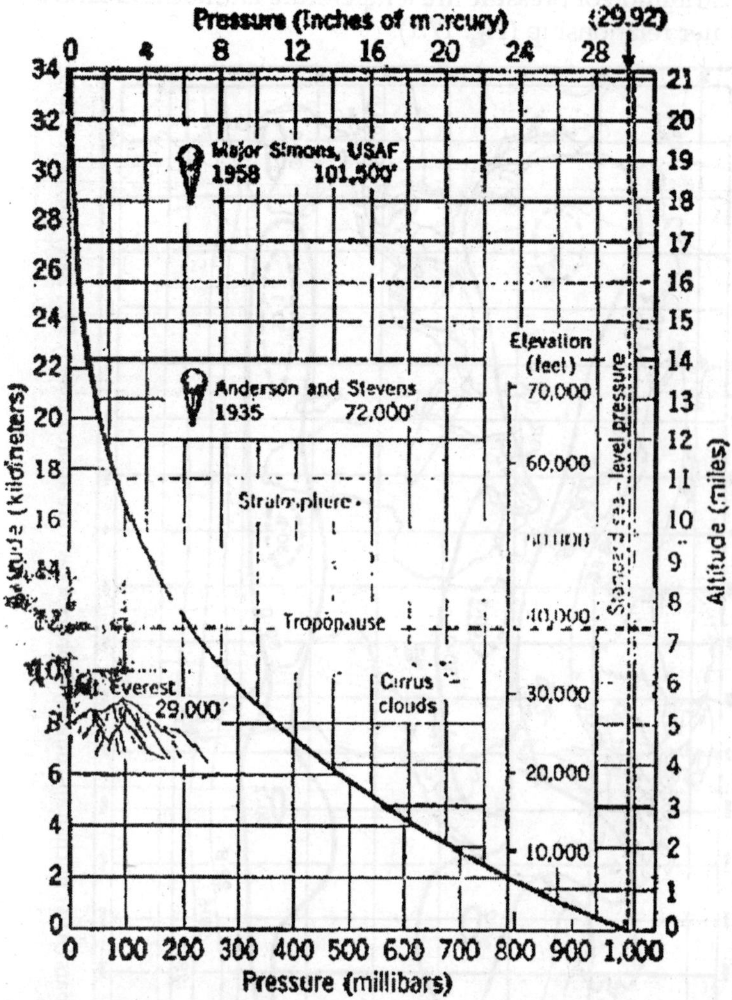

Fig. 17.2: Decrease of air pressure with elevation. (Data from Humphreys, Physics of the Air)

Along the equator lies a belt of low pressure known as equatorial low or doldrums. In the cold polar latitudes a high pressure zone exists which is called polar highs. Sub-polar low pressure belt exists at about 60° to 70° south latitude. Subtropical high pressure belts exist between 25° to 35° north

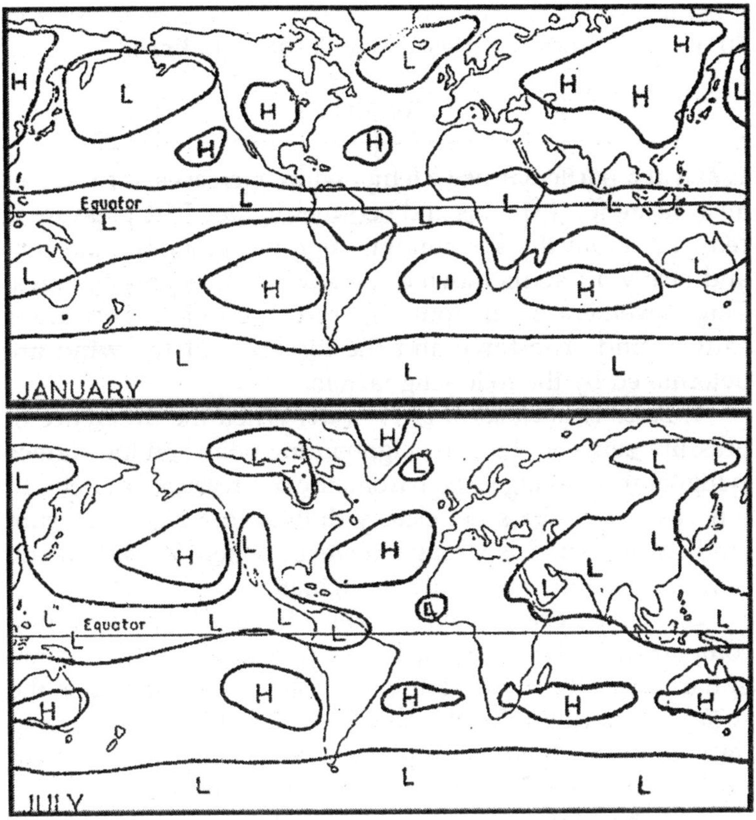

Fig. 17.3: World Pressure 'Cells' in January and July.

and 25° to 35° south latitude. These pressure belts are generally associated to temperature. However, it will not be wise to regard temperature as a sole cause, because wind plays an important role in the redistribution of temperature. These pressure belts are also affected by the shift in the position of the sun from north to south and *vice versa*. The distribution of land and water also affects the pressure (Fig. 17.3). During winter season the low pressure is mostly over the water bodies, whereas the land surfaces have high pressure. North Atlantic ocean is under the dominance of high pressure in summer and low in winter. On the other hand, southern hemisphere which has more homogeneous surface, mostly water, does not show great seasonal changes. For example, the isobars at 40° and 70° south latitude are merely parallel straight lines.

WIND

When at a particular place temperature increases, the mass of air is heated. It expands and becomes lighter. The pressure is reduced. From the surrounding high pressure area the cold and heavy air starts moving towards the low pressure area. This movement of air from high pressure to low pressure is called wind. The force and the direction of the wind are determined by the following factors:

1. *Barometric Gradient or Pressure*: When the difference of pressure between the two neighbouring points on the surface is higher it is having steep barometric or pressure gradient. In this case the isobars are closely drawn. In a steep pressure gradient the wind is strong. However, when the barometric gradient is gentle the winds are lighter.

2. *Geotropic Force or Corrial is Force*: It refers to the deflecting component in the centrifugal force produced by the rotation of the earth. Due to the earth's rotation a body moving on the surface in northern hemisphere is deflected towards its own right. In the southern hemisphere is deflected towards its left. As a result of this force the winds move in clockwise direction around the high pressure and anti-clockwise around low pressure in northern hemisphere. In southern hemisphere it moves anti-clockwise around high pressure and clockwise around the low pressure.

3. *Friction*: The power of deflection is reduced by ground friction. This results in the modification of the wind direction of a slight angle.

Types of Wind

Local Winds: Local winds may be classified as slope winds, coast winds and mountain winds. Any wind phenomena which cannot fit into any simple classification may be considered in local winds.

Slope Winds: Most familiar of all the winds are the gentle down slope winds. These winds occur in the inland parts on clear night during quiet weather. During the quiet weather the surface is cooled by radiation. The air along the slope becomes dense whereas the air in the valley at the same level is relatively warmer. The denser air close to the slopes sinks gently into the valley and the warmer air is slowly lifted up. These cool slope winds have a speed of 2 or 3 miles per hour. In high mountain areas its speed may be 5 to 10 miles. These are well known in Switzerland. The wind that flows in wisper valley opening into Rhine is called wisper wind. It is quite common over the glaciers. Air descending along the glacier is continuously cooled resulting in an increase in the speed.

Land and Sea Breeze: During the day the land becomes considerably warmer than the ocean due to isolation and radiation. The air above the land is heated. Due to increase in temperature the air expands and becomes lighter and starts rising. Since the sea is relatively cool the air above the sea is also cool and heavier. The heavier air starts moving from the sea towards the land. This flow of gentle breeze from sea to land during the day is called sea breeze. On the other hand, the land becomes cool quickly during the night by radiation, whereas the sea is relatively warmer. The air on land becomes cool and dense whereas the air above the sea is still warm and lighter. The cool dense air starts flowing gently from land towards the sea. This gentle breeze of air from land to sea at night is called land breeze.

The sea breeze starts in the mid-morning, is maximum in the prenoon session and gradually vanishes towards the end of the day. The land breeze develops in the early hours of the night, reaches maximum before midnight and fades out after midnight.

Bora or Coastal Slope Winds: Where a cold unstable air mass flows into a region of high snow covered plateaus or mountains descend into relatively warm seas a very strong wind develops over the coast known as coastal slope wind or Bora. When the resultant offshore breeze reinforces a general flow the wind over the sea near the coast becomes very strong with squall and gale posing danger to shipping. Notorious Bora affecting the northern shores of Adriatic is a best example. Its direction is usually northeast or east-east north. They also occur on north Caucasian shore of Black Sea and some arctic coasts. In southern California it is known as Santa Ana.

Mountain-Gay Winds/Mistral: It is a cold and dry wind flowing in northerly direction down the Rhone valley and into the warm waters of the Gulf of Lyons. It is associated with the breakthrough of cold air masses from the north whose path to Mediterranean is obstructed by the Alps and the Pyrenees. It streams through the gaps or passes wherever possible as in the case of Gibraltar strait. It is an example of mountain gap wind.

Foehns: In all mountainous regions the winds which blow from the ridges to the leeward low lands are local developments. They are known as Foehns. They are very pronounced in European Alps. In North America, east of the rocky mountains they are known as chimook winds. In Argentina near the Andes they are called zonda winds. Foehns are dry and warm winds which occur in all season.

Planetary Winds

The difference in pressure from one latitude to another leads to the development of planetary winds. These winds are of permanent nature blowing throughout the year in one direction. The two most important types of planetary winds are the tropical easterly winds and the westerly winds. (Fig. 17.4)

Tropical Easterly Winds/Trade Winds: The equatorial belt of calms and variable winds lies over the equatorial trough of low pressure between 5°N and 50°S latitude. North and south of the doldrums are the trade wind belts covering the earth surface between 5° to 30° North and South latitudes. They are initiated as a result of pressure gradient from the subtropical belts of high pressure to the equatorial trough of low pressure. In the northern hemisphere wind is deflected towards the south-west due to corrialis force. Thus the prevailing wind is

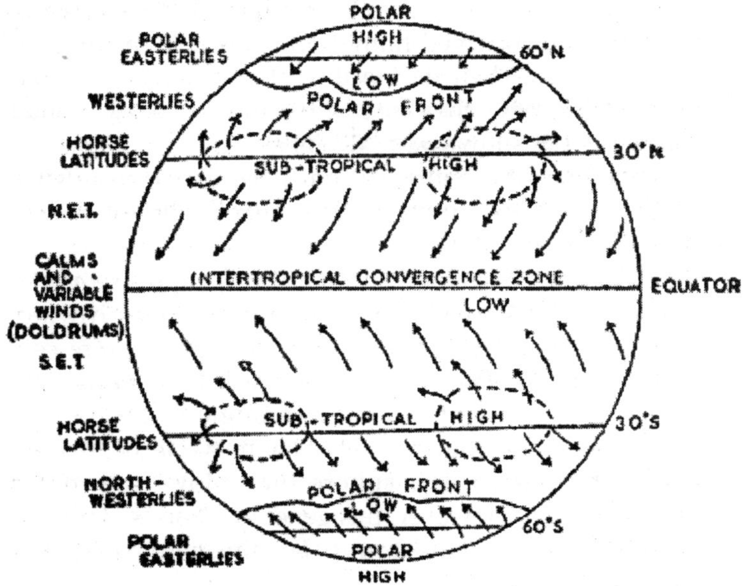

Fig. 17.4: The Planetary System of Pressure and Winds at the Equinoxes.

from the north-east and the winds are named as north-east trade winds. In the southern hemisphere it is deflected towards its left and hence it takes a direction from the east. Here it is named as south-east trade wind.

The trade winds blow continuously in the same direction throughout the year. They are noted for their steadiness. The system of doldrums and trade wind belt shifts seasonally north and south of the equator through several degrees of latitude. In the northern hemisphere these belts are shifted much farther because it is a land hemisphere. In the southern hemisphere the shift is less marked because it is a water hemisphere. Trade winds develop perfectly well over the Atlantic and Pacific oceans but are retarded in the Indian ocean due to the proximity of huge Asiatic land mass.

During the days of sailing vessels trade winds provided excellent avenue for westward travel. Since these winds are associated with steadiness in direction and flow and generally clear weather, it was the favourite zone of the mariners. However, they have contrasting nature at different places. They are stable and descending near their origin. As they reach the equator they are warmed-humid and unstable leading to a copious rainfall. When the north-east and south-east trade winds meet each other they are uplifted and give heavy rainfall. The eastern parts of the trade winds are dry as compared to the western parts. Secondly, the trade winds are more stable and steady in the eastern parts of the oceans, whereas they are variable in western parts.

The Westerlies: Between 35° and 60° north and south of the equator lies the belt of westerlies or prevailing westerly winds. These are initiated from the subtropical high pressure cells, also known as horse latitude, blowing towards the poles. They also blow in a constant direction throughout the year. In northern hemisphere they blow from south-westerly direction, whereas in southern hemisphere they blow from north-westerly direction. However, this generalization is somewhat misleading because winds from polar direction are frequent and strong.

An indepth study reveals that within the westerlies winds blow from any direction but westerly components are predominant. Storms are common, cloudy days are frequent, precipitation is high and the weather is highly changeable.

Westerlies are stronger and persistent in southern hemisphere between 40° to 60° latitude whereas these are disrupted by the land masses in the northern hemisphere. In the southern hemisphere they are known as 'Roaring Forties' around 40° latitude "Furious Fifties" around 50° latitude and the "Screaming Sixties" around 60° latitude. The westerly winds carry a number of west to east moving temperate. Cyclones embedded in them. This belt was extensively used by sailors to sail eastward from the South Atlantic Ocean to Australia, Tasmania, New Zealand and the southern pacific islands and further east around the world.

Atmospheric Disturbances

Atmospheric disturbances occur due to the horizontal movement and vertical uplift of the air columns. Both horizontal movement and vertical uplift are associated with air pressure. The air moves from high pressure areas towards the low pressure areas. The disturbances taking place due to variations in pressure have different nature and frequency in various latitudinal zones. Cyclones, anticyclones, thunder storms, planetary winds, lead to disturbances in the atmospheric conditions.

Tropical Cyclones

Tropical Cyclones occur over all the oceans except the South Atlantic Ocean. They are known as hurricanes in West Indies, Typhoons in China Sea and Philippines, Cyclones in Indian Ocean and Willy-Willies near North Western Australia. In the eye the Barometric Pressure may drop by 40 millibars in a few hours. Eye of the tropical cyclone is about 20 kilometres across. Around this calm eye the speed of the wind can be as high as 270 kilometres per hour.

In the initial stage a depression is formed by the convection of strong rising moist air from the islands. Under special circumstances a low pressure is established by the sheer of the fronts at higher level. It further deepens to give rise to strong winds converging from all sides.

The violent turbulent disturbances in the tropical zone are called tropical cyclones. The tropical cyclones are more characteristic in parts of the oceans. They are turbulent whirlwinds with a diameter varying from 80-400 kms. The Isobars on the maps are drawn in series of close concentric circles. Even if their diameter is small the pressure at the centre is very low, usually 960 mbs. or 28.35 inches.

Majority of the tropical cyclones occur towards the end of the hot season in both the hemispheres. They are observed first between 6°-20° north-south of the equator. Their track is normally around the southern and western borders of the great subtropical high pressure cells. They become weaker beyond 30° north-south of equator. The speed of their west-ward movement and pole-ward movement varies from 16 to 24 km. per hour.

The tropical cyclones are strongest over the oceans and as soon as they cross over to land they immediately lose their energy. On the land their movement is slow and has very poor cloud and rain potential. Hence the devastation and damage resulting from the extremely violent winds, associated with tropical cyclones, are limited to certain coastal areas and groups of islands in tropical oceans and seas.

However, tropical cyclones are absent in south Atlantic Ocean. It is because the belt of calms never moves further south than the equatorial region.

(1) *Distribution and Frequency*: In West Indies, Caribbean Sea, and Gulf of Mexico they are known as hurricanes. In this region they occur during the months of August, September and October. Their frequency is low as compared to typhoons. The hurricanes occur four or five times in a year. Some of them are also noticed towards the west of Mexico.

(2) *East Indies*: The tropical cyclones which develop over the sea east of Philippines are locally known as typhoons. They move westwards and then take a turn towards the north up to the coast of China. The majority of these typhoons occur from June to October. On an average between 20°-25° such typhoons have been reported from some parts of this large area.

(3) *Indian Ocean*: In the southern part of Indian Ocean, tropical cyclones occur during the months of January, March and April. About six depressions are noticed in this region in one year. They cause damage in Mauritius and Madagascar.

(4) Similar turbulent and violent cyclones occur in the Bay of Bengal. They occur either during the relatively calm hot weather in April and May or at the end of the monsoon in October. All of them are not as violent as the typhoons.

(5) *North Western Australian Coast*: A few depressions occur along the north-east coastal area of Australia. But they are better known to the north-western Australian Sea coast where they are known as Willy Willies.

General Behaviour

The behaviour of all the tropical cyclones is more or less erratic. Their route of travel is not always regular. Sometimes, when a cyclone shows the signs of intensity it may get rejuvenated as it moves onwards.

From their diameter, frequency and rather erratic behaviour it is clear that many tropical Islands and coasts may be fortunate enough to escape serious damage over many years.

Sometimes, when a tropical cyclone reaches higher latitude, e.g. on the east coast of the United States or southern Japan it generally loses its intensity and may attain the characteristic of a temperate cyclone.

Approaching warm front is preceded by continuous rain, heavily overcast sky. The wind moves in an anti-clockwise direction before the arrival of the warm front but it changes into clockwise direction at the arrival of the front. On the cold front temperature falls sharply resulting in heavy showers of rainfall—sometimes accompanied by thunders. Sometimes, exceptionally stormy conditions called line-squall associated with the cold front. It is also associated with gale, thunder and heavy downpour of hail.

The warm sector is gradually lifted off the ground by the cold air front known as occlusion. If the overtaking cold air is colder than the air in the front it is called cold occlusion and if it is not as cold it is warm occlusion. After the formation of occlusion the cyclone dissipates.

The temperate cyclones normally move parallel to the isobars in the warm sector.

Temperate Cyclones

Temperate cyclones are local low pressure systems in the temperate latitudes. A Norwegian Scientist postulated a wave theory about their origin in 1914 and 1918. The theory postulates that two adjacent air masses of cold polar air and warm tropical air are separated by a frontal zone. This frontal zone may be as wide as 80 km. The warm air is flowing towards the east whereas the cold air is flowing towards the west more or less parallel. Waves are formed when the warm air mass bulges into the cold air mass, forming a kind of salient, at the northern tip of which a low pressure centre is developed. When the pressure in the centre is very low than that on the margins, a deep depression is formed. If the pressure gradient is less the depression is weaker in the bulge sector of the cyclone. The eastward edge moving in the general direction of advance is called warm front along which the warm air is rising above the cold air.

With the further development of the cyclone the cold air moves under the warm air from the back forcing the warm air upwards and forming a cold front.

As the cyclone advances the approach of different fronts is associated with different types of weather.

Anticyclones

In temperate latitudes a system of closed isobars may have a high pressure in the middle forming an anticyclone. Anticyclones move at a slow and in an irregular manner. The slow speed of the anticyclones reveals that the weather was stabilized for some time. In anticyclones the pressure declines from the centre. Outwards as a result of which winds move outwards from the centre. The winds take a clockwise direction in the northern hemisphere. The weather is associated with light and variable winds. An anticyclone presents a picture of extremely slow descending air current.

During the summer season weather in an anticyclone tends to be dry warm and sunny. The winter is having an 'Anticyclones Gloom' depicting cloudy and overcast sky. The winter weather may also have clear sky days and sharp frosts by night and in industrial centres fogs are also common.

AIR MASSES AND FRONTS

Air Masses

A body of air in which the upward gradient of temperature and moisture are fairly uniform over a large area is known as an Air mass. In its area an air mass may extend over a subcontinent and in its vertical dimension it may extend through the troposphere. An air mass has a special combination of temperature, lapse rate, relative humidity etc. Hence the air masses vary in temperature from very warm to very cold and in moisture content from very dry to highly moist.

Air masses can be divided into different groups on the basis of their origin, trajectory, temperature and humidity. On the basis of temperature air masses can be of two types, i.e. warm and cold air masses. Polar air mass can be of two types, i.e. maritime air mass and continental air mass. Tropical air masses are also of two types. If we combine the above criteria we get four main types of air masses as Trewarths puts them:

(i) Polar Maritime Air mass.
(ii) Tropical Maritime Air mass.
(iii) Polar Continental Air mass.
(iv) Tropical Continental Air mass.

The polar continental air mass has its origin from the continental interior of North America, Asia, Greenland, forming masses of cold dry air.

As the polar continental air masses spread out towards the coastal areas over north Atlantic they are warmed from below leading to an increased water content forming polar maritime air mass. This air mass is unstable in winter and gives showers.

The air over the tropical high pressure cells is called tropical continental air mass and when moves either towards the poles or towards the equator is called tropical continental warm air mass.

Tropical Maritime Air mass is associated to the maritime climatic conditions in the tropics. Thus they can be divided into tropical maritime cold air mass and tropical maritime warm air mass. In the first case it is moving towards the warmer areas and the later is moving towards the cooler areas.

Trewarths had a similar scheme of air masses for Eurasia. Eurasia is a transition zone where invading air masses are modified. The air masses controlling weather of western Europe resemble to those of western North America, i.e. Polar Maritime Air mass. Tropical Maritime Air mass in Europe resembles to Pacific Maritime in North America but is cooler, drier and stable.

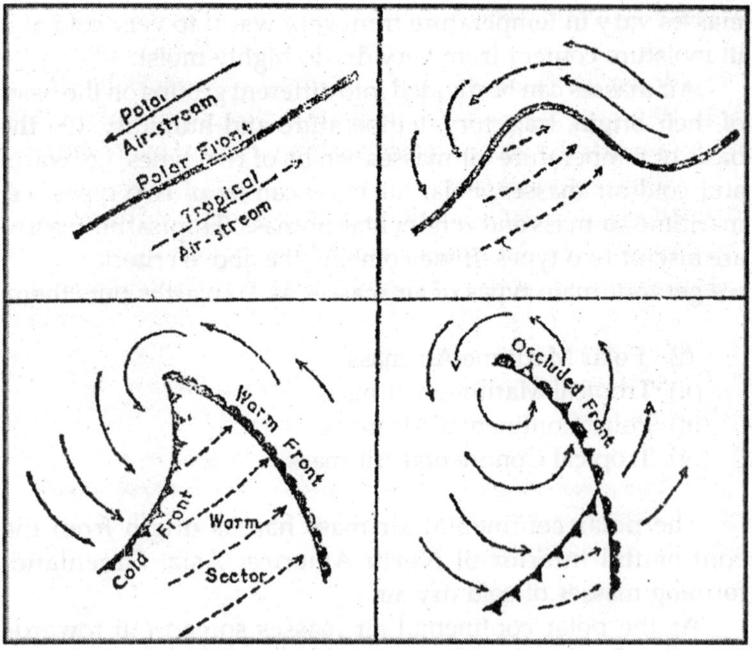

Fig. 17.5

The Mediterranean borderlands in winter are a region of Air mass convergence. Maritime polar and cold polar air from Europe and tropical continental air from North Africa get mixed up and modify the temperature and humidity resulting in instability leading to cyclonic storms.

Air Front

The boundary between two Air masses of different nature may be very sharp. This in fact is a surface of discontinuity between the two air masses and is usually termed as Air front. An air front can be defined as an axis which is being pushed by an air mass. If the air mass behind this line is cold it is named as cold air front and if it is being pushed by a warm air mass it is named as warm air front. The fronts separating air masses of varying characteristics cannot remain in a stationary position for a very long period. Normally, one is more active than the other and moves into the area of the other (Fig. 17.5).

Properties of Warm Air Front

(1) Warm air front is pushed by warm air mass behind it.
(2) When the warmer air mass is more active it moves towards the cold air mass pushing its front over wedge of the cold air mass.
(3) Warm air front is relatively unstable than cold air front.
(4) Warm air front is associated to heavy rainfall.
(5) Whenever there is a conflict between the warm and cold air fronts, warm front always overrides the cold front.
(6) Vertical movement in warm front is more than that of cold front.
(7) Warm air front is usually found in a straight line.
(8) Horizontal movement of warm front is slower than that of cold front.

Properties of Cold Front

(1) It is pushed by the cold air mass behind.
(2) It is more stable than warm air front.
(3) Cold air front always pushes the warm air front upwards leading to condensation.
(4) The cold air front near the surface is bulging upwards and captures the entire surface area.
(5) Sometimes cold front is obstructed by the ground friction leading to piling up of cold air mass. Cold air then overrides the warm air leading to violent convectional overturning.
(6) The horizontal movement of the cold front is more than that of the warm front.
(7) Vertical movement of cold front is very less as compared to the warm front.

Occlusion Front

When the two fronts, i.e. warm air front and cold air front meet each other either the warm air front overrides the cold front or cold front under runs the warm front. The line along which they mix experiences a change in temperature. This part of the front is called occlusion front.

Occlusion front is highly unstable. It is associated with heavy rainfall, thunder storms, thunder showers and hail storms. If the overtaking cold air is colder than in the front the occlusion is known as cold occlusion. If the overtaking cold air is not as cold as in the front it is called warm occlusion.

18

ATMOSPHERIC MOISTURE

The Global Hydrological Cycle

The global hydrosphere can be regarded as a series of reservoirs interconnected by water cycling in various phases. These reservoirs are the oceans; ice sheets and glaciers; terrestrial water (i.e. rivers, soil, moisture, lakes and ground water); the biosphere (i.e. water in plants and animals); and the atmosphere. The oceans, with a mean depth of 3.8 km and covering 71 per cent of the earth's surface, hold 97 per cent of all the earth's water ($1.31 \times 10^{18} m^3$). Approximately 75 per cent of the total fresh water is locked up in ice sheets and glaciers, while almost all of the remainder is ground water. It is an astonishing fact that at any instant rivers and lakes hold only 0.33 per cent of all fresh water and the atmosphere a mere 0.035 per cent (about $12 \times 10^{12} m^3$) (Figure 18.1) water cycling is accomplished by evaporation, the transport of water vapour in the atmosphere, condensation, precipitation and terrestrial run off. The average residence time of ware within these reservoirs varies from hundreds or thousands of years for the oceans and polar ice to only about 10 days for the atmosphere.

Because of its large latent heat, the global occurrence and transport of water is closely linked to global energy. Atmospheric water vapour is responsible for the overriding per cent of total global energy lost into space by infrared radiation. Over 75 per cent of the energy input from the surface into the atmosphere comes from the liberation of latent heat by condensation and, principally, the production of rain fall. The average shortage of water vapour in the atmosphere given

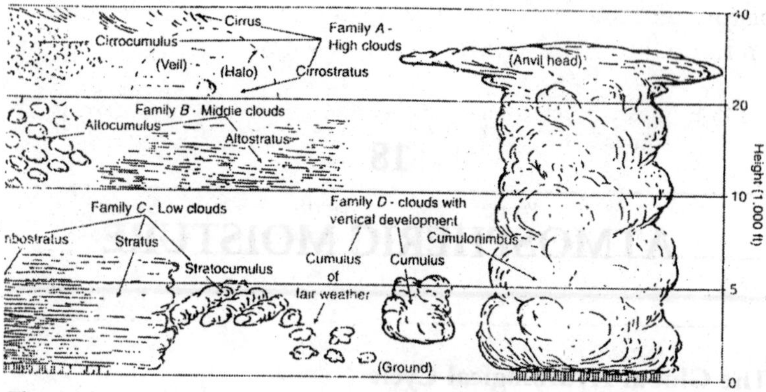

The ten basic cloud groups classified according to height and form. Edited after Strahler, 1965

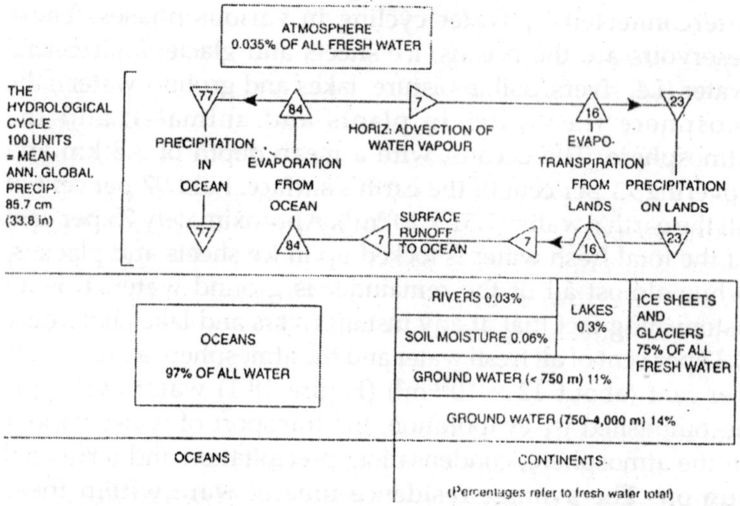

Fig. 18.1: The hydrological cycle and water storage of the globe. The exchanges in the cycle are referred to 100 units, which equal the mean annual global precipitation of 85.7 cm. The percentage storage figures for atmospheric and continental water are percentages of all *fresh* water. The saline ocean waters make up 97 per cent of *all* water. The horizontal advection of water vapour indicates the *net* transfer.

Source: From More 1967.

in Table 18.1 termed the perceptible water content (about 2.5 cm), is sufficient for only some 10 days supply of rain fall over the earth as a whole. However, intense (horizontal) influx of moisture into the air over a given region makes possible short-term rainfall totals greatly in excess of 3 cm.

Table 18.1: Mean water content of the atmosphere (in cm of rain fall equivalent)

	World	*North Hemisphere*	*South Hemisphere*
January	1.9	2.5	2.2
July	3.4	2.0	2.7

Source: After Sutcliff, 1956.

Atmospheric Moisture and Precipitation

Water is found in three forms, i.e. solid, liquid and gaseous. It occurs simultaneously in liquid form as running water, water in oceans etc. ; it occurs in solid form as snow and ice at higher altitudes and in gaseous form as water vapours suspended in the air. According to an estimate about 100000 cubic miles of water is evaporated every year. The water vapour which remains suspended in the air forms atmospheric moisture measured in grammes at a given temperature per unit area. This atmospheric moisture is known humidity which can be expressed either in terms of absolute humidity or relative humidity.

Absolute Humidity

Absolute Humidity is the actual amount of water vapour (moisture) present in per unit area of air at a given temperature. It is normally measured in grammes per cubic metre of air. Air at a particular area with a particular temperature and pressure can hold water vapour to a certain limited quantity. When this limit is reached (when as many molecules of water enter the air as are leaving it) the air said to be saturated. For example, an air at place (A) at 50°F temperature can hold only 9.14 grammes of water vapour per cubic metre, while at the same place at 68°F it can hold 17.117 grammes. If a fraction of a fraction is added in it, it will condense in the form of water.

This level beyond which the air cannot hold the vapours is called saturation level. The difference between the absolute humidity and the saturation level is called saturation deficit.

Absolute humidity is highest near the equator throughout the year because evaporation is very high. It is lowest in the high pressure belts near the poles. Water vapour also have weight which is maximum when the air is saturated. It is called vapour pressure which is 30 mbs at the equator.

Relative Humidity

When the actual amount of water vapour in the air is expressed as a percentage of the water vapour, where the air saturated, it is called relative humidity. We have seen that a saturated air at 50°F contains 9.14 grammes of water vapour per cubic metre. If, for example, air at that temperature actually have 6 grammes of water vapour per cubic metre the Relative humidity would be [(6 × 100) ÷ 9.41] which is approximately 63.7 per cent. As the temperature rises, the relative humidity falls because the carrying capacity of the air is increased. Conversely if the temperature of the air falls, the relatively humidity increases till it reaches the saturation beyond which any further rise in water vapour leads to condensation.

Specific humidity

It is the weight of the water vapour per weight of given mass of air including the water vapour (usually expressed as grammes of water vapour per kilogram of air). More than 50 per cent of the atmospheric moisture is below 850 mb (approximately 1,450 m) and more than 90 per cent below 500 mb (5,575 m). It is also apparent that the seasonal effect is most marked in the lowest global distribution of atmospheric vapour content in January and July illustrated in Figure 18.2. Mean perceptible water generally decrease from the equator to the poles due to direct relation between atmospheric moisture and temperature. There is especially low value over deserts, where there is strong air subsidence. The greatest

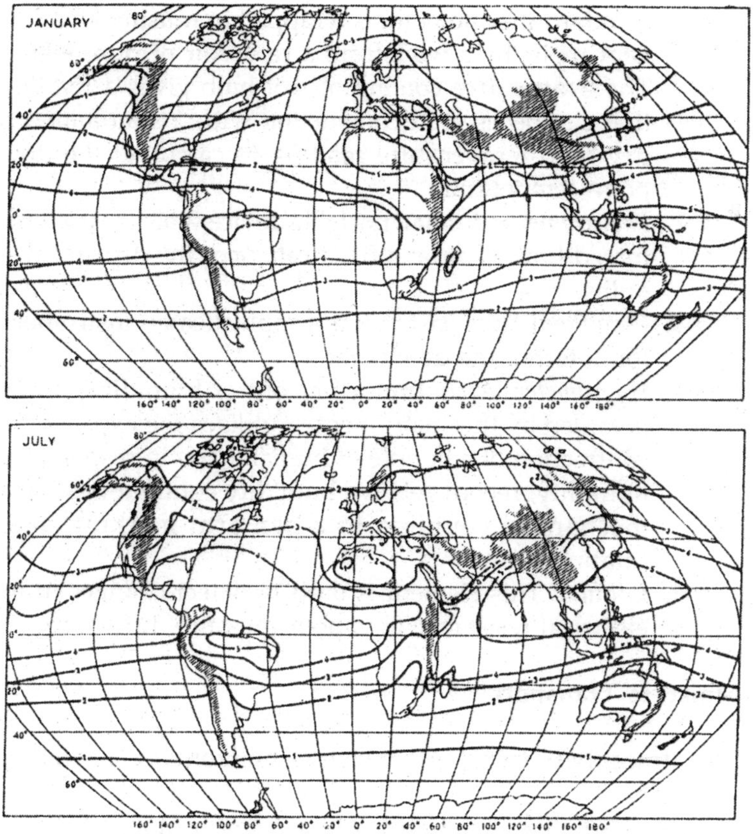

Fig. 18.2: Mean atmospheric water vapour content in January and July, 1951-55, in cm of precipitable water.
Source: After Bannon and Steele, 1960 (Crown Copyright Reserved).

seasonal variations occur in the belt 20-30° N, the location of the Asian and African monsoon.

Measurement of Atmospheric Moisture

The atmospheric moisture can be measured by at least five types of instruments:

1. *Wet-bulb thermometer*: It is installed in a louvered instrument shelter (Stevenson screen). The bulb of the

standard thermometer is wrapped in muslin, which is kept moist by a wick from a reservoir of pure water. The evaporative cooling of wet bulb gives a reading that can be used in conjunction with a simultaneous dry bulb temperature reading to calculate the dew point temperature.

2. *Psychrometer*: A small portable device which is used for forced flow of air at a fixed rate over the dry and wet bulbs.
3. *Dew Point Hygrometer*: It is a sophisticated instrument for determining dew point.
4. *Hygrometer*: A continuous record of humidity can be obtained by this instrument. It utilizes the expansion/contraction of a bundle of human hairs in response to humidity, record relative humidity continuously by a mechanical coupling to a pen arm marking on a rotating drum.
5. *Lithium*: This element is used to detect changes in its electrical resistance to vapour pressure difference.

Moisture Transport

The atmosphere transports moisture horizontally as well as vertically.

There are two vital aspects of the transfer of moisture:

(a) *Land-Sea moisture exchange*: There is continuous exchange of moisture taking place between the continents and the oceans. The amount of moisture gain through precipitation on the continents is more than that lost through evaporation. Oceans on the contrary, lose more moisture through evaporation than what they receive through precipitation. Winds carry a part of moisture from over the oceans to the continents where water vapour condenses and falls as rain or snow. However, part of this precipitation returns to the oceans through surface run off or groundwater seepage. The remaining part is also brought back to the oceans by off-shore winds laden with land-evaporated moisture. As a matter of fact, direct precipitation over oceans brings back a larger part of the moisture evaporated to the atmosphere.

(b) *Meridian moisture exchange*: It involves the transfer of moisture between different latitude belt on the earth's surface. This also involves a transfer of energy. In certain latitude zones, there is a marked excess of precipitation over evaporation. On the contrary, there are certain zones where a evaporation exceeds precipitation.

Evaporation

This is a process by which the water is changed into gaseous form from the liquid and solid states. The effectiveness of rain-fall is usually taken to be the difference between the actual rain-fall and total possible evaporation.

The rate of evaporation is mainly dependent upon a number of factors, e.g.:

(1) Temperature,
(2) Movement of the air,
(3) Relative humidity, and
(4) Nature of the surface configuration.

Rate of evaporation is low in equatorial region and cold areas.

Condensation

The process by which water vapour changes into water is called condensation. If the temperature at the time of condensation is below freezing point the vapour directly changes into solid form, the process is called sublimation.

Condensation, the direct cause of all the forms of precipitation, occur under varying conditions, which in one way or the other are associated with change one of the linked parameters of air volume, temperature, pressure and humidity. Thus the condensation takes place:

1. When the temperature of the air is reduced but its volume remains constant and air is cooled to dew point.

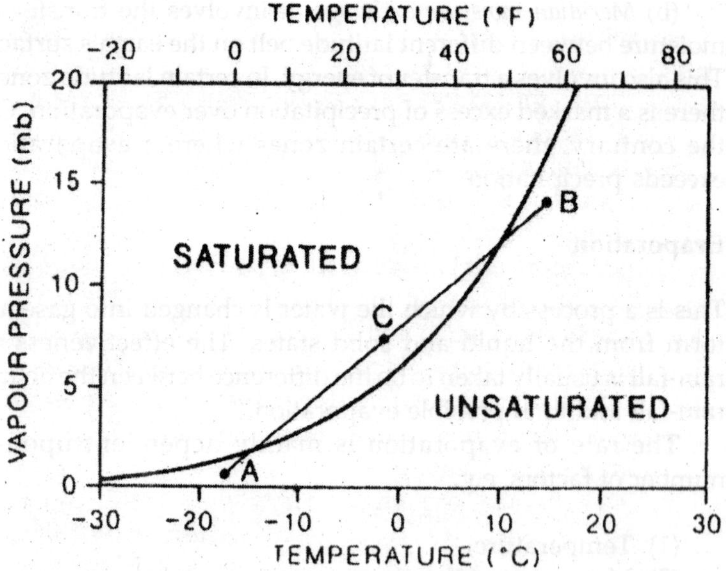

Fig. 18.3: The effect of air-mass mixing. The horizontal mixing of two unsaturated air masses A and B will produce one supersaturated air mass C. The saturation vapour pressure curve is shown (*cf.* Figure 1.14B, which is a semi-logarithmic plot).

Source: From Petterssen 1941 (1969).

2. If the volume of the air is increased without addition of heat, this cooling occurs because adiabatic expansion causes energy to be consumed through work.
3. When a joint change of temperature and volume reduces the moisture holding capacity of the air below its existing moisture content.
4. By evaporation adding moisture to the air.

The key to the understanding of condensation clearly lies in the fine balance that exits between these variables. Whenever, the balance between one or more of them is disturbed beyond a certain limit, condensation may result.

The most common circumstances favourable to the production of condensation are those producing a drop in a air temperature, namely contact cooling, mixing of air masses of different temperature and dynamic cooling of atmosphere.

Contact cooling is produced for example, within warm, moist air passing over a cold land surface. On a clear winter night, strong radiation will cool the surface very quickly and this surface cooling will gradually extend to the moist lower air, reducing the temperature to a point where condensation occurs in the form of dew, fog or frost, depending on the amount of moisture involved the thickness of cooling air layer and the dew point value. When the latter is below 0°C it is referred to as the hoar frost point, if the air is saturated with respect to ice.

The mixing of the different layers within a single air mass or two different air masses can also produce condensation. Figure 18.3 indicates how the horizontal mixing of two air masses (A and B) of given temperature characteristics may produce an air mass (C) that is oversaturated at the intermediate temperature and consequently forms cloud. Vertical mixing of an air layer can have the same effect. Fog or low stratus, which drizzle—known as 'crachin' which is common along the coasts of south China and gulf of Ton kin in February–April can develop as a result of either air mass mixing or warm advection over a colder surface.

The addition of moisture into the air near the surface by evaporation occurs when cold air moves out over a warm water surface. This can cause steam fog. However, the most effective cause of condensation is the dynamic process of adiabatic cooling.

Types of Condensation

(1) *Silver Thaw: Super Cooling-Glazed Frost*: Sometimes, droplets of water remain suspended in the air and remain in liquid form even when the temperature is below freezing, as long as the air is undisturbed. This phenomenon is known as super cooling. A similar condition is known as Glazed Frost or Silver Thaw in America.

(2) *Dew*: It occurs after a warm day of higher evaporation to increase the humidity of the air. The evening must be calm and a clear sky at night due to which back radiation will be higher. The bodies such as leaves, hard surfaces, grass etc. are

cooled by natural radiation to a temperature below the dew point. The air above these bodies is warm and moisture laden. As it touches these nocturnally cooled surfaces the water vapour condenses to form dew.

(3) *Hoar Frost*: When the dew-point is below freezing point, the vapour directly turns in ice spicules on the cooled bodies.

(4) *Mist*: Mist is also formed as a result of condensation of water vapour. It is notable for reduction in visibility. When the visibility through the condensed air ranges from 1 to 2 kms it is called mist.

(5) *Haze*: When the visibility is impaired due to dust or smoke particles ranging from 1-2 kilometres it is called haze.

(6) *Fog*: When the visibility is less than 1 km it is known as fog. During winter months the fog is so thick at Batote or higher altitude areas, that one cannot see beyond a few metres even. There are five major types of fog:

(a) *Radiation Fog*: In Great Britain, during the months of spring and autumn, during a calm and clear period the surface of the earth is rapidly cooled by radiation and hence the layer of air resting upon it is also cooled. The cooled dense air in hilly areas flows by gravity towards the hollows. As it comes into contact with the cold earth the water vapours condense producing a shallow horizontal layer of white radiation fog.

(b) *Advection Fog*: It is formed when a current of warm moist air moves over a cold land or sea surface. The term advection is used for the horizontal transfer of air. This fog occurs over New Found Land Island for 70 days in a year and for 100 days in the Strait of Belle Isle.

(c) *Cyclonic Fog*: It is associated with the passage of warm front of a cyclone. It is shortlived and occurs in the form of very fine drizzle forming gloomy foggy conditions impairing the visibility. It is also called frontal fog.

(d) *Steam Fog—Arctic Smoke*: In high latitude areas when cold air passes over a much warmer water surface, the moisture of the air is converted into ice crystals for a short period only. In low latitudes, the water below appears to steam known as steam fog.

(e) *Hill Fog*: When the cloud, in higher altitude areas comes into contact with the surface, it is known as hill fog.

(7) *Rime*: When a slight wind drives the fog comprising super cooled droplets, against prominent objects as poles, telephone wire, trees, etc. the drops freeze on them in the form of Rime. The temperature of the droplets as well as the objects is below freezing. Rimes have a white opaque appearance as the ice particles contain some air.

(8) *Clouds*: Clouds are another form of condensation in which very small particles of ice or water are floating in the air at various altitudes up to a height of 40,000 feet. When a cloud comes in contact with the earth surface it is termed as fog. Cloudiness may vary from area to area and lines of equal cloudiness are called Isonephs.

On the basis of their height the clouds can be divided into three types, i.e. High clouds, Medium clouds and Low clouds. According to their form also they can be divided into three types:

(1) Feathery or Fibrous type of clouds such as "Cirrus clouds".
(2) Globular of heaped type clouds such as "Cumulus clouds".
(3) Sheet or Layer type clouds such as "Stratus clouds".

However, ten basic cloud forms can be identified as under:

(1) *Cirrus Cloud*: It is a feathery high level cloud containing tiny ice spicules. It indicates approaching cyclone.
(2) *Cirrostratus*: Cirrostratus is a layer of white clouds at higher levels. Through these clouds sun shines with a distinct halo.
(3) *Cirrocumulus*: Small globular clouds with rippled appearance found in a series of lines are called Cirrocumulus and the phenomena are also referred as 'Mackerel Sky'.

(4) *Altostratus*: These are much denser sheet-clouds of grayish colour. Sun can be seen through these clouds. These clouds indicate the approaching warm front.
(5) *Altocumulus*: It is found in bands of globular patches separated by blue sky. They are fleecy and cellular in appearance. Sometimes bring heavy rain and thunder.
(6) *Stratus*: A grey uniform sheet of clouds is called Stratus.
(7) *Nimbostratus*: It is heavier, darker giving heavy rain continuously at the warm front, it is called Nimbostratus.
(8) *Stratocumulus*: It is still darker, lower and heavier type of cloud which is usually seen in winter.
(9) *Cumulus*: Cumulus is the convection cloud which is formed by the uplift of air to a level of condensation. They are usually fair weather clouds which die out at the end of the day.
(10) *Cumulonimbus*: If a cumulus cloud keeps on rising up to 6 or 7 miles and result in heavy downpour, they are called cumulonimbus. These cumulonimbus clouds are associated with heavy rain, hail and thunderstorm.

Precipitation

When the condensed water vapour taking the shape of a droplet or an ice spicule becomes so heavy that the air is unable to carry its load, it falls down on the surface of the earth in the form of precipitation. Precipitation may take the form of rain, snow, sleet or hail.

Snow: If the condensation of water vapours takes place when the temperature is below freezing point, the vapour directly changes into solid state forming spicules of ice! The small ice spicules unite to form snow flakes of different shapes and sizes. The snow spicules are rarely flat they are usually in the form of hexagonal plates or hexagonal prisms revealing infinite beautiful symmetrical patterns. Where the temperature of lower layer of the air is sufficiently low the flakes reach the surface in solid form.

Sleet: Sleet is a mixture of snow and rain. The British scholars say that it is a state of partially melted snow, whereas, the Americans believe it to be a condition of partially frozen raindrops.

Hail: Formation of hail is associated with the conditions of extreme instability. When droplets are formed in the lower part of a cloud they coalesce into larger drops. If the ascending air is strong it may carry the drop upwards to a point where it freezes in a small grain. Further drops of super cooled air collide with this ice grain and freeze around it in the form of different layers. When it becomes heavier it falls down in the form of hailstones. Hailstones can be of varying sizes. In Great Britain the largest size recorded is 2″ in diameter. These are common in spring and early summer in mid-latitudes.

Types of Rain

Water vapours condensed in the form of droplets remain suspended in the air in the form of clouds. If these droplets coalesce they will form a bigger drop. The weight of the drop becomes so heavy that the rising air is unable to bear its load, it falls on the surface in the form of rain drop. The rainfall can be of three types:

(1) *Convectional Rainfall*: When a mass of air is warmed by conduction from heated land surface it is uplifted and expands as a result of which gets adiabatically cooled and condensation takes place resulting in convectional rainfall.

It is mostly a local phenomenon. Normally, it can be easily observed in those areas which are surrounded by high mountain walls. The evaporation takes place due to local heating increasing the moisture in the air. The air mass further gets warmed by conduction. As it rises it expands. It further rises because there is no possibility of horizontal drift. As it attains sufficient altitude it condenses and result in rain.

It is mostly associated to calm atmosphere. Convection rain occurs on the equator daily throughout the year. In temperate zone convectional rainfall occurs in early summer when the upper atmosphere is still cool after winter.

(2) *Orographic Rainfall*: This type of rainfall is associated to the relief features of an area hence it is sometimes referred to as Relief Rainfall. A heap of moisture laden air is forced to ascend along the side of a mountain. The monsoon winds moving from the south-west are pushed over the Western Ghats.

The vapours condense at higher level and coalesce in drops of rain. Such rainfall occurs in many parts of the world where the mountain ranges are parallel to the coast.

(3) *Cyclonic Rain or Frontal Rain*: This type of rainfall is associated to the cyclones as the name suggests. It occurs along the frontal zones of inter-tropical convergence zone and the polar parts. In case of polar fronts associated with cyclones travelling from west to east in mid-temperate latitudes the rainfall is continuous drizzle in a large belt along the warm front. With the cold front associated a squall shower at the head of the occlusion front. The cyclonic rain also occurs when warm front overrides the cold front pushing the warm air upwards. The occluded front is marked with heavy rainfall if the temperature is sufficiently low.

World Distribution of Precipitation

A.N. Strahler divided regions of the globe into 7 precipitation:

(1) *The Wet Equatorial Belt*: It is the region of very high rainfall. The average annual rainfall exceeds 200 cms. It extends along the equator including Amazon River Basin in South America, Congo River Basin of Africa, African Coast from Nigeria to New Guinea and East Indies. Thunderstorms are frequent in this belt.

(2) *Trade Wind Coasts*: Trade wind coasts are associated with high rainfall. These are narrow coastal belts with average

Precipitation

Sl. No.	Name of Region	Range of Latitude	Continental Nature Annual	Locational Precipitation	Precipitation (in cms.)
I.	Wet Equatorial Belt	10°N-10°S	Interior and Coasts	Rain	+200
2.	Trade Wind Coasts	5°-30°-N&S	Narrow Coastal Zones	-do-	+150
	MOISTURE				177
3.	Tropical Deserts	10°-35°N&S	Interior and West Coasts	Rain Below	25
4.	Mid-latitude Desert and Steppes	30°-50° N&S	Interiors	Rain Snow	10-50
5.	Humid Sub-Tropical Region	25°-45°N&S	Interiors	Rain Snow	100-150
6.	Mid. Latitudes West Coasts	35°-65°N&S	West Coasts	Rain Snow	+200
7.	Arctic and Polar Deserts	60°-90° N&S	Interiors and Coasts	Rain Below Snow	30

Source: A.N. Strahler.

annual rainfall ranging from 150 to 200 cms. It extends over eastern coast of Brazil, Central America, Madagascar and North-Eastern Australia. These are the trade wind coasts or windward tropical coasts. The rainfall is orographic.

(3) *Tropical Deserts*: There are two zones of large tropical deserts approximately along the tropic of cancer and tropic of Capricorn. Average annual rainfall is less than 25 cms. and in many places even less than 5 cms. These deserts are located along the west coasts of the continents. Rainfall is convective in nature. It includes deserts of Australia, South Africa, South America, North Africa, etc.

(4) *Mid-latitude Desert and Steppes*: In the interior of Asia and North America between 30° to 50° North latitude are located the great mid-latitude continental deserts and semi-arid grasslands known as steppes. Average annual

precipitation ranges between 10-50 cms. These arid lands are located in the rain shadow zones.

(5) *Humid Sub-Tropical Region*: The average annual precipitation in this region ranges from 100-150 cms. It extends over south-eastern sides of the continents of North America and Asia in 25°-45° latitudes. Tropical cyclones bring heavy rainfall in this region.

(6) *Mid-latitude West Coasts*: This is a region of prevailing westerly winds. Where the coasts are mountainous the rain-fall is over 200 cms. Areas of Alaska, British Columbia, Patagonia, Scotland, Norway and South Islands of New Zealand are included in it.

(7) *Arctic and Polar Deserts*: North of 60°N. Latitude annual precipitation is largely in the form of snow but is below 30 cms. per annum.

EL NINO AND THE SOUTHERN OSCILLATION

El Nino is a remarkable disturbance of ocean and atmosphere phenomena that begins in the eastern Pacific Ocean and spreads its effects widely over the globe for more than a year's time, bringing with it unseasonable weather patterns with abnormalities in the form of drought, heavy rainfalls, severe spells of heat and cold or a high incidence of cyclonic storms. In 1890, it was reported that Peruvian fishermen were using the expression "Corriente del Nino" (Current of the Christ Child) to describe an invasion of warm surface water that occurred once every few years around Christmas and greatly depleted their catch fish. The occurrence of El Nino phenomenon was noticed in 1891, 1925, 1940-41, 1965, 1972-73 and 1982-83.

In the Pacific the cold current or Peru current flows northward off the South American coast, then, about at the equator, it turns westward across the Pacific as the south Equatorial current. The Peru Current (Humboldt current) is characterized by upwelling of cold, deep water which carries nutrient ions that serve as food for plankton. Fish feed on this high concentration of plankton. The anchoveta, a small fish used commercially to produce fish meal for animal feed,

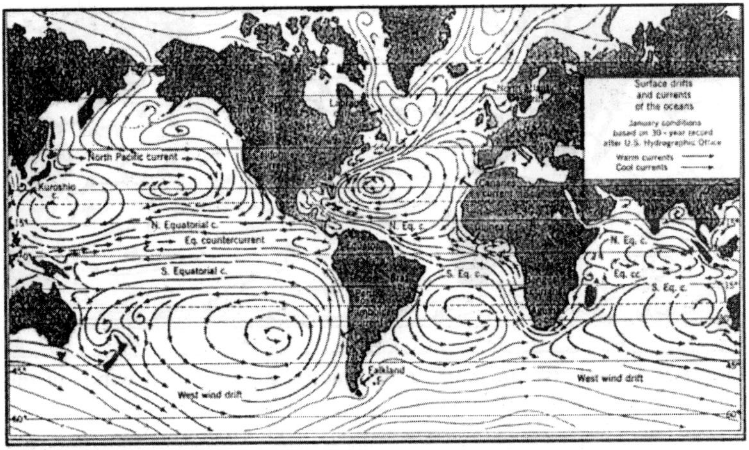

Fig. 18.4: Suface drifts and currents of the oceans in January. (Based on data of U.S. Navy Oceanographic Office).

flourish in great numbers is harvested by the Peruvian fishermen. With the onset of El Nino upwelling stops, the cool water replace by warm water from the west and the plankton and their anchoveta predators disappeared. Vast numbers of birds that feed on the anchoveta die of starvation. To understand what sets El Nino in action (Fig. 18.4). Review the January average conditions of pressure and winds in the equatorial and subtropical zones of Pacific. Note the position of Intertropical convergence zone ITC along the axis of the equatorial trough of low pressure. Note the pattern of easterly winds or trades. Figure 18.5 review the January average conditions of pressure and winds in the equatorial and subtropical zones of Pacific. Note the position of Intertropical convergence zone ITC along the axis of the equatorial trough of low pressure. Note the pattern of easterly winds or trades. Figure 18.5 shows how the south equatorial current is related to the pressure and wind system. Note especially that low pressure along the ITC decreases steadily from east to west and that starting at about 160 W, the low pressure trough broadens greatly in latitude and deepens over a large area that includes Indonesia, New Guinea and northern Australia. This area in January (high sun season in the southern hemisphere)

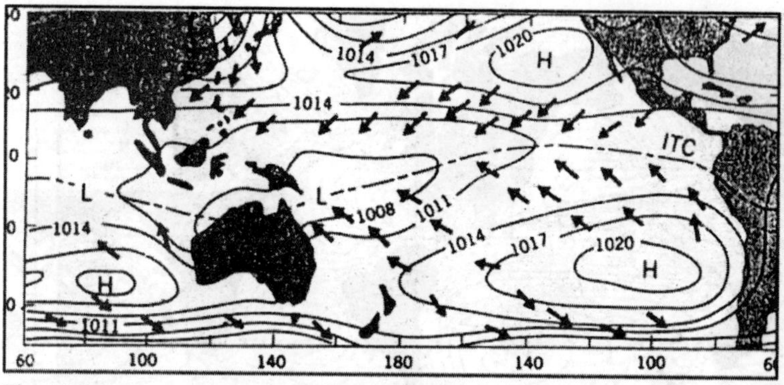

Fig. 18.5: January average surface barometric pressure and winds of the Pacific equatorial zone.

normally experiences heavy rainfall. For northern Australia this is the rainy or monsoon season of tropical wet-dry. As for the Peruvian coast, a desert climate prevails throughout the year and January is usually almost rainless.

In 1904, Gilbert Walker, was observing an occasionally strange reversal in barometric pressure system. In some years, the Indonesian/Australian low pressure region was replaced by high pressure while low pressure deepened over eastern Pacific portion of the equatorial trough. Walker called this phenomenon the Southern Oscillation (SO). He observed that the onset of high pressure brought drought to the western regions while the eastern Pacific region experienced abnormally high rainfall.

Since Walker's time, his observation has been confirmed as precursory to the onset of El Nino. Two observation stations have been singled out as end points on this meteorological station: Darwin, on the northern coast of Australia at point long 130° east, and the Island of Tahiti, at about long 150° west. In Figure 18.6, a special map is showing the level of correlation in barometric pressure reading over the oceanic region. At any given point over the entire map area, pressure during SO event is compared with pressure at Darwin to obtain a pressure anomaly. Lines with large numbers (6 to 8, positive or negative) indicate a high level. Level of correlation between Darwin and Tahiti, the former having high pressure and the latter having

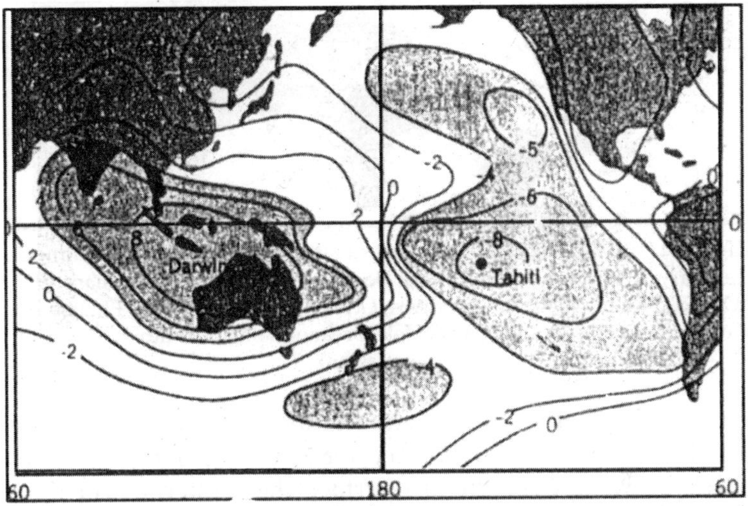

Fig. 18.6: Correlation of annual mean sea-level pressures with the pressure at Darwin, Australia. Isopleths give correlation coefficients. (From Eugene M. Rasmusson, 1985, *American Scientist,* Vol. 73, p. 169; Figure 2. Used by permission)

correspondingly low pressure. Today the SO combined with El Nino "EN" under the short form ENSO.

The onset of SO pressure anomaly in May 1982 was soon followed by two important charges in surface environment. First, the trades stopped to blow; second, sea surface temperatures rose sharply just south of the equator in the eastern Pacific, the temperature rise started along the south American coast in October 1982, when upwelling stopped. The warm zone then extended progressively further west between the equator and latitude 5° south, reaching almost to long 180° by December 1982 and persisting through February 1983, the sea temperature has risen to more than 2°C (3.6F) above the normal value in what is referred to as the "mature face" of the ENSO, lasting from December 1982 through February 1983.

Cessation of the trades beginning in the previous June (1982) was followed by a curious event, the onset of low level of equatorial westerly winds about in mid-Pacific in the longitude ranging from 160° east to 150° west. At the same time, the precipitation zone of south-western zone shifted far

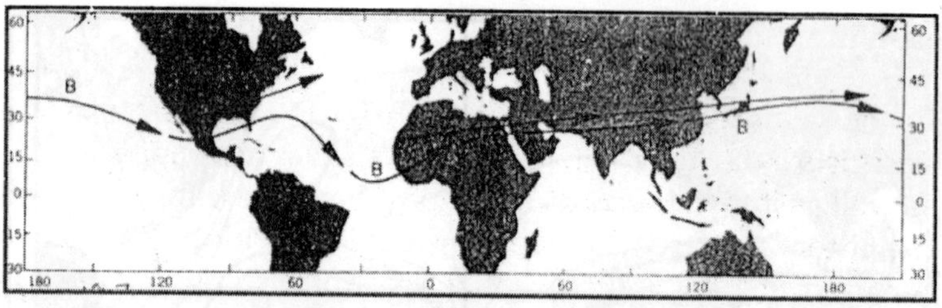

Fig. 18.7: (A) The typical prevailing jet stream pattern for December through February. (B) Pattern during ENSO December 1982 through February 1983. (Data of Eugene M. Rasmusson, 1984, *Oceanus*, Vol. 27, No. 2, p. 12, Figure 9. Copyright © 1984 by the Woods Hole Oceanographic Institution. (Used by permission)

eastward into the area of equatorial westerly, reaching its maximum intensity in January 1983 between long 140°-160° west. Blowing over the sea surface the westerlies set up a succession of eastward moving. Kelvin waves, which drove warm water in flows to the Pacific shores of South America in what can be thought as a kind of sloshing effect raising sea level along those coasts. A consequence of the warm water zone off the south America was the occurrence of torrential rainfalls, causing devastating floods in the Andean high lands of Peru, Bolivia, and Colombia. The warm surface layer spread northward along the Central American and North American coasts, latitude as far north as Oregon. Associated with mature face of ENSO was an important change in the pattern of northern hemisphere polar jet as shown in the Figure 18.7.

It is noticed that ENSO jet swept across Mexico and the Caribbean and continue strong across the Atlantic to enter North Africa and continue in strength across the Arabian Peninsula and north India. One effect far southward position of Polar jet over North America was to bring intense and frequent invasions of cyclonic storms. These dumped heavy snowfall over the western mountain ranges, as well as heavy rains over the Gulf Coast and Florida.

Early in 1983, the dry high pressure area of the western Pacific moved slowly northward to cover Philippines and spread westward to include southern India and Sri Lanka. Meanwhile, a severe drought was occurring in southeast Africa.

La-Nino

In the years following the great 1982-83 El Nino, research scientists looked for signs of a recurrence of the phenomenon, as well as into the records of the past for more information on the interval of recurrence. A weakly developed El Nino occurred in 1987 and served as a test of a computer models devised for prediction. Perhaps the most significant development has been the recognition of a distinctive set of contrasting or complementary condition that prevails between El Nino events, known as La Nino (the girl), it is a period during which sea surface temperature in the central and eastern Pacific Ocean fall to lower level, e.g. below 25°C(77°F). The south Pacific subtropical high becomes very strong during the high sun-season of that hemisphere and causes abnormally strong south-east trade winds. The force of these strong winds drags warm surface water westward, exposing cooler water at the surface. La-Nino conditions were recognized during 1988 and were suspected of having been casually related to a severe drought experienced over North America in summer of 1988.

Climate Change

Al Gore and UN's Intergovernmental Panel on Climate Change jointly won the 2007 Nobel Peace Prize for their efforts to spread awareness of manmade climate change and to lay the foundation for fighting it. At the United Nations climate change talks held at Bali, Indonesia, India along with 190 countries agreed on road map to fight against climate change. In 2012, the first commitment period of the Kyoto Protocol (36 industrial nations have to cut greenhouse gas emissions by 5 per cent lower than 1990 ends. Over the next two years the "Bali road map" will guide the next stage of the global fight against climate change.

According to IPCC to prevent global warming from eventually exceeding 2°C by 2050, the world will need to cut its emission to roughly 15 per cent of the volume in 2000. IPCC has given global figure for carbon dioxide production in 2000 and divided it by current population (2007). This gives a base-line figure of 3.58 tonnes of carbon dioxide per person. An 85 per cent cut means that (if the population remains constant) the global output per head should be reduced to 0.537 tonnes by 2050. The U.K. currently produces 9.6 tonnes per head and US 23.6 tonnes. Reducing these figures to 0.537 means a 94.4 per cent cut in the UK and 97.7 per cent cut in the US but the world population will rise in the same period. So it is not possible to reduce carbon dioxide with present model of the Bali conference. The real issue in Bali was not technical or economic. The crisis we face demands a profound philosophical discussion, a reappraisal of who we are and what progress means.

EVAPORATION

Evaporation (including transpiration) provides the moisture input into the atmospheric part of the global hydrological cycle, of which the oceans provide 84 per cent and the continents 16 per cent. The highest values (150 cm) averaged zonally around the globe, occurs over the tropical oceans, associated with trade wind belts and over equatorial land area in response to high solar radiation receipt and luxuriant vegetation growth. (Fig. 18.8). The largest evaporation losses in winter, for each hemisphere represent the effect of outflows of continental air over warm ocean currents in the western north Pacific and North Atlantic and strong trade winds in cold season of southern hemisphere. The relationship of saturation vapour pressure and temperature means that evaporation process limit low-latitude ocean surface temperature (i.e. where evaporation is at a maximum) to value of 29°C. This factor plays an important role in regulating the temperature of ocean surface and overlying air in tropical latitudes. Evaporation occurs whenever energy is transported to an evaporating surface if the Evaporation pressure in the air is below the saturated value.

The rate of evaporation depends on a number of factors. The two most important are the difference between the saturation vapour pressure at the water surface and vapour

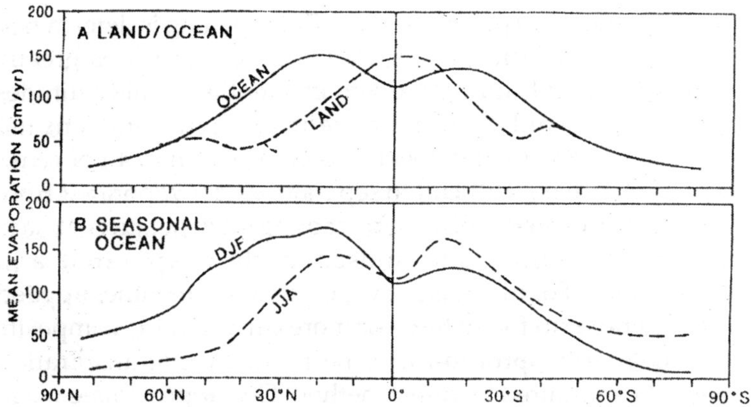

Fig. 18.8: Zonal distribution of mean evaporation (cm/year) (A) annually for the ocean and land surfaces, and (B) over the oceans for December to February and June to August.

Source: Peixoto and Oort 1983. From *Variations in the Global Water Budget.* A Street-Perrott. M. Beran and R. Ratcliffe (eds) 1983, Fig. 22. Copyright © D. Reidel, Dordrecht, by kind permission of Kluwer Academic Publishers. Also partly from Sellers 1965.

pressure of air and the existence of a continual supply of energy to the surface. Wind velocity also affects the evaporation rate, because the wind is generally associated with importation of fresh saturated air, which will absorb the available moisture.

Water loss from plant surfaces, chiefly by leaves, is a complex process termed **transpiration.**

It occurs when the vapour pressure in the leave cell is greater than the atmospheric vapour pressure and vital as a life function in that causes a rise of plant nutrients from the soil and cool the leaves. The cells of the plant roots exert an osmotic tension within them increases. If the tension of the soil films exceeds the osmotic root tension, the continuity of the plant's water supply is broken and wilting occur. Transpiration is controlled by the atmospheric factors that determines evaporation as well as by plant factors such as the stage of plant growth, leaf area and leaf temperature and also by the amount of soil moisture. It occurs during the day, when the stomata (i.e. small pores in leaves) through which transpiration takes place, are open. This opening is determined by the light intensity. Transpiration naturally varies greatly with season, and during the winter months in mid-latitudes conifers lose only 10-18 per cent of their total annual transpiration losses and deciduous trees less than 4 per cent.

In practice it is difficult to separate water evaporated from the soil **intercepted moisture** remaining on vegetation surface after precipitation and subsequently evaporated and transpiration. For this reason evaporation is sometimes applied as a general term for all these or more correctly, the composite term **Evapotranspiration** may be used. Evapotranspiration cannot be measured by direct method. One approximate mean of indirect measurement is based on the moisture balance equation:

Precipitation = run off + evapotranspiration+ soil moisture storage change.

If the soil block is regularly irrigated so that the vegetation cover is always yielding the maximum possible evapotranspiration, the water loss is called the **Potential Evapotranspiration (PE).** More generally, PE can be defined as the water loss corresponding to the available energy. Potential evapotranspiraton forms the bases for the climatic classification developed by the C.W. Thornwthwaite.

19

WORLD CLIMATES

CLASSIFICATION OF CLIMATE

Different combinations of climatic controls produce many variations in climate from place to place resulting in a large number of climatic types. The fundamental problem in classifying climates is concerned with the search for the factors which constitute distinctive climatic types. A "single climate" is the synthesis of all the climatic elements in a unique combination. The purpose of climatic classification is summarized by C.W. Thornthwaite in following lines: "The purpose of a climatic classification is to provide a concise description of the various climatic types in terms of truly active factors so far as possible. The scheme should not only differentiate between them. It should be able to characterize the climates of the whole earth. At the same time, it should supply the framework for a differentiation of the innumerable microclimates that make up a climatic type."

There are two well known types of classifications—one suggested by Thornthwaite and the other by Koeppen.

Koeppen's Classification

Wladimir Koeppen (1846-1940) was a German biologist who devoted most of his life to climatic problems. He used world vegetation map of de-Candolle a French Plant physiologist and devised his first classification in 1900 largely on the basis of vegetation zones.

Again in 1918 revised it with greater attention to temperature, rainfall and their seasonal characteristics.

Koeppen divided world climates into five major categories which are represented by capital letter. For further subdivision he has used small letters. The following are the main classifications of the climate:

(A) Tropical Forest or Humid,
(B) Dry Climate,
(C) Warm Temperate Rainy/Mesothermal,
(D) Cold Forest/Humid Microthermal, and
(E) Polar climate.

Sub-types of the Koeppen's Classification

Af = Hot; rainy in all seasons; A = Hot, f = rain in all seasons.
Am = Hot; seasonally excessive rainfall; m = Excess rain in rainy season.
Aw = Hot; winter dry; 'W'—stand for dry winter.

(B)

Bsh = Semi-arid (steppe); hot.
Bsk = Semi-arid (Cooler) cold.
Bwh = Desert; hot.
Bwk = Desert; cool or cold

(C)

Cfa = Mild winter; moist all seasons; hot summer.
Cfb = Mild winter; moist all seasons; warm summer.
Cfc = Mild winter; moist all seasons; short cool summer.
Cwa = Mild winter; dry winter; hot summer.
Cwb = Mild winter; dry winter; short warm summer.
Csa = Mild winter; dry summer; hot summer.
Csb = Mild winter; dry summer; short warm summer.

(D)

Dfa = Severe winter; moist all seasons; long hot summer.
Dfb = Severe winter; moist all seasons; short warm summer.
Dfc = Severe winter; moist all seasons; short cool summer.
Dfd = Severe winter; moist all seasons; short summer extremely cold winter.
Dwa = Severe winter; dry winter; long hot summer.
Dwb = Severe winter; dry winter; cool summer.
Dwc = Severe winter; dry winter; short cool summer.

Dwd= Severe winter; moist; summer; short summer extremely cold winter.

(E)

ET = Polar climate; very short summer.
EF = perpetual ice and snow.

Koeppen has used small English letter for sub-division of climate

a = Mean temperature of warmest month about 22°C (71.6°F).
b = Mean temperature of warmest month under 22°C, at least four months above 10°C (50°F).
c = Only one to four months above 10°C coldest month above –38°C (–36.4°F).
d = Temperature of coldest month less than –38°C.
f = Constantly moist (precipitation adequate in all seasons).
g = Ganges type of temperature trend, maximum before the summer-rainy season.
h = Hot annual temperature above 18°C (64.4°F).
i = Isothermal difference between warm and coldest months less than 5°C(14°F).
j = Cold winter annual temperature less than 18°C warmest month above 18°C.
k = Same as K, but warmest month less than 18°C.
l = Mild all months 10° to 22°C.
m = Monsoon regime of precipitation, short dry season compensated for heavy rains during rest of year.
n = Frequent fog.
o = In frequent fog, but high humidity, low rainfall and relatively cool, the summer temperature below 24°C (72.2°F).
p = Same as in except summer month 24°C to 28° (28.4°F).
p = Same as in above 28°C.
s = Dry season in summer.
u = Coolest month after summer solstice.
v = Warmest month in autumn.
w = Dry winter not compensated for by total rain in rest of year.
w = Rainy season in autumn.
w = Two distinct rainfall maximum, separated by two dry seasons.

x = Maximum rainfall in spring or early summer, dry in late summer.

x = Same as above but with infrequent heavy rain in all seasons.

The Tropical Humid Climates

This type of climate is found in an area on both sides of the equator up to 26°N and south. The maximum pole-ward extension of this climate occurs along the eastern or wind-ward sides of the continents.

In the interior and towards the western sides of the continents, however, the humid topics are bounded by the dry/ B climates.

Rainfall is relatively abundant, well above 30 inches. Much of the precipitation is convectional in origin, the heavy showers often being accompanied by severe thunder and lightning.

The two principal climatic types within the humid tropic are distinguished from each other on the basis of their seasonal distribution of precipitation. The tropical wet climates have ample rainfall throughout the year. The other is tropical wet-and-dry climate. There is a distinctly wet and distinctly dry season.

Af Type: Tropical wet climate (Af) is associated with uniformly high temperatures and heavy precipitation distributed throughout the year. The latitudinal spread may be increased to 15° or even 25° along the wind-ward margins of the continents. But mostly we consider its spread up to 10°N both sides of the equator. The Amazon basin in north, south America, Congo basin and Guinea coast in West Africa are the two largest contiguous areas with tropical wet climate. A third extensive but not contiguous area is composed of the East Indies, the Philippine Islands, western Colombia, Malaya peninsula, low-lands and slope sections of eastern Brazil and eastern Madagascar. The yearly average temperature of this region ranges between 77° and 80°F. The annual temperature range or difference between the warmest and coldest months is usually less than 5°. The hottest month at Belem (Amazon basin) and at Akassa (Niger Delta) experience 80°F temperature. The daily range of temperature is usually 10° to 25° or several times greater than the annual range.

During the afternoons the thermometer ordinarily rises to temperatures varying from 85° to 93°, and at night sinks to 70° or 75°.

Rainfall is both heavy and distributed throughout the year. There being no distinctly dry season. Taken as a whole tropical wet climate is coincident with the belt of the world's heaviest precipitation. Ward estimates the average rainfall of the doldrums belt to be in the neighbourhood of 100 inches, with less over the continents and more over the oceans. This is no distinctive seasonal rainfall regime characteristic of the rain-forest type of climate. In the rainy periods precipitation falls on a large majority of the days, although there are few days without rainfall. Fewer rainy days and less rain on each day are characteristic features of the less wet season. Belem, in the Amazon valley, receives 94 inches of rain. There are 6 rainy days in March and 10 times more rain than what it is in November. The month of November has 10 rainy days in addition to the heavy local showers associated with afternoon thermal convection. Rainfall accompanies the passage of extensive atmospheric disturbances of the cyclonic or the wave type.

Cloudiness is cumulus in character, relatively high in the doldrums, averaging in the neighbourhood of 58 per cent. At Manaos in Amazon valley, cloudiness varies between 6/10 (for total sky number denoted by la) and 7/10 for each month. At Belem it is 4/10 in the driest month and 8/10 in the rainiest. The whole region is poorly ventilated. The importation of cooler air from the sea during the heat of the day is greater boon to residents along the littoral area causing tropical coasts to be much more livable than the interiors.

Aw Type: The tropical wet-and-dry climate (Aw) differs in two principal respects from tropical wet climate:

(a) It usually has less total precipitation, and
(b) Rainfall is unevenly distributed throughout the year.

There being a distinctly wet and a distinctly dry season. The typical latitudinal location of the wet-and-dry climate is about 5° to 20° and it may extend still farther pole-ward on the eastern or wind-ward side of the continents. The Ilanos of the Orinoco Valley (Colombia and Venezuela) and adjacent parts of the Guiana, highlands of north and south America, Sudan and Congo basin of Africa and northern parts of Australia have 'Aw' type of climate; whole of Asia is included in this type of climate.

The temperature elements in tropical wet and dry climate and in tropical wet climate have smaller variation.

In general, however, yearly ranges are somewhat greater than in typical rainforest regions, usually over 5° but seldom exceeding 15°.

Since temperatures do not differ much within the tropics, rainfall becomes the more critical element in setting apart the various climatic types of low latitudes characteristically; the total amount of the rainfall of the wet-and-dry type is less than that of the tropical wet climate being 40 to 60 inches. This contrast between the two types is principally due to their latitudinal location. The tropical wet type is almost constantly within the inter-tropical convergence where there is a large scale ascent of warm, humid, unstable air masses whereas the wet and dry type is on the margins of the I.T.C.

During its rainy season the weather of wet-and-dry climate closely resembles that tropical wet climate at its worst.

Am Type: Am type of climate is associated with high temperature and excessive rainfall. The only difference in this type is the dominance of monsoon regime of precipitation, short dry season, compensated by heavy rains during rest of the year. The basic characteristic is the series of monsoon winds which experience a complete reversal of the direction between the summer and winter season. The annual average temperature is about 26°C and the annual range is about 3°C. The type of climate occurs typically in south-eastern Asia, western Guinea, coast of Africa, eastern Amazon valley, West Indies and Guiana coast.

Dry Climates-B

The important feature of a dry climate is the potential evaporation from the soil and vegetation exceeding the average annual precipitation. As a result of rainfall deficiency there is no surplus of water to maintain a constant groundwater supply. Hence the permanent streams cannot originate within such areas.

Four commonly recognized subdivisions of dry climate are:

1. *Low latitude desert* (*Bothnia*): The heart of the tropical dry climates is in the vicinity of 20-25° N and south latitude. The low latitude deserts probably are the most nearly rainless regions on the earth. This type of climate occurs in the Sahara, the Thar, Arabia, south-western U.S.A. and desert of Australia. Exact amount of rain can be accepted as defining the outer, or humid margins of the hot deserts. Over much of the Sahara precipitation is under 5 inches. Similar conditions prevail in Kalahari desert, Atacama desert of Peru, Australian desert, Thar desert of India, Somora desert of Mexico and south western parts of U.S.A. At Cairo the average annual rainfall is 1.2; at Lima Peru 2 ; William Greek Australia 5.4; Yama Ariz 3.3; and Port Nolloth in Africa 2.3 inches. Many parts of north Chile do not experience rainfall for 5 to 10 years in succession.

In this region most of the precipitation comes in violent convectional showers. On the pole-ward margins of low latitude deserts there are occasional widespread rains of a less violent nature associated with fronts of middle latitude cyclones. Skies are clear so that the sunshine is abundant. Over much of the Sahara (December and January) have a cloudiness of only 1/10 while from June to October it drops to about 1/30. Because of strong surface heating, due to the intense insolation and the nearly bare ground must give rise to vigorous convectional currents, but the whole mass of air is to stable and has low relative humidity to allow there rising air currents. Evaporation, due to high temperature and low relative humidity, is excessive after being 20 or more times the precipitation.

Relative humidity as low as 2 per cent have been recorded in Egyptian Sahara. Hot weather temperature averages between 85° and 95°F (Yuma 90°, Timbuktu 94°; Nullagin Australia, 90°) and mid-day readings of 105 to 110°F are common. During the period of low sun the days still are warm, with the daily maximum usually averaging 65 to 75°F. Annual ranges of temperature in the low latitude deserts are larger than in any other types of climate within the tropics, 20 to 30°C being temperature of 59°F in January and 93°F in July, resulting in an annual range of 34°C. Daily ranges average 25 to 45° and in rare cases may reach 50° to 60°. Along the pole-ward margins of the deserts the uniformity is reduced because of the promote of the mid-latitude.

2. *Low-Latitude Steppe* (*Bsh*): This climate surrounds the low latitude deserts except possibly on their western sides.

This climate occurs in southern Australia, Mesopotamia, Arabia, north-western India. They are more on the margins of the dry settling tropical air masses associated with subtropical highs and trades. They are closer to the humid climates than the deserts. The areas are encroached upon, for a short period of the year, by rain bearing winds and their associated storms.

Rainfall is meagre, variable and undependable. Koeppen further divided steppe type of climate in two classes: (a) steppes with low-sun-rainfall (Bshs where smalls stand for summer drought). There are those belts of steppe climate found on the pole-ward sides of the tropical deserts. They are fairly close to Mediterranean climate and have nearly all the year, however, there steppes are dominated by dry settling air masses associated with the subtropical anticyclones. (b) Steppes with high sun-rainfall (Bshw) climate found on the equator-ward margins of the deserts. Here the temperature gradient towards the adjacent areas is small.

3. *Middle-Latitude Deserts (Bwk) Climate*: Because of their generally lower temperature and less evaporation the humid boundaries of the middle-latitude deserts have lower rainfall than those of tropical desert. Some amount of precipitation does occur in this climate. Characteristically this subtype of the middle-latitude dry climate occupies the low latitude basin

like areas in the continental interiors. Tarim, the Gobi desert, Dzungaria, Russian Turkistan and central Bran are all surrounded by highland rims. Hence there regions develop a character of rain shadow zone with descending currents leading to excessive aridity. A further consequence of enclosure, when combined with low elevation, is the excessively high summer temperature with warm month averages in some cases even approaching those of tropical deserts. In Central Asia in July, average temperature is 90°F while the daily maximum for that month often rises to 110°F. Patagonia and Argentina narrow land mass with cool water offshore, temperatures are more marine than continental.

4. *Middle Latitude Steppe (Bsk)* : The general characteristics of these continental steppe have already been analyzed. Middle latitude steppes, like their counterparts, the semi-arid lands of the tropics, occupy transitional or intermediate positions between deserts and the humid climates. Because of the greater precipitation than in desert the steppes are somewhat better fitted for human settlements. Maximum temperature is recorded in the month of June 63°F. December and January are coolest months of the region. Years with below normal rainfall are more numerous than those in which it is above.

Warm Temperate Rainy Climates

The Humid Mesothermal Climates: This group of climates is characterized by a very definite seasonal rhythm in temperature conditions. Thus temperature becomes equal determinant with rainfall for determining the various types of middle latitude climates. Winter in this type of climates is relatively mild, as a consequence, restricted either (a) To the equator-ward margins of the middle latitude continents where latitude ensures winter mildness or (b) To marine locations usually on the windward or western side of continents away from the equator. Two of the mesothermal climates, e.g. dry summer subtropical and humid subtropical, occupy the first type of locations while the third of the marine climate is found in the second.

Dry-summer sub-tropical climate is subdivided based principally upon degree of location. Two subdivisions of this climate are recognized:

I. (*Csa*) *An Interior Sub-type with Hot Summer*: Interior from the coast a short distance, however, Mediterranean climate has a more continental temperature regime. Winters, of course, are still mild for the latitude is distinctly sub-tropical in most regions, mountains or water bodies protect against strong importations of low temperatures from higher latitude. This type of climate occurs in California, Sacramen to valley, Mediterranean Europe, North Africa, Rome and northern parts of Italy. January temperature of Red lands an interior station of California is 51°F. Sacramen to valley has a July temperature of 81.5°F. Mediterranean Europe has an average hot month temperature approximating 73°F. Those of north Africa are in the neighbourhood of 80°F. Summer days are charmy and dry. The diurnal range of temperature is 30° to 40°F in July. Interior locations have average cold month temperatures ranging from 10 to 20°F above freezing. The growing season is impeded during the winter because frost occasionally occurs during the winter months. Rain-fall is generally less than moderate 15 to 25 inches. Snow-fall is in the lower altitudes, whereas highlands may have a moderate to heavy snow cover. This snowfall in highlands provides an invaluable source of irrigation for the adjacent drier low lands. Periods of brilliant sun shine, extremely meagre precipitation, nearly cloudless skies and desert like relative humidity are the main features. Thunderstorms are rare in summer.

II. (*Csb*) *Marine or Coastal Type with Cooler Summers*: Littoral locations are likely to have somewhat modified Mediterranean conditions. Summer are usually cool, owing partly to general marine location, but this condition many times is accentuated by the cool ocean currents offshore. Such conditions prevail over the parts of Morocco, Chile and Australia. The average cool month temperature at Valparaiso, Chile and Perth (Australia) is 55°F, at San Francisco it is 50°F. In such marine locations the annual temperature range is uncommonly small approximately 9° at San Francisco and 110°F at Valparaiso. Daily ranges are likewise small in humid subtropical climate.

There are three principal differences between the humid subtropical and the dry summer subtropical climates: (a) The former is characteristically located on the eastern rather than on the western sides of continents; (b) It has a more abundant precipitation; and (c) This precipitation is either well distributed throughout the year or else concentrated in the warm season. This climate characteristically extends from latitude 25° pole-ward to 35°-40°. The parts of North America, China, Asia, Argentina, Australia and American states experience this type of climate.

Average hot month temperature of 75 to 80° is characteristic. Along the immediate coasts, especially of the smaller southern hemisphere land masses, they are as often below as above 75°. But everywhere summers are distinctly warm to hot, and this is particularly true to North America and Asia. Winters are, of course, relatively mild in these sub-tropical latitudes. Cool month temperatures usually average between 44°and 55°F. Thus Montgomery has an average cool month temperature of 49°F; Shanghai, China 38°F; Buenos Aires, Argentina, 50°F, and Sydney, Australia 52°F. Annual ranges are usually small although there is considerable variation depending upon the size of the continent and the latitudinal location of the stations. The mid-day temperatures in winter are likely to be warm, the thermometer usually rising from 50°F to 60°F. Frost occurs for seven months in a year. Rainfall is relatively abundant ranging from 30 to 65 inches within the humid sub-tropics, but still there are considerable differences within several regions. On the land-ward frontier of this type, where it makes contact with steppe climates rainfall reaches the lowest totals. In general there is no marked drought season. Summer usually has more precipitation than winter. In the southern hemisphere where the monsoon tendency is weak, the seasonal accent is less marked.

China and India, by contrast with the best developed monsoons and the fewest cyclonic interruptions in winter, have the greatest rainfall contrast between summer and winter. A considerable part of the summer rainfall at low elevations originates in convectional storms, many of them accompanied by thunder and lightning.

A considerable amount of rainfall is brought by the week cyclonic storms. During the late summer and early fen tropical hurricanes also bring adequate rainfall.

Marine Climate: Marine climate is the result of the land and sea breezes. The temperature is brought to mildness by sea breezes and land breezes. This type climate occurs along the water bodies. The marine type of climate characteristically occupy positions on the western or windward side of middle latitude continents, pole-ward from about 40° where the onshore westerly winds import to them conditions from the oceans. This climate extends in areas of narrow belt of Tasmania, Islands of New Zealand, Great Britain and parts of Australia and Africa. However, marine climate may not be limited completely to the western margins. On the equatorial margins this climate makes contact with the dry summer sub-tropical type.

On the pole-ward side marine climate extends into the higher middle latitudes, where it is eventually terminated by either the sub-arctic or tundra type. In the North and South America and Scandinavia marine conditions are confined to relatively narrow strips of territory seaward from the highlands. Temperature conditions are temperate. Summers are moderately cool and more or less ideal for human comfort. Seattle wash, has a mean July temperature of 63°F; Dublin, Ireland 60°F; Bergen, Norway 58°F; and Paris, 66°F. Winters, on the whole are more abnormally, mild for the latitude than the summer are cool. The most marine parts of Western Europe are 20° to 30°F warmer for their latitude in January. Winter Isotherms tend to parallel these coasts rather than follow the lines of latitudes. January average of 35° to 50°F in Western Europe is matched by others of 0° to –40°F in the continental climates of interior Asia in similar latitudes. Frosts are more frequent as well as more severe.

Nevertheless, the growing season is usually long for the latitude 180 to 210 days. Unusual cold spells in these marine climates are caused by importations of cold continental air from the anticyclones.

These are humid climates with adequate rainfall at all the seasons. The total amount, however, varies greatly from region to region, depending mostly upon the character of the relief. Where lowlands predominate, as they do in parts of Western Europe, rainfall is only moderate, usually 20 to 35 inches. But on the other hand, where west coasts are elevated and border by mountain ranges, as is the case in Norway, Chile and western North America, precipitation may be excessive, even reaching such totals as 100 to 150 inches. Marine climate is one of the cloudiest climates of the earth.

The Humid Micro Thermal Climate

Colder winter, a durable snow cover a longer frost season, and larger annual ranges of temperature distinguish the severe micro thermal climates from the mesothermal types. Emphatically micro thermal climates are land controlled and are, therefore, distinctly continental in their characteristics. It is because they are land controlled, being associated with large continents in higher middle latitudes, that they are confined exclusively to the northern hemisphere. Only Eurasian and North America are able to produce them. Of the southern hemisphere continents south America alone extends pole-ward sufficiently far to permit of severe climates but the narrowness of that land mass south of latitude 35° prevents genuinely severe conditions in spite of the latitude.

Micro thermal climates are excluded from the western or windward coasts because of the dominance thereof maritime air masses. Micro thermal groups differ substantially from one another only in degree and that chiefly in one element temperature.

Polar Climate

Because of a wide latitudinal spread, there are marked temperature contrasts within those regions classed as micro thermal. Of the two extreme seasons, it is the winter cold, rather than the summer heats which is most characteristic and distinctive. Nevertheless, summers are warm for the latitude. Not only are the seasons extreme but they are like-wise variable in temperature from one year to another. There is a effect of a snow cover upon temperature. Once a region is overlain by such a white snow mantle the ground itself ceases to have much influence upon air temperature. Moreover, although loss of energy by earth radiation goes on very rapidly from the top of a snow greatly to retard the flow of heat from the ground below to replace that which is being lost. Obviously, the effect of a snow cover is markedly to reduce winter temperatures.

Although winters are not without precipitation, summer is normally the season of maximum. The seasonal distribution is related to the following conditions: (a) the specific humidity or reservoir of water vapour in the atmosphere is much less over. The continents during the cold winter than it is in summer when the temperature much higher. (b) During winter the settling air in the continental seasonal anticyclone is likewise conducive to specific humidity. This same subsidence makes for increased stability of the atmosphere. (c) The continental anticyclone which develop over the colder, more northerly parts of the land masses in winter, are areas of converging air currents a condition that is antagonistic to the development of fronts and cyclones. In summer, although cyclones may be fewer and weaker, they can nevertheless penetrate deeper into the continents. (d) Convection is maximum during the warm summer months for at that season the warm land surface has a tendency to make unstable the air masses moving over it. In winter on the other hand, the cold snow surface tends to increase the stability of air masses. (e) Consequent upon the seasonal extremes of temperatures and hence of pressure a tendency towards it monsoon system of winds is developed, which leads to an inflow of tropical maritime air with high rainfall potentialities in summer and to an outflow of dry cold air in winter.

In the micro thermal climate, however, where the severe winters create a completely dormant season for plants, it is highly essential that period of sufficient heat and sufficient rainfall coincide.

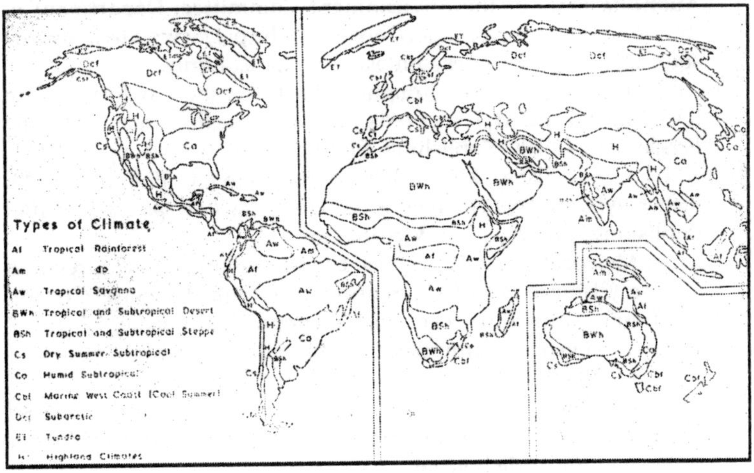

Fig. 19.1: Koppen's World Map of Climates.

REFERENCES

Allan, R.J., 1985, 'The Australasian summer monsoon, teleconnections, and flooding in the Lake Eyre Basin', *South Australi in Geographical Papers*, no. 2, 1985, Royal Geographical Society of Australasia (S.A. Branch).

Allan, R.J., 2000, 'ENSO and climatic variability in the last 150 years', in HF Diaz and V. Maɪkgraf (eds.) *El Niño and the Southern Oscillation: Multiscale Variability, Global and Regional Impacts*, Cambridge University Press, Cambridge, pp. 3-55.

Alves, O., Wang, G., Zhong, A., Smith, N., Tzeitkin, F., Warren, G., Schiller, A., Godfrey, S. and Meyers, G., 2003, 'POAMA: Bureau of Meteorology operational coupled model seasonal forecast system', in R. Stone and I. Partridge (eds.) *Science for Drought, Proceedings of the National Drought Forum*, Brisbane, April 2003, Queensland Department of Primary Industries, pp. 49-56.

Anon, 1901, *Royal Commission to Inquire into the Condition of the Crown Tenants,* Western Division of New South Wales, Government Printer, Sydney.

Beeadle, NCW, 1948, *The Vegetation and Pastures of Western New South Wales with Special Reference to Soil Erosion,* Government Printer, Sydney.

Beer, T., 2006, 'Atmosphere', theme commentary prepared for the 2006 Australia State of the Environment Committee, Department of Environment and Heritage, Canberra, <http://www.deh.gov.au/soe/2006/commentaries/atmosphere/index.html>.

Botterill, L.C., 2003, 'Beyond drought in Australia: the way forward', in L.C. Botterill and M. Fisher (eds.) *Beyond Drought: People, Policy and Perspectives,* CSIRO Publishing, Collingwood, Victoria, pp. 197-207.

Botterill, L.C. and Fisher, M., 2003, Introduction in L.C. Botterill and M. Fisher (eds.) *Beyond Drought: People, Policy and Perspectives,* CSIRO Publishing, Collingwood, Victoria, pp. 1-7.

Bureau of Meteorology, 2002, *Annual Climate Summary 2002,* prepared by the Climate Analysis Section, National Climate Centre, Commonwealth Bureau of Meteorology. Melbourne, <http://www.bom.gov.au/announcements/media_releases/climate/change/20030106.shtml>.

Bureau of Meteorology, 2005, *More than 150 high temperature records set in January and April,* media release 2 May 2005, Australian Bureau of Meteorology, Melbourne.

Bureau of Meteorology 2006, *Special Climate Statement 9: An exceptionally dry decade in parts of southern and eastern Australia: October 1996-September 2006.* Issued on 10 October 2006, National Climate Centre, Australian Bureau of Meteorology, Melbourne.

Calder, Nigel (1975). *The Weather Machine.* New York: Viking.

Callendar, G.S. (1938). "The Artificial Production of Carbon Dioxide and Its Influence on Climate." *Quarterly J. Royal Meteorological Society* **64:** 223-40 Online here.

Dahan-Dalmedico, Amy (2008). "Climate Expertise: Between Scientific Credibility and Geopolitical Imperatives." *Interdisciplinary Science Reviews* **33:** 71-81 [doi: 10.1179/030801808X259961].

Dahan-Dalmedico, Amy (2001). "History and Epistemology of Models: Meteorology (1946-1963) as a Case Study." *Archive for History of Exact Sciences* **55:** 395-422.

Eady, E.T. (1957). "Climate." In *The Earth and Its Atmosphere,* edited by D.R. Bates, pp. 113-29. New York: Basic Books.

Easterling, David R., et al. (1997). "Maximum and Minimum Temperature Trends for the Globe." *Science* **277:** 364-67.

Easterling, David R., et al. (2000). "Observed Climate Variability and Change of Relevance to the Biosphere." *J. Geophysical Research* **105**(D15): 20, 101-20, 114.

Economist (2000). "Beyond the Hague." *Economist*, 2 Dec., pp. 19-20.
Faegre, A. (1972). "An Intransitive Model of the Earth-Atmosphere-Ocean System." *J. Applied Meteorology* **11:** 4-6.
Faegri, Knot, et al. (1964). *Textbook of Pollen Analysis.* New York: Hafner.
Gribben, John (1975). "Cause and Effects of Global Cooling." *Nature* **254:** 14.
Haag, Amanda (2006). "Church Joins Crusade over Climate Change." *Nature* **44:** 136-37.
Kutzbach, John E. (1976). "The Nature of Climate and Climatic Variations." *Quarternary Research* **6:** 471-80.
Lacis, Andrew A., and James E. Hansen (1974). "A Paramaterization for the Absorption of Solar RAdiation in the Earth's Atmosphere." *J. Atmospheric Sciences* **31:** 118-33.
Maddox, John (1983). "Great Greenhouse in the Sky?" *Nature* **306:** 221.
North, Gerald R., et al. (1981). "Energy Balance Climate Models." *Reviews of Geophysics and Space Physics* **19:** 91-121.
Pearson, Ronald (1978). *Climate and Evolution.* London: Academic Press.
Rind, D., et al. (1991). "Positive Water Vapour Feedback in Climate Models Confirmed by Satellite Data." *Nature* **349:** 500-03.
Shaw, Gleen E. (1983). "Bio-Controlled Thermostasis Involving the Sulfur Cycle." *Climatic Change* **5:** 297-303.
Urey, Harold C., et al. (1950). "Paleotemperatures of the Upper Cretaceous." *Science* **111:** 462-63.
Zeng, Ning (2003). "Drought in the Sahel." *Science* **302:** 999-1000.

20

OCEANOGRAPHY

'The word ocean came into English from the French word 'ocean' which in term was derived from the Latin 'Ocean us' and Greek 'Okeanos'. The ocean serves as a store house of heat and water and affects mankind in many ways. It has got profound influence on the climate. Hence the study of oceans is as important as any other subject. Scientists have been studying the ocean waters from time immemorial. The ocean has sometimes been worshipped as a god. Men have feared the ocean.

A scientific study of oceans covering all the aspects is known as oceanography. Although it may be regarded as a separate science it actually is a common meeting ground of four sciences. Oceanography studies the following aspects of the oceans:

(i) The physical study of water and wave movements in oceans,
(ii) The geologic study of the form of the ocean basin and the characteristics of the sediments, laid down on the ocean basin,
(iii) The chemical study of the water and dissolved substances, and
(iv) Biological study of plants and animal life in the sea.

When such a study is conducted for the lakes and fresh water bodies it is called limnology.

Movement of Ocean Water

The physical study of water and wave movements involves the movement of water from one ocean to the other and from latitude to latitude. The movement of the water is carried on by drifts, creeps, tides, waves and currents:

(i) *Ocean Drifts*: It is a movement of water which is hardly discernible. It is a slow sweep of water having broader courses.
(ii) *Ocean Creeps*: These are sub-surface movements of water. Near the poles the cold water sinks and a slow current of cold water starts moving below the surface, towards the equator.
(iii) *Tides*: The regular and periodic rise and fall of the ocean waters is referred as tide.
(iv) *Ocean Currents*: They are those powerful movements of the ocean water that proceed regularly in a definite direction. The movement is clearly visible. They move like streams.
(v) *Waves*: When the wind blows on the ocean it causes a drag on the water surface by friction, the water rises and falls in crests and troughs forming waves.

The geological study of the form of the ocean basin includes the marine relief or the features on the ocean floor. This study also involves the assessment of the characteristics of the deposits found on the ocean floor. The configuration of the ocean floor is an important area of the study of oceanography. Movement of the vessels in the ocean and search for future sources of economy such as animal food plant life, rubies etc. can be planned on the basis of the careful study of the configuration of the ocean floor.

Configuration of Ocean Floor

The surface of the ocean floor is having features similar to those found on the surface of the land. If we start moving from the land into the ocean along the bottom we find that the nature and shape of the floor varies to a considerable degree. The ocean floor can be divided as under:

1. Continental shelf,
2. Continental slope,
3. Deep sea plain, and
4. Ocean deeps and trenches.

1. Continental Shelf: From the coast to about 100 fathom (600 feet) mark lies a shallow platform known as continental shelf. From this shallow platform rise the islands. The slope of continental shelf is gentle towards the sea. It slopes at an angle of less than one degree. The continental shelf is well developed off in Western Europe where it extends up to 200 miles from the lands at some places. Near the coasts where the old mountains run parallel and close to the sea the continental shelf is altogether absent. In most of the oceans of east Pacific coast the continental shelf is narrower off the northern coasts of Siberia, it is about 750 miles wide.

It has been noticed that the valleys of many rivers seem to continue across the continental shelf.

The continental shelf can be subdivided into zones:

(i) *Littoral Zone*: This is an area which extends from the tide mark to about a depth of 200 feet. This portion of the continental shelf is very shallow and comprises shingle sands and shelf sands.

(ii) *Shallow Zone*: Away from the littoral zone extends the shallow water zone with a maximum depth of 600 feet. This lies between the continental slope and littoral zone comprising of sand, mud and corals.

2. Continental Slope: The ocean floor towards the sea from the continental shelf is having considerably steep slope known as continental slope. The continental slope extends from 100 fathoms (600 feet) to 2000 fathoms (12000 feet). The degree of slope varies from 2° to 5° although some extremely steep slopes exceeding 15° have also been noticed. The continental slope is

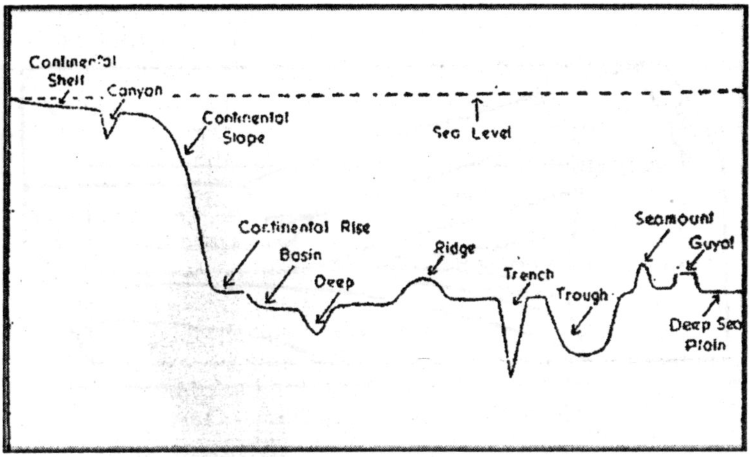

Fig. 20.1a: Topographical Features of the Sea Bottom.

covered by mud of different colours and origin. Off Ireland coast the continental slope falls an angle of 5° but off the coast of Spain it is much steeper. Near Cape Tarinana it is in one place as steep as 36°, the continental slope area is called Bathyal zone. It is a transition or link between the shallow zone of continental shelf and deep sea plain.

***3. Deep Sea Plain*:** The greater part of the ocean floor is formed of the deep sea plain. About two-thirds of the entire ocean floor lies at depths between 2000 to 3000 fathoms forming an undulating deep sea or 'abyssal' plain. It is not a flat surface as its depth varies from area to area but the slopes are imperceptible. (Fig. 20.1a)

The deep sea plain is interrupted by long curving ridges and large submarine plateaus. Sometimes, one finds volcanic peaks rising steeply from the deep sea plain which sometimes reach above the surface of the water to form islands. Sometimes, the flat topped wave truncated mountains are also found under the water known as Guyot.

The deep sea plain is entirely free of sediments brought by the stream. However, the surface of the deep sea plain is not formed of hard rock. A large part of the deep sea plain is covered with mud of different colour and origin. The mud is called ooze consisting of shells and tests of minute animals and plants which when alive float on the surface of the water.

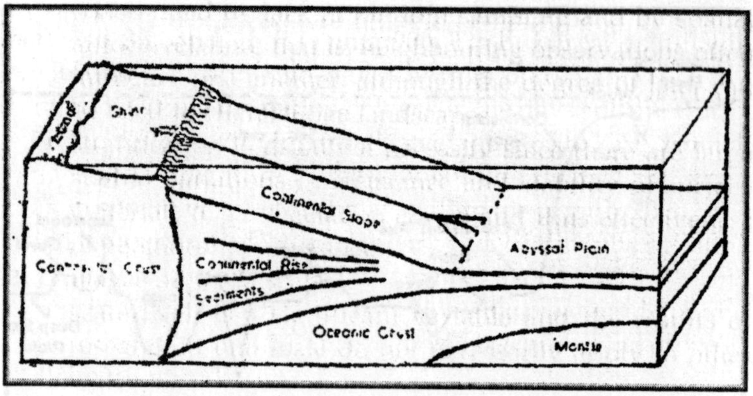

Fig. 20.1b: Continental and Oceanic Crust.

In the deeper parts of the deep sea plain one finds red clay mud formed mainly from fine ash thrown by the volcanoes.

Continental Rise: Between the continental slope and deep sea plain lies a very gently sloping area known as Continental Rise. This is a transitionary zone which is raised above the level of the flat ocean floor. Continental rise is formed by the deposition of river sediments at the base of the continental slope.

***4. Ocean Deeps and Trends*:** These are the depressions in the ocean floor deeper than the deep sea plain. These deeps cover smaller areas and their sides are very steep. Generally, these deeps are found on the margins of the oceans along the outer edges of the island areas. It has been found that most of them occur near the coasts where volcanoes are still active and earthquakes are common, for example, the shore of Pacific, Atlantic and Indian Oceans.

The deepest deep in the world is located near Philippine island off Mindanao in Pacific Ocean which is 36,198 feet deep. In Atlantic ocean the deepest deep is about 8,374 feet deep. In Indian ocean the deepest deep is 25,344 feet deep. In the Caribbean sea it is 24,720 feet. In all other seas and oceans the maximum depth is less than 20,000 feet. The deeps are associated to major trenches, e.g. Mindanao trench, Mariana, trench, Japan trench, Peni trench in pacific, Caribbean trench in Atlantic and Sunda trench in Indian Ocean etc. (Fig. 20.1b)

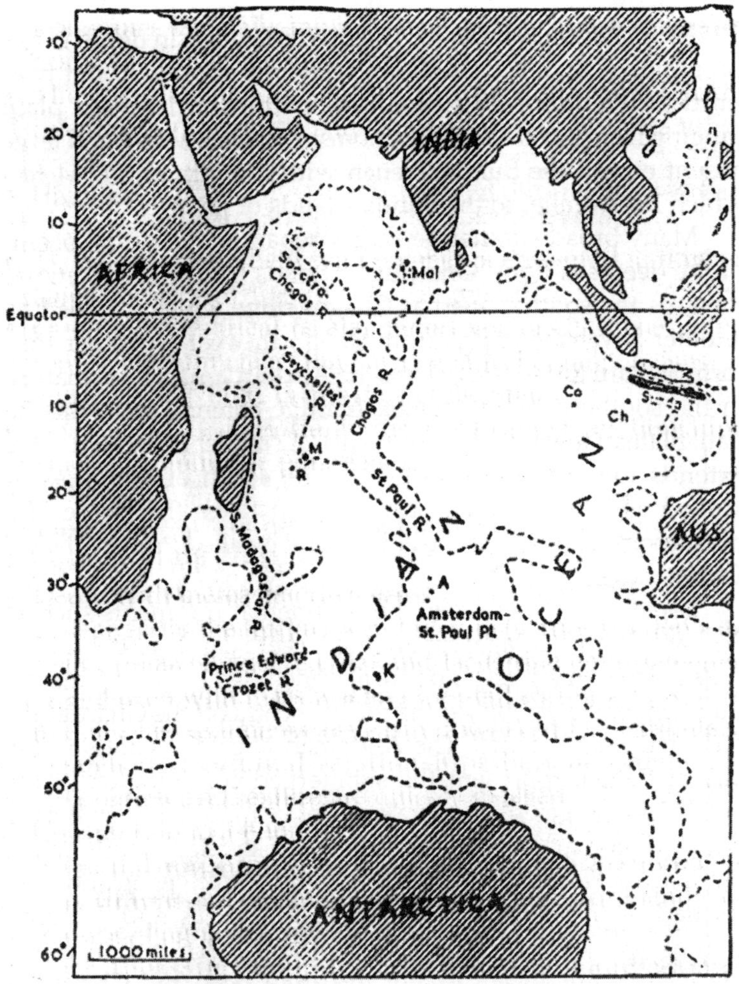

Fig. 20.2: The configuration of the Indian Ocean.

Configuration of Indian Ocean: The Indian Ocean is smaller than the Pacific and Atlantic oceans. This ocean is landlocked in the north. It extends roughly between the tropic of Cancer in the north and extends up to the Antarctica in the south. Indian Ocean shores consist of the ancient land masses of Africa, Australia and Antarctica. Towards the northeast it is bounded by the island of East Indies. (Fig. 20.2)

Floor of the Indian Ocean: The diversification in depth is less pronounced than the Pacific and Atlantic oceans. About two-third of the area of the ocean floor forms a deep sea plain ranging from 2,000 to 3,000 fathoms. Trenches are by and large absent except the Sunda Trench with a depth of 25,344 feet which lies parallel to the Sunda islands of Java.

Many broad submarine ridges separate individual basins of the deep sea floor. One ridge extends from Cape Comorin towards the south named as Chagos ridge which is joined by another Seychelles ridge from the north-west. Together they form St. Paul Ridge which joins the Amsterdam St. Paul plateau in the southern Indian ocean. A ridge extends from this St. Paul plateau towards the south-east. A smaller elongated ridge extends towards the south from the south of Madagascar known as South Madagascar ridge. It takes the shape of a mushroom cap towards the bottom and is called Prince Edward Crozet ridge.

Islands of Indian Ocean: The continental fragments such as Sri Lanka and Madagascar are the larger islands of the Indian ocean. Some smaller islands are also included in the continental fragment category such as the rocky Socotra off the Guardefuie, Zanzibar and Comoro. The Andaman Nicobar islands in the Bay of Bengal represent the submerged continuation of the outer fold ranges of the Arakan Yoma mountains in Burma (Now Myanmar). There are a number of smaller islands on the Submarine ridges. The Laccadive and Maldives islands are formed of the coral accumulation. Mauritius and Reunion islands are volcanic in origin. In the eastern Indian ocean the islands are absent except for tiny Cocos and Christmas islands.

Configuration of the Atlantic Oceans: The Atlantic ocean covers a little less than one-sixth of the total area of the world. In other words, it is about half of the area of the Pacific Ocean. Its general shape resembles the latter 'S'. The Atlantic Ocean is relatively narrow near the equator. The northern basin of the ocean is enclosed by Greenland and Iceland, whereas the southern basin opens broadly into southern ocean. (Fig. 20.3)

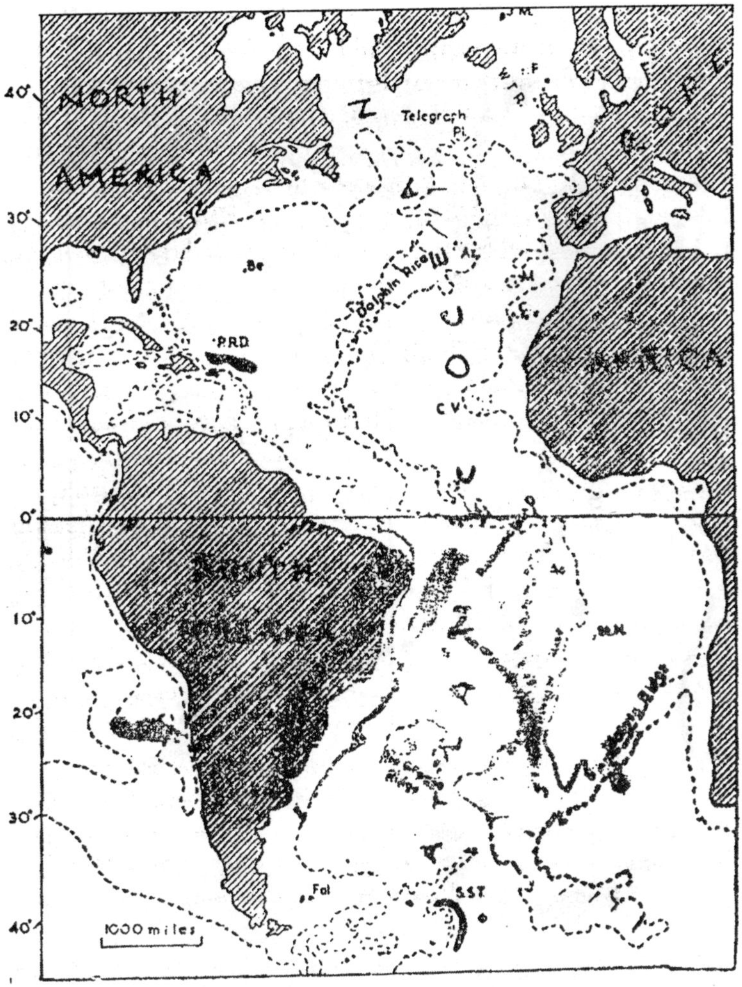

Fig. 20.3: The configuration of the Atlantic Ocean.

Floor of Atlantic Ocean: The presence of a mid-Atlantic ridge running in north-south direction is the most important feature of the Atlantic ocean. The vast rise of sea floor south of Greenland is called Telegraph Plateau. From the Telegraph Plateau a submarine ridge extends southward known as Dolphin Rise. This ridge goes on becoming narrower towards

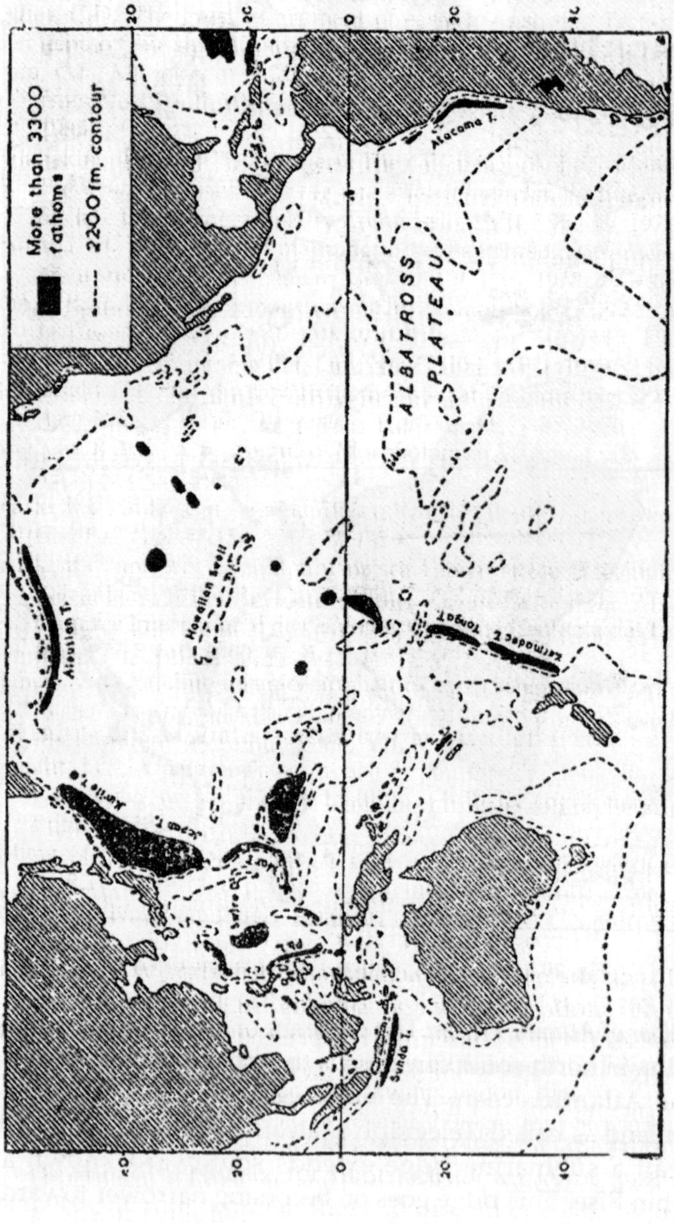

Fig. 20.4: The configuration of the Pacific Ocean.
Mar. T. = Mariana Trench; Ph. T. = Philippine Trench. Small island groups are omitted.

the south. In the middle of the North Atlantic it turns towards the south-west and becomes very narrow near the Caribbean sea, south of the Caribbean sea it turns towards the south-east. Near the equator it turns due south where it is called Challenger Rise. At the base of the Challenger Rise it is joined by another ridge called Walvis Ridge extending from south-west Africa shelf. Rio Grande Ridge is located between the south Challenger ridge and the south Brazilian shelf. Weyville Thomson Ridge extends in northern Atlantic between Scotland and Iceland.

Linear deeps or trenches are not common in the Atlantic ocean except for Puerto Rico trench and south Sandwich trench in the extreme south. The continental shelf is of considerable extent in north Atlantic but less in south Atlantic except the Falkland rise.

Islands of Atlantic Ocean: In the North Atlantic Ocean Great Britain, Ireland, Greenland, Iceland, New Found Land and West Indies are the major islands. In the south Atlantic the islands are of much smaller size such as Falkland, south Orkneys, south Shetlands, Georgian and Sandwich islands. The true oceanic islands project from the Mid-Atlantic-Ridge such as the Azores, Ascension and Tristan da Cunha whereas St. Helenalies just east of the ridge. The coral islands of Bermuda are built on volcanic cones. Madaira, off the Moroccan coast is almost entirely built of volcanic material.

Distribution of Temperature in the Ocean

Daily Variation of Temperature in the Ocean: The mean daily variation in oceanic surface temperature approaches the last value, namely 0.2° to 0.3°. The mean maximum daily variations of 0.24° in southern North Sea with a water depth of 50 m for instance, compare with in Finnish offshore waters of the Gulf of Bothnia with a depth of approximately 6 metres.

Annual Variations: In contrast to the daily variation, the annual variation significantly influences the physical, chemical and biological processes in the ocean. The annual variation of the surface temperature depends on different factors such as annual variation in radiation, ocean currents, and the prevailing winds. In the equatorial and polar regions, the annual variation remains small, namely less than 2°. The maximum value of approximately 7° has been obtained in latitudes of 30-40° where the radiation is maximum. However, the north Atlantic and north Pacific oceans are exceptions, especially the western parts. There, the prevailing westerlies bring very cold continental air masses over the oceans in winter decreasing the surface temperature appreciably resulting in an annual variation of 18°C. In Mediterranean sea and marginal seas, which frequently remain under continental influence during the whole year, the corresponding variation in water temperature follows variation of air temperature. In the Baltic Sea, the Black Sea and Persian Gulf, it reaches more than 14°C.

The penetration of heat represents a much more complicated process in the ocean than on the land where only the process of physical heat conduction is involved. During heating in spring, the temperature increase is restricted to only a thin layer which becomes less dense. The Austausch process is almost exclusively responsible for the transport of heat to the deeper layers. Further heating results primarily from the storage of heat in the near surface layers, and the thermo cline intensifies and moves slightly downward.

The summer thermo cline is a typical phenomenon of an area with large yearly variations in surface temperatures. Conditions concerning the penetration of heat are more complicated in the marginal seas. The stratification in the lower layers as well as the thermo cline will be destroyed by the frictional turbulence as a result of strong tidal currents. The conditions in the Bornholm Basin in Baltic sea represent a typical case of this kind. In winter, the surface water becomes denser, and a vertical thermal convection becomes effective. In time, the thermo cline will diminish until the temperature differences disappear which is rather typical for the winter conditions. This winter circulation determines the distribution of vital gases and nutrition in the ocean and has extreme biological importance.

Long Range Variation: The annual variation and, therefore, the annual mean of the surface temperature are subject to long-range variations. This is true for the geological past indirectly confirmed from long bottom cores showing the various compositions of the sediments with depth of core. It also applies to the current past for which observations of the surface temperature are available. The observation of North sea shows the deviation of monthly means from the 30-year mean. The greatest temperature increase 0.6° to 0.7° can be found at the extension of Gulf Stream system near the Faro Shetland area.

The recent climatic changes also affect the long range variation of temperature. The effect of the present climatic change is demonstrated by various phenomena on the earth especially since the changes restricted not only to the temperature and wind conditions but also to precipitation and evaporation.

Distribution of Salinity in the Oceans and Seas

Most of the naturally occurring elements have been identified in dissolved form in the ocean. On an average, 1000 grams of sea water contains 35 grams of dissolved inorganic salts.

The salinity of the ocean is thus frequently expressed as 35 per cent, although it actually varies between 30 per cent to 38 per cent. Chlorine and sodium together account for over 86 (or about 30 parts per thousand) of the dissolved material in sea water. Below given table shows important constituents of sea water:

Sl. No.	*Constituents*	*Ion*	*Percentage of all dissolved material*
1.	Chlorine	Cl^-	55.04
2.	Sodium	Na^+	30.61
3.	Sulphate	SO_4^{--}	7.68
4.	Magnesium	Mg_2^+	3.69
5.	Calcium	Ca_2^{--}	1.16
6.	Potassium	K^+	1.10
7.	Bicarbonate	HCO^{---}	0.41

The salinity of the ocean varies from place to place, although the proportion of the major components of the salinity remains constant. However, numerous cases of salinity differences are encountered in horizontal as well as in vertical direction. These are the results of different physical processes in the oceans. When ocean water evaporates or freezes, most of the salt remains in the water resulting in an increase of salinity near the surface. Conversely, precipitation, river discharge from the continents and the melting water decrease the surface salinity. Circulation of ocean waters decrease the salinity not only at the surface but through all layers.

Isoclines (line joining the places having equal salinity value) can be drawn to show the salinity at the surface or at any intermediate depth. In the Atlantic Ocean the areas of higher salinity lie near the tropics (about 37%) where the clear skies, constantly high temperatures and brisk trade winds maintain evaporation. Salinity decreases towards the equator where it is 35 per cent where rainfall is heavier and evaporation less because of the higher relative air masses of the doldrums there is also a decrease towards the poles (less than 34") due to the result of melting ice and decreasing evaporation. In the areas where big rivers like Amazon and Zaire join the sea the salinity is less. Variations in salinity are most marked in partially or wholly enclosed seas. The Baltic sea for example reveals decreasing salinity with decreasing distance from the North Sea. Off the south coast of Sweden it is about 11 per cent, off Bornholm about 8 per cent, and it decreases to about 2 per cent at the head of Gulf of Bothnia. This sea receives water from a number of rivers such as the Oder and the Vistula and much melt water while evaporation is low. The Black sea also has many large rivers (including the Danube, Dnester, Dneper and Don) hence its salinity is between 18 per cent and 20 per cent. The Red sea, which receives no rivers and experiences great evaporation, has a summer salinity exceeding 40 per cent. The land seas and lakes have a much higher salinity than the open sea, for while their rivers may bring down only a small quantity of salt. The Great Salt Lake in Utah has a salinity of 110 per cent, the Dead sea of 238 per cent and Lake Van in Asia minor of 330 per cent. Salinity may vary locally even within a lake or inland sea. The northern part of the Caspian sea into which flow the Volga and Ural rivers has a salinity of only 13 per cent, while the Gulf of Kara-Bagaz in the south-east, which is almost separated from the sea by land spits, has a salinity of 195 per cent.

Ocean Deposits

Ocean deposits include all those materials which are accumulated on the ocean floor. The sediments brought by a number of streams are finally deposited in the oceans layer after layer. Here these materials are mixed with the local material found in the oceans such as the remains of plants and animals. Numerous animals and plants either float in the ocean or live on the ocean floor. When they die they also accumulate in the form of fossils on the sea floor. Chalk is an example of such deposits. It consists of the great thickness of the remains of Foraminifera deposited in clear water which was free from other sediments such as mud or sand. Millstone Grit is another example which consists of consolidated beds of coarse sands and gravels laid down in shallow waters near the margin of the land in a series of deltas.

Types of Ocean Deposits

On the basis of the location the ocean deposits can be divided into four types:

(1) *Littoral-Deposits:* These are found in the upper part of the continental shelf near the shoreline shingle sands and shelly sands are common.
(2) *Shallow-Water-Deposits* occur from low water mark up to the edge of the continental shelf, i.e. up to 100 fathoms. The commonest deposits in the shallow water are the sands, muds and corals.
(3) *Bathyal Deposits* generally occur on the continental slope. They comprise mud of differer t colours such as blue mud, red mud, green mud and corals.
(4) *Abyssal Deposits* occur on the deep sea plain and in ocean deeps. These deposits consist of two types of oozes, e.g. calcarious oozes and siliceous oozes.

(A) Calcarious oozes are of two types e.g. Pteropod oozes and Globigerina oozes:

(i) Pteropod oozes are calcarious in nature. They are thin and fragile conical shells.

(ii) Globigerina oozes consist of various foraminifera.

(B) Siliceous oozes are of three types:

(i) Radiolarian oozes are the skeletons of foraminifera. They have a lattice like structure. They generally occur in the deep water.

(ii) Diatom oozes are the skeletons of the Diatoms like Microscopic plants flourishing in cool waters of the oceans.

(iii) Red clay is a hydrated silicate of alumina. The redness is due to the Iron oxide. It is largely formed by volcanic and meteoritic dust which settles in the deep sea plain. They also consist of iceberg debris, insoluble relics of organic life, e.g. shark teeth.

Coral Deposits

Corals are mostly found in tropical oceans. Some corals known as polyps live as individual animals but most of them are joined to each other. They are generally found in the form of colonies. These are microscopic sea animals of many colours and shapes. They may be smooth and round, or like masses of sponge, or occur in branched clusters or even like pieces of lace. All of them have hard skeleton of calcium carbonate. As each coral dies these skeletons accumulate to form white coral limestone. In addition to corals many other organisms help in building a coral reef such as calcareous algae act as cement between the two dead corals, remains of molluscas and echinoderms (star fish, sea-urchins and sea-cucumbers).

Corals cannot survive in water where temperature is below 60°F. That means they are highly concentrated in and around

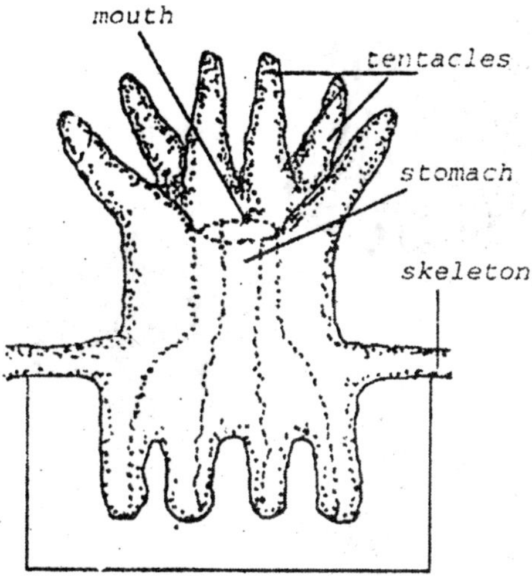

Fig. 20.5: Coral polyp.

the tropics within 30° North and South latitude. However, they may extend in some cases such as Bermudas that too because of the Gulf Stream. Sometimes because of cold currents even the tropical coasts, especially the western sides, are free of corals. Corals are rarely found above low tide level, whereas they rarely survive at depths exceeding 30 fathom.

A coral is not one coral but a number of them growing together. If you look at a coral piece closely you will find numerous pores on it. Each one is the place where one individual polyp (Figure 20.5) lived. Every coral colony begins as a minute skeleton of one single polyp and then keeps on growing by adding more and more individuals. According to their genes there are four common forms of coral colonies. (Figure 20.6):

1. Massive (stone like)
2. Foliaceous (leaf like)
3. Branching (tree like or flower like)
4. Encrusting (coating like)

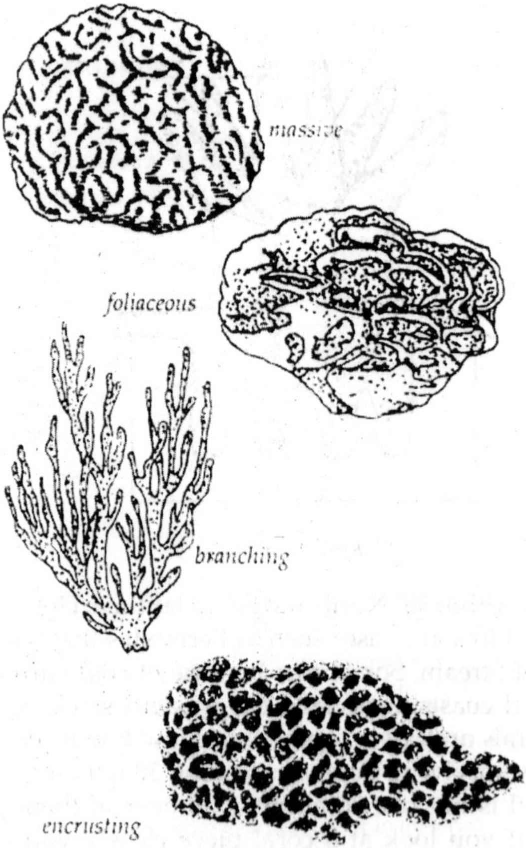

Fig. 20.6: Coral colonies.

At times, water depth and light may also control the shape. At greater depths, the water pressure is high and coral skeleton tends to grow flatter. The growth forms also change because the corals are continuously on the look out for light. All corals that we see on a reef have microscopic plants (called Zooxanthellae) living inside their cells. These algae have chlorophyll pigments and produce organic matter by photosynthesis. For this they need sun light. So the coral has to remain near the surface where the sunlight is abundant.

The colour of the corals, in fact comes from the algal pigments. If the algae were not there, then the coral tissue would be a transparent layer through which we could see the skeleton. When the algae live with the animals, both are benefited. The algae produce organic matter by photosynthesis but do not use all of it. Some portion of this is passed on to corals for their nourishment. The corals like all animals excrete nitrogen, phosphorous salts and carbon dioxide. All these are needed for photosynthesis by algae and are readily taken up by them.

Food of the corals: In fact, corals are basically carnivores. They catch tiny animals from seawater and swallow them. That is their main food. What they get from the algae is only supplementary. Some corals can also feed on detritus (dead organic matter) and some other can take up dissolved organic matter directly from the seawater. Corals are thus versatile in their feeding habits. This is what helps them to survive and thrive in ocean deserts.

Natural enemies to corals: Parrotfish feed on corals. These have powerful teeth with which they can snip off pieces of coral skeleton along with polyp tissues and ingest them. The butterfly fish also feed on corals but since their teeth are not so strong, they just nibble the corals. The crown-of-thorns starfish feeds on corals. Black band and white band diseases are known so far to effect corals.

Species of corals: There are somewhere between 800 and 1000 known species of the corals. In India 252 species of corals are known from Indian reefs, with a majority of them occurring in the Andaman and Nicobar islands. Some corals species such as *Porites mannarensis* are unique to India. There are other corals that are different from the ones we see on a reef. They resemble the true corals but do not have zooxanthellae. They occur sometimes in profusion in deeper pats of the reefs. The fire corals, tree corals, wire corals, sea fan and sea fern are some such corals. (Fig. 20.7)

Coral resources : Reef abound with food fish. Most of them live in the lagoon. Some fish like tuna and sharks live outside the reef but come closer to the reef to feed on the small fish. The biological productivity of the reefs is very high. A substance used as bone transplant should have nearly the same chemical

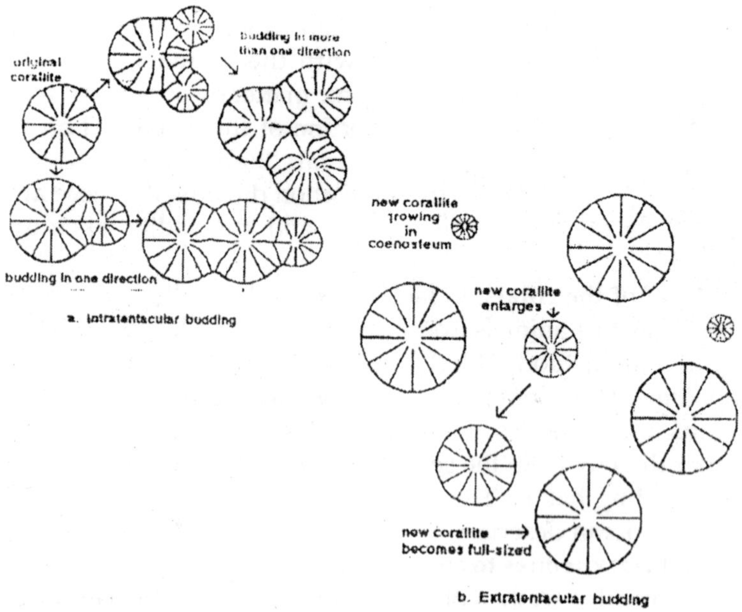

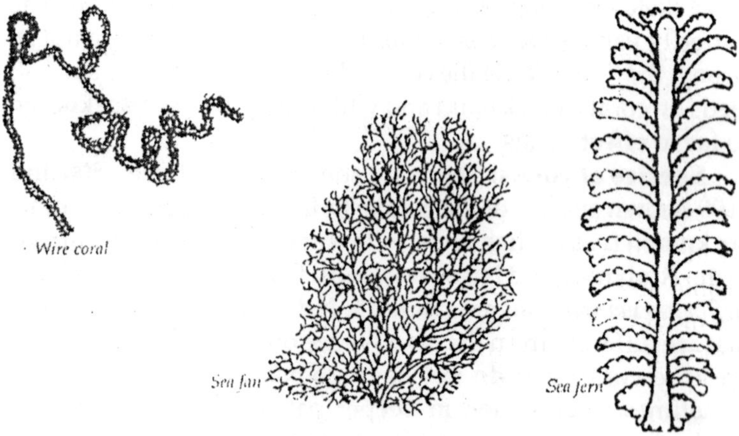

Fig. 20.7: Species of Corals.

composition as the bone should be tough to give mechanical strength and at the same time porous enough to allow the fine blood vessels to pass through. Among the several biomaterials, corals like the massive *Goniopora* are the best choice. Many medicines preparations includes them.

Coral Reefs

Dead corals accumulating layer after the layer form a wall in the sea sometimes parallel to the coast or around an island. They also accumulate around the volcanic islands. They also form low coral islands for example Gilbert Islands, Ellice Islands and the Marshal Islands in the Pacific. Western and central Pacific have maximum concentration of corals. It is also found in Indian ocean as in the Laccadives, west of Srilanka, the Andaman, Seychelles and Mauritius. In Atlantic oceans west Indian archipelago is the only coral belt. Coral reefs are of three types:

1. Fringing Reef,
2. Barrier Reef, and
3. Atolls.

Fringing Reef: Fringing reef is located at some distance from the mainland. This is an uneven platform fringing the coast formed by accumulation of dead corals. There is a shallow lagoon between the reef and the mainland. The lagoon is narrow and calm.

Barrier Reef: Barrier reef is separated from the mainland by a much deeper and wider channel. The barrier reef is hardly visible above the water surface. They pose greatest threat to the moving ships. Ships can move across the barrier reef where they are broken. The greatest barrier reef in the world is known as the great Barrier Reef of Queensland near Australia.

Atoll: An Atoll is a coral reef which forms a lagoon. The shape of an atoll may be circular, elliptical or like a horse shoe. The reef may either be forming a perfect enclosure or may be broken at many places. The atoll does not have an island in the centre. (Fig. 20.8)

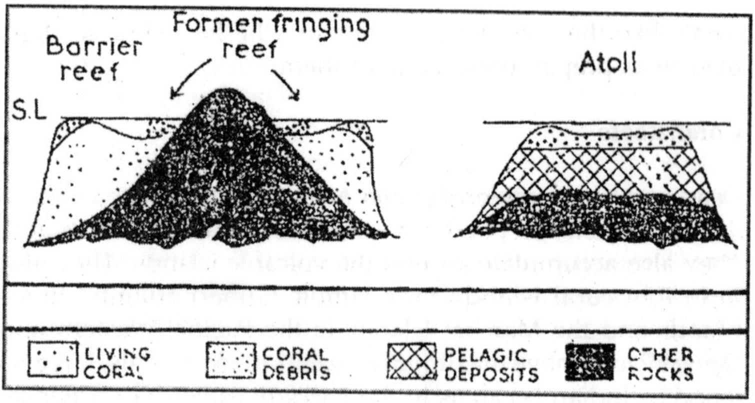

Fig. 20.8: Atoll.

Theories About the Origin of Barrier Reefs and Atolls

The depth of the water enclosed by the barrier reefs needs explanation. Important theories are as follows:

I. Darwin's Theory

According to him the Atolls and Barrier Reefs have grown in a gradual process, mainly subsidence. A fringing grows around a volcanic island. As this slowly subsides, the coral continues to grow upwards keeping pace with the subsidence. Coral growth is more vigorous on the outer side of the reef forming a higher rim, whereas the inner part has a deep and wide lagoon. Ultimately, the island in the centre is completely submerged for a ring of the coral reef known as Atoll.

He has been supported by Dana and others.

II. Murray's Theory

In 1872 Sir John Murray voyaged in the Challenger and found that in most cases subsidence was not involved in Atolls. According to him the base of the reef consists of a submarine hill or plateau rising from the ocean floor. He also writes that subsidence is not involved in the formation of barrier reefs. He explains that as a fringing reef grows masses of the coral fragments gradually accumulate on the seaward side of the reef and become cemented into solid bank by the pressure of the waves. Coral generally grows on the seawards side farther and farther on its own debris.

Ocean as Storehouse of Resources for the Future

The Ocean plays an immense role in the life of humanity. Since ancient times it has been the source of fish, salt and pearls. Water covers almost 70.8 per cent of the earth's surface and the world ocean constitutes over 94 per cent of the hydrosphere. The ocean fulfils function of importance for human life. Its plants thanks to chlorophyll, collect the energy of sunshine and by means of it form molecules of sugar (glucose) and free oxygen molecules from molecules of carbon dioxide and water. Their process known as photosynthesis enables plants together with inorganic salts to form protein and fats and to saturate the atmosphere with oxygen. The annual release of oxygen by marine plants provides 70 per cent of the oxygen in the atmosphere. The organic substance of plants in turn is a great kitchen for species living near the surface of the water and they feed organism living deeper.

The moisture of the sea, evaporating through the effect of solar heat, forms stores of fresh water in the atmosphere that later fall as precipitation. Every year 4,47,900 cubic kilometres of water are evaporated, 90 per cent being returned to the ocean in rain, while 10 per cent falls first on land, irrigating it, and then returning in the discharge of rivers to the larger hydrosphere, which serves as an accumulator and regulator of solar heat on our plant.

The ocean's reserves of minerals, in the most general term have been theoretically estimated at 60 × 100 (60,000 million) tonnes. The ocean is rich in zinc, bismuth, strontium, barium sulphate etc. Magnetite and titano-magnetite placers are mined along the coasts of Japan and other countries. Titanium containing sands are found along the coasts of Florida, India, Srilanka and Australia. Magnesite sands containing thorium occur in the zone of Brazil and India, and the sands with a high content of iron and chromite of Alaska.

Beds of phosphorite modules occur at depths from 60 to 80 to 300 metres in the boundary zone between the continental shelf and continental slope. Deep water silts contain calcium and silicon.

The ocean's resource provides raw materials for chemical industries. Many valuable elements are dissolved in the sea water. Every cubic kilometre of sea water contains about 37.5 million tonnes of solids. The greater part of the material in sea water (30 million tonnes) in common salt, 4.5 million tonnes magnesium and the remainder potassium, calcium, bromine, fluorine, iodine, manganese, nickel, cobalt, copper, iron, thorium, molybdenum, etc.

Each kilogramme of water contains, on average 35 grammes of salt. This is equivalent to saying that the total volume of salt in the ocean is 48 × 10 tonnes. Eighty per cent of it is common salt (sodium chloride). Common salt is important as a food product and as an industrial component.

The ocean contains 3,00,000 million tonnes of heavy water or around 1/5000 of its total volume 20 of the heavy water now used throughout the world is of marine origin.

Recent cold fusion experiment increased the demand of heavy water; sea water also contains 400 million tonnes of uranium. There are 2,00,000 million tonnes of lithium in sea water, a metal viewed as one of the source of energy in twenty first century.

The increasing need of water can be fulfilled by ocean water. Many countries are already using desalinization plants for ocean water. Experimental irrigation using sea water is being carried on in certain areas, namely in the Russia, USA, Italy, Germany and Tunisia. There are a number of ways of utilizing energy of seas and oceans, namely by building of hydro technical structures, tidal power stations and sub-marine equipments to extract solar energy. At present continental shelf is a source of mineral oils. The urgency of exploiting the food resources of the ocean is underlined by the fact that by the year 2010 the world's population will require approximately more than 210 million tonnes of protein annually. This demand of protein could be achieved by extending fishing zones.

The geography of fish catches is currently quite uneven. 35 per cent is taken in northern hemisphere, 19 per cent in the tropical zone and 28 per cent in the southern hemisphere. As regards the intensity of fishing per unit area of Atlantic Ocean comes first followed by the Pacific and then by the Indian Ocean.

There are considerable possibilities for the exploitation of sea weeds. They contain nutrients, sometimes more valuable than wheat, meat, and potatoes and cheaper to produce. A mass of 15 tonnes of sea weed can be taken from the one hectare of the sea bed, while only four tonnes of grass can be harvested from the same area of hay meadow. Marine plantations moreover can yield more than ten harvests a year without any outlay of ploughing, harrowing and watering. Around 70 species of edible seaweeds are known.

There are many medicinal plants life in the ocean. The transport importance of the ocean for man is immense. Today marine transport carries more than 80 per cent of world cargoes.

REFERENCES

Impact from the Deep: October 2006; *Scientific American* Magazine; by Peter D. Ward; 8 Page(s).

Tom Garrison. "Oceanography: An Invitation to Marine Science", 5th edition. Thomson, 2005. Page 4.

FURTHER READING

Hamblin, Jacob Darwin (2005), *Oceanographers and the Cold War: Disciples of Marine Science.* University of Washington Press. ISBN 978-0295984827.

Steele, J., K. Turekian and S. Thorpe (2001), *Encyclopedia of Ocean Sciences.* San Diego: Academic Press. (6 vols.) ISBN 0-12-227430-X.

Sverdrup, Keith A., Duxbury, Alyn C., Duxbury, Alison B. (2006), *Fundamentals of Oceanography,* McGraw-Hill, ISBN 0072826789.

21

ENVIRONMENTAL POLLUTION AND ITS CONTROLS

Environmental Pollution

Pollution can take many forms. The air we breathe, the water we drink, the ground where we grow our food, and even the increasing noise we hear everyday all contribute to health problems and a lower quality of life.

Environmental pollution is on the increase. Exhaust emissions from vehicles are adversely affecting the air quality in cities. Toxic elements have started entering the food chain. Inefficient water management is leading to increase in various communicable and water-borne diseases. Solutions devised to protect and regenerate the environment have not shown any tangible results so far. Industries, power generation plants, automobiles are the major cause of air pollution. Some of the highly polluting industries are metal smelting, cement, dyes, fertilizers, steel, leather, pesticides, petrochemicals, pharmaceuticals, paper, and refineries

Serious health problems are caused by the presence of flyash in the air. Since flyash production is only going to increase rapidly, the situation is only going to get worse. Flyash is responsible for air pollution and wastage of land that could be used to grow crops. This same flyash can be put to good use for building roads and pavements. It can be used to fill mines and make cement. Although technology is available to put flyash to good use it is hardly being used for this purpose. Small scale industries located in residential areas are also major causes of pollution and related health risks. The government should have an incentive scheme for industries that install pollution control devices. Certain areas have crossed the maximum limit for allowable pollution while some areas can bear an additional pollution load. The government should identify such areas and carry out pollution carrying capacity studies of these areas.

Major Forms of Pollution and Major Polluted Areas

The major forms of pollution are listed below along with the particular pollutants relevant to each of them:

Air pollution is the release of chemicals and particulates into the atmosphere. Common substances that cause air pollution include CO_2, SO_2, CFCs (chloro fluro carbon), and NO_x produced by industry and motor vehicles. Photochemical ozone and smog are created as nitrogen oxides and hydrocarbons react to sunlight.

Water pollution is mainly due to surface runoff, leaching to groundwater, liquid spills, wastewater discharges, eutrophication and littering.

Soil contamination occurs when chemicals are released by spill or underground storage tank leakage. The most significant contaminants are hydrocarbons, heavy metals, herbicides, pesticides and chlorinated hydrocarbons.

Noise pollution, includes roadway noise, aircraft noise, industrial noise, high-intensity sonar etc.

Light pollution, includes light trespass, over-illumination and astronomical interference.

Visual pollution refers to the presence of overhead power-lines, motorway bill-boards, scarred landforms, open storage of trash or municipal storage waste.

Thermal pollution is a temperature change in natural water bodies caused by human influence, like the use of water as coolant in a power plant.

Pollution is an undesirable change in the Physical, Chemical or Biological characteristics of air, water and land that may or will harmfully affect human life. It is a manmade problem mainly of developed countries. Environmental pollution can be classified as under:

1. Air pollution in the form of smoke and gases.
2. Water pollution—Discharge of industrial waste in water bodies.
3. Soil pollution—By dumping of industrial waste and acid rain.
4. Noise Pollution—Sounds of auto-vehicles, aero planes, loudspeakers etc.

All these are related to scientific development. Man, at every step of scientific development, is depriving mankind and other living beings of a part of their life span. Every step of technological development causes irreparable loss to the environment we are living in. Knowingly or out of ignorance man is slow poisoning himself by destroying his physical environment.

Geographers have always projected their studies at the relationship between man and his environment. The environment that we keep in mind includes all those conditions which surround the man, may be physical, cultural, economic, social, and political. In the early stages of civilization these conditions were purely natural, i.e. climate, vegetation, soil, relief etc. With the development or the rise of civilization man created artificial surroundings which became a part of his environment. Today a city dweller thinks about his environment in terms of concrete buildings, jam packed roads, zooming traffic, crowded buses, smoke spitting chimneys, over-crowded hospitals etc. The homo sapiens have certain environmental needs which are highly specific. It becomes difficult to survive in conditions with variations from the normal. We cannot reduce the level of oxygen or increase the level of hydrocarbons. Similarly, man cannot live without water. In other words, man needs a well balanced ecosystem for survival.

Man has been worried about his environment since the dawn of civilization. Hippocrates, presenting a case for relationship of man with airs, waters and places, advocated the need for a cordial relationship between man and nature as early as 5th century B.C. Man has always tried to change or modify the natural environment according to his needs. It seems as if in his challenge to nature man has eroded the natural environment out of innocence and ignorance in the early stages of development but at present he is doing the damage knowingly but helplessly. He knows that the development of high technology increases the proportion of danger to his environment everyday but he cannot afford to stop. He would allow his environment to be injured rather his false ego of being left behind. Girosimov and Pulyarkins observed that "the danger to environment at present is growing through the uncontrolled techno-economic impact on the nature leading to the violation of ecological balance."

With the aggravation by man himself how long will this environment support human life on this planet called earth? Not very long. Never before in the history had the inhabitants of this tiny little habitat had been worried so much about the physical environment as they are today. Continuous growth of population is creating imbalances in the production-consumption patterns on the one hand and increasing environmental pollution on the other. Today environmental pollution and slow growth of pollution absorb ants are causing concern on this globe. Environmental pollution can be of various types, e.g. air pollution, water pollution or noise pollution. Air and water pollution may lead to a number of physical deformities whereas noise pollution may lead to mental infirmities.

The smoke released by the motor vehicles and factory chimneys is the cause of the poisonous gases in the air. Apart from carbon monoxide different sprays such as D.D.T., perfumes, insecticides etc. also pollute the atmosphere.

Pollution of the water being carried on by the man is not less dangerous. Quite a number of streams are being used as repositories for disposal of sewage, industrial garbage and other waste material. The presence of Hydrogen Sulphide gas in the campus of major hydro projects is also dangerous. In the hydro power stations like Bhakra, Sabragiri and Kundhra following effects were noticed immediately after their commissioning: presence of foul smell, corrosion and tarnishing of metallic parts, damage to structures and pipe fittings, overheating of machine parts and diseases to the personnel of the project. It seems as man has forgotten that he is not the only claimant of water, there are other living beings who have better claims on water than man, e.g. fish and other water animals. By polluting water with poisonous elements is interfering in the life of water animals. We must protect water animals for ourselves because 5-6 of our protein requirements come from water animals.

Do we have a solution in hand? There is no direct answer to this big question. If the present state of affairs is continued in the development of high technology the answer may not be very much optimistic, but there is no need to be disheartened. We have to think about the pollution absorbants and noise absorbants. A major absorbant known since time immemorial is the forest cover. The trees absorb most of the injurious gases spread in the air and give us in return in the form of food products, oxygen, blanket for soil erosion, controls of temperature, energy resources and fodder. In other words, our life is by and large dependent on trees. Professor Edward Ayensu, Director of Biological Conservation for the Smithsonian Institute, Washington, opined that "world media gives more importance to political events neglecting the serious issues of environmental concern and remarked 'it is the content of the forests, that are going to keep us alive not Khomeinies'." A study of the forests conducted by the World Health Organization in 1975 showed that during past 150 years the tropical forests of the world had dwindled 50 per cent. Tropical countries of the world have the largest reserves of renewable fuel and the forests in this area have the greatest regeneration capacity. Here again the developing countries have suffered from forest the destruction with growth of the economic development. About 90 per cent of the energy needs of the developing world met by the fuel wood. Unfortunately, the people of the developing world do not realize the ecological damage taking place due to removal of forest cover and the case is applicable to India also.

In India during last thirty years a number of measures have been taken towards aforestation in the form of social forestry, farm forestry, *van mahotsava, chipko* movement etc. indicate that our people have started realizing the importance of this self-regenerative renewable resource. Even then several people who live adjacent to nature need more education about preservation of the forest resources.

Controls

Following measures are to be taken to check environmental pollution:

1. Aforestation.
2. Protect existing forest cover.
3. Protection of wild life.
4. Reduction of smoke exhaust from auto-vehicles.
5. Produce smokeless chimneys for large industrial units.
6. To use better silencers in auto-vehicles and aeroplanes.
7. Impose restrictions on the use of loudspeakers.
8. To educate the people to protect their environment.

REFERENCES

Edelson, (Ed.), *Clean Air,* New York: Chelsea House, 1992.

Gay, Kathlyn, *Air Pollution,* New York: F. Watts, 1991.

Lucas, Eileen, Helen J. Challand, and Harriet Stubbs, *Acid Rain,* Chicago: Children's Press, 1991.

Oliver, Dennis J., "Killer Air Plagues Bay Area", *Oakland Tribune,* May 9, 1996.

Revkin, Andrew, *Global Warming: Understanding the Forecast,* New York: Abbeville Press, 1992.

Rock, Maxine, *The Automobile and the Environment,* New York: Chelsea House, 1992.

Sandak, Cass R., *A Refcrence Guide to Clean Air,* Hillside, NJ: Enslow, 1990.

Tyson, Peter, *Acid Rain,* New York: Chelsea House, 1992.

22

SOIL GEOGRAPHY

Soil geography is that field of earth sciences that deals with soil as very slowly renewable natural resources on the earth's surface. It evolves "the study of soil formation, classification and mapping, the physical, chemical, biological and mineralogical properties of soil from microscopic to macroscopic scales and behaviour of soil systems and their use and management." It is an integrative discipline that interlinks knowledge of atmosphere, biosphere, lithosphere and hydrosphere. Soil geography provides knowledge to understand the causes and consequences of spatial variability of the soil and to take a more holistic approach to the dynamic processes affecting ecosystems.

Soil geography focuses primarily on near-surface processes that govern the quality distribution of soil resources relative to landform evolution (geomorphology), geochemical environment and organism habitat (ecology/biology). These pedogenic processes are interactively conditioned by lithology, climate and relief through geologic time. Soils are welded together into landscapes like chains; processes which perturb and impact higher topographic surfaces directly affect processes on adjacent lower surface.

Soil geomorphology is the scientific study of the origin, distribution and evolution of the soil, landscapes and surficial deposits and the processes that create and alter them. As a science, soil geomorphology is directly linked to Pedology, Geology, Hydrology, Archaeology, Geomorphology, Physical Geography and Geotechnical Engineering.

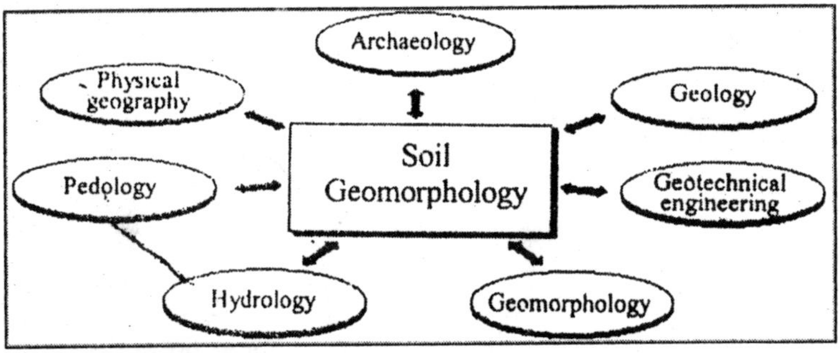

Fig. 22.1: Conceptual diagram showing the relationships of Soil Geomorphology to other sciences.

A major focus of soil geomorphology is the movement of water through soil bodies and the way that a landscape sheds or concentrates water. This focus is the result of the impact of surficial fluvial processes on major agricultural areas and population centres. Consequently, terms have been evolved to describe portions of the Earth's surface that share a common location, form and geomorphic process and practical emphasis on surficial water movement. Soil geomorphology has close relationship with other sciences Fig. 22.1 reveals the relationship of soil geomorphology to other sciences.

Functions of Soil Geomorphology

Soil geomorphology serves two important functions: (1) It provides basic principles for understanding the geomorphic history of landscape, e.g. spatial and time relationships among soils, landscapes and surficial sediments. (2) Geomorphic-based landscape models segregate the soil continuum into meaningful soil bodies with a minimum of effort and can explain soil distribution from local hill slope.

Soil and Geomorphic Description of Landscapes

1. Hill Slopes

Soil patterns or sequences on hill slopes will generally follow a catenaries (a repeating sequence of soils that occurs from the

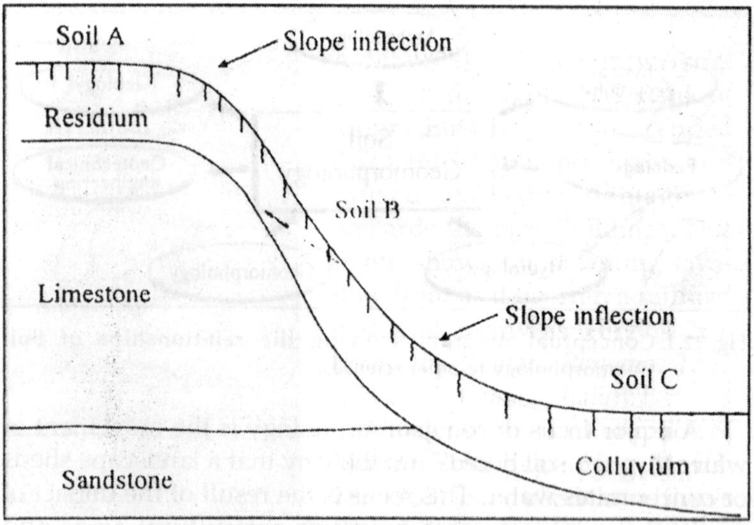

Fig. 22.2: Diagram of natural landscape boundaries.

top of a hill slope to the adjacent valley bottom) relationship inherent in the catena (It is a fundamental concept that explains the pattern of soils on hill slopes) can be used to predict and describe soil occurrence. A simple set of geometric or morphometric descriptors can be used to define hill slopes and therefore, associated soil patterns. These descriptors include slope gradient, aspect, shape, complexity and position and geomorphic position.

(a) ***Slope Gradient*:** Slope gradient, which is measured along the vertical profile of a slope, is the angle of inclination of the ground surface from the horizontal plane. Commonly, slope gradient is expressed in degree from the horizontal plane (e.g. 45 degree) or as a per cent, which for a 45 degree slope is one unit of drop or rise per unit of distance (slope gradient equals 100%). Slope gradient is a proxy expression of potential energy that drives mass movements and the erosive force of surface runoff on a slope. Soil survey map prepared by vanWambeke and Forbes in 1986 used Inflection points or changes in slope gradient that

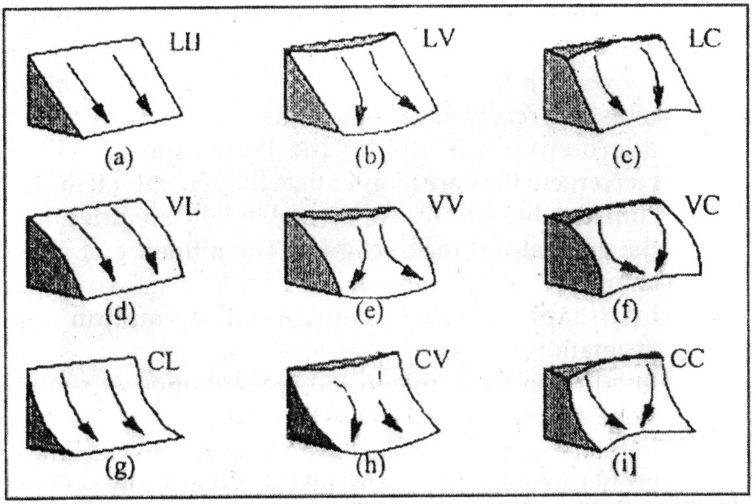

Fig. 22.3: Three-dimensional diagram of the nine hill slope shapes with flowlines shown; L = linear, V = convex, C = concave [Schoeneberger and Wysocki, 1997]

repeat on a landscape are readily discernible, and generally correspond to differences in the internal structure (underlying lithology), past erosion events (stream incision) or contact between landforms or sediment bodies. Inflection points on slope, therefore, often mark natural boundaries between soils. (Fig. 22.2)

(b) ***Slope Shape*:** Slope shape is the three dimensional geometry of a slope. The geometric form is obtained by combining the shapes of both the vertical (up and down slope) and the elevation contours (across slope). A two dimensional shape is either linear or curved; the shape can be convex or concave. This yield nine possible geometric forms to describe all slopes. (Fig. 22.3)

Slope shape is a property of hill slopes that strongly influences the lateral movement of the water across the surface as overland flow and internally as through flow. For example, a slope that is linear in both vertical profile and contour shape creates parallel, linear flow (Fig. 22.3), whereas a slope that is convex in both profile and contour causes divergent flow (Fig. 22.3) and a slope concave in both profile and contour causes convergent flow (Fig. 22.3). Slope shape redistributes moisture received by precipitation, creating distinct micro-environments on the landscape. Areas of convergent flow are moister than the general soil micro-climate whereas areas of divergent flow are drier than the general soil micro-climate. The influence of slope shape on water movement and soil moisture on landscapes, in turn, control soil formation and vegetation.

In addition to its role in the redistribution of runoff, slope shape is a visual clue the internal structure of a hill slope. Hill slopes that have a steeply convex vertical profile are inherently resistant to erosion. Convex slope occur predominately in landscapes where erosion is controlled by resistant bedrocks. Soils on these slopes tend to be shallow and usually display limited horizon development. In contrast linear or concave slopes generally occur on unconsolidated material or weekly resistant rocks. Soils on these slopes to be thicker and display greater horizon development.

Concave slope denote a decrease in slope gradient. Both potential energy and the velocity of the overland flow decrease at the inflection of a concave slope. An accumulation of colluviums or sediment derived from slope wash often occurs on concave slopes. Soils on these slopes often form by cumulic processes and commonly have thicker and less poorly developed horizons compared to soils on adjacent uplands.

(c) ***Slope Complexity*****:** Slope Complexity is a simplistic explanation of the ground surface in respect to down slope path of overflow. Complex slopes contain substantial irregularities in slop conditions. Surface water flow is appropriate to be interrupted and largely non-parallel with considerable changes in flow velocity and subsequent erosion, sediment transport capacity and deposition. (Fig. 22.5a)

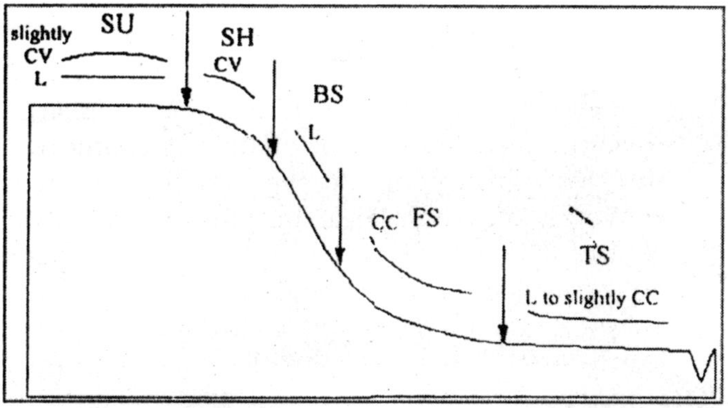

Fig. 22.4: The profile elements and associated slope shapes, SU = summit, BS = backslope, FS = footslope, TS = Toeslope, CV = convex, CC = concave

(d) ***Slope Position:*** Slopes can be divided into segments or elements along a two dimensional cross-sectional profile based on slope shape, the degree of erosion or deposition, the presence or absence of sediment, and the nature of the sediments.

Wood (1942) identified four segments across a fully developed slope for a bedrock controlled landscape. These elements are the waxing slope, free face, debris slope and pediment.

King (1957) elaborated on geomorphic processes active on these four slope elements. The waxing slope is the convex crest of a hill slope dominated by chemical weathering rather than erosional removal. The free face is an outcrop of bare bedrock on the upper reach of the hill. Erosion is most active on this element. The debris slope occurs below the face and is composed of material erode and transported from the free face. Below the debris slope is the pediment, which is an inclined ramp extending from the hill slope base to an alluvial basin. Ruhe (1960) modified Wood's two dimensional hill slope elements and applied them to the study of soil landscapes. His hill slope elements, which includes the summit, shoulder, backslope, footslope and toeslope can be distinguished by inflections in slope gradient and line segment shape. (Fig. 22.4)

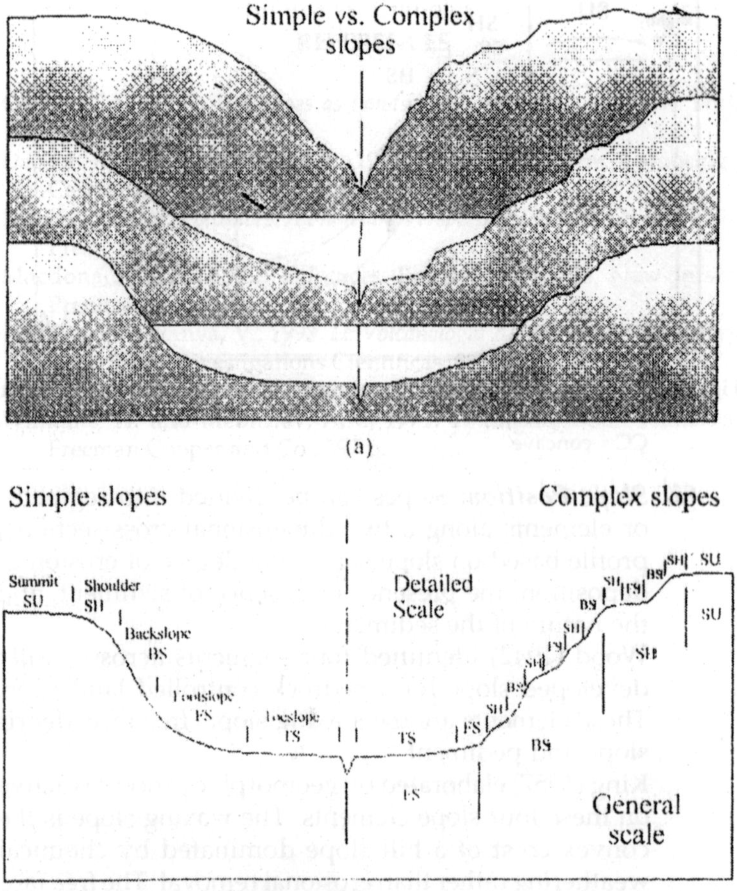

Fig. 22.5a: Two-dimensional diagram of simple *vs.* complex slopes.

The summit is relatively level, uppermost portion of hillslope profile. It is the most geomorphically stable and least erosive part of hillslope. (Fig. 22.5a)

2. *Mountains*

Mountains represent a unique geomorphic setting due to their scale and characterized by comparatively short, simple slopes composed of bare rocks, residuum or short transport (angular)

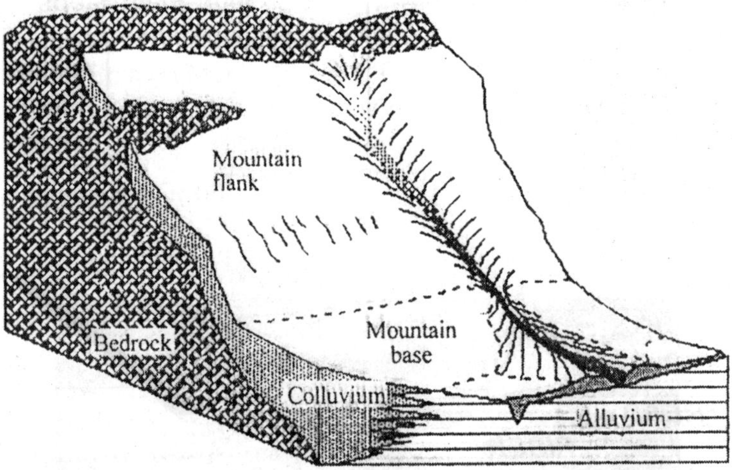

Fig. 22.5b: Geomorphic components for mountain landscapes.

colluvial sediments. In humid climates, soil on mountaintop can be quite thick. The side of mountain, the mountain flank, is characterized by comparatively long, complex slope dominated by layer of long (subangular) colluviums. Residuum, if present is usually buried by 1 to 2 m of colluviums. Incised drainage ways, structural benches and mass movement can be common. The mountain base is an apron of colluviums at the bottom of the mountain slope and analogous to the base slope in hills. The mountain base is characterized by a thick layer of colluvium and commonly, contain a comparatively high percentage of course rock fragments. (Fig. 22.5b)

3. *Terraces*

Stream terraces and flood-plain steps originally developed by alluvial processes and sediments. The tread is comparatively broad, generally level part of a terrace or flood-plain step. Treads can lateral for many kilometres. Treads are level or gently inclined (low slope gradient) and underlain by alluvial, lacustrine or sediments. In India Zanskar terraces are the best example of such types of terraces.

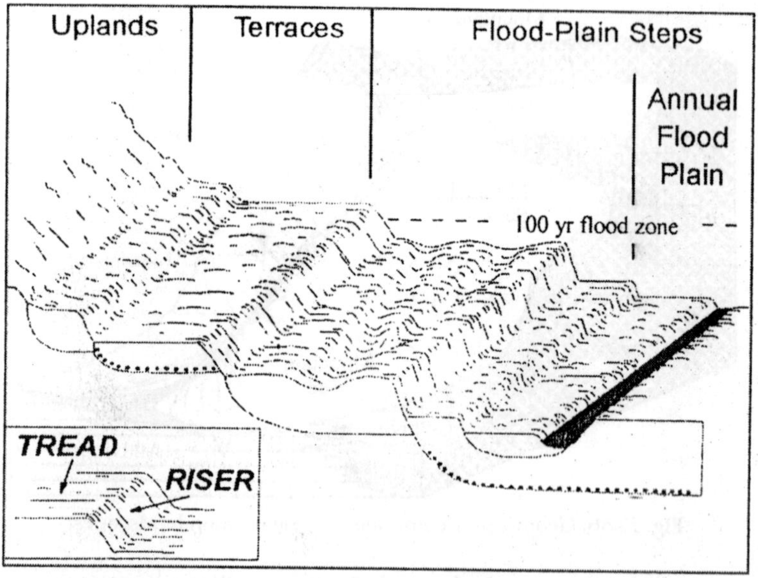

Fig. 22.6: Geomorphic components for terraces and flood plains.

There is a distinction between stream terraces and flood plain steps are an arbitrary break in a natural band. Terraces are defined as a part of the fluvial system that no longer actively experiences fluvial modification, e.g. no longer aggrandize, no longer experiencing significant flooding. Flood plain steps are morphometrically similar to terraces except that they occur within the fluvial active portion of a flood plain and are subject to relatively frequent modification. Terraces commonly have more extensively develop soils. Flood-steps receive fresh sediment. (Fig. 22.6)

4. *Flat Plains*

Flat plains are subdued landscapes dominated by broad areas with low slope gradient (e.g. 0—1%) called Talf and closed depressions, with only nominal changes in local relief (i.e. micro high and micro lows). (Fig. 22.7) A rise is a slightly elevated area (i.e micro high) which tend to be broad with low slope gradient (e.g., 1—3%). The low slope gradient of flat

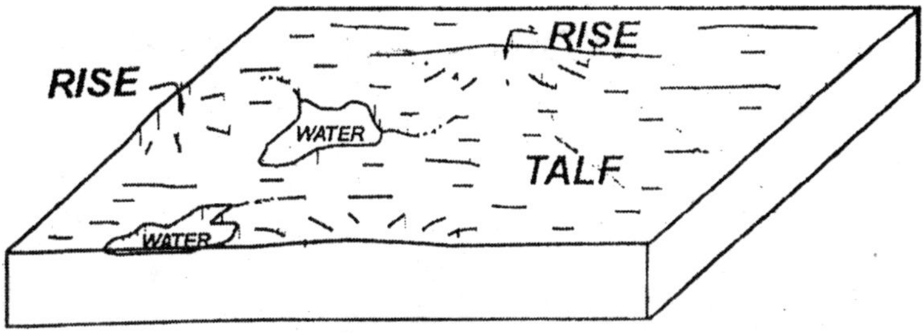

- very low gradients (e.g., slope 0-1%)
- deranged, non-integrated, or incipient drainage network
- "high areas" are broad and low (e.g., slope 1-3%)
- sediments commonly lacustrine, alluvial, or till

Fig. 22.7: Geomorphic components for flat low relief plains.

plains result in minimal erosion by running water, especially if vegetated. Consequently, fluvial drainage network tend to be poorly developed, e.g. non-integrated, incipient or deranged. Precipitation tends to pond locally and lateral transport is slow both above and below ground. These conditions favour accumulation of organic matter and retention of water and fine earth sediments.

Soil Hydology

The basic purpose of soil hydrology is to understand how water moves through landscape; where the water goes, so goes soil development. A variety of terms has been developed or adopted which partition water flow in soils systems. These terms are placed in idealized schematic to demonstrate relationships. (Fig. 22.8)

These terms depict the potential fate of precipitation into, and through soil. The emphasis is on water at or above the permanent groundwater table. Water dynamics and aquifer conditions below the water table generally are the purview of groundwater hydrology. Water flow can be predominantly downward, lateral, or upward towards the ground surface, depending upon prevailing energy dynamics. This can be

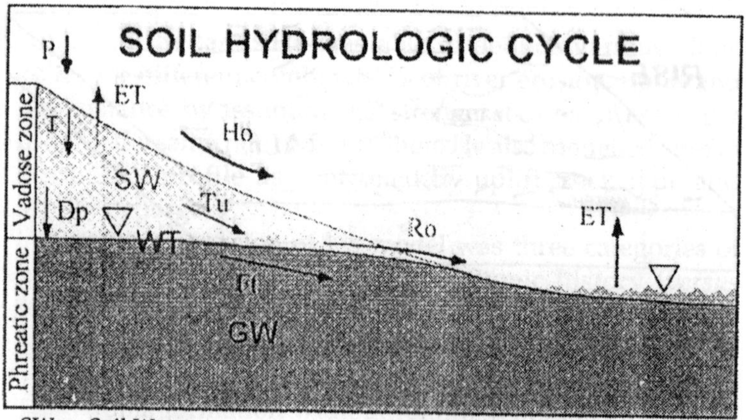

SW =	Soil Water	I =	Infiltration
WT =	Water Table	Dp =	Deep percolation
GW =	Ground Water	Tu =	Throughflow
P =	Precipitation	Ft =	Flowthrough
ET =	Evapotranspiration	Ro =	Reflow
Ho =	Overland flow (or runoff)	∇ =	Zero tension water surface

Fig. 22.8: Soil hydrologic cycle with terminology.

demonstrated by looking at near-surface water movement patterns in different climatic setting.

In humid environments, water tables and flow paths tend to loosely mimic the local topography with water generally infiltrating (recharging) in topographically high areas, flowing to and discharging from lower lying areas. (Fig. 22.9a)

In sub-humid environments behave as tension or transition zones between these two extremes. These transition zones temporarily follow either pattern depending upon short term climatic conditions (e.g. seasonal climatic variations or annual cycles). (Fig. 22.9b)

In arid to semi-arid environments, conditions are reversed, with recharge site dominant in low areas and sub-surface water moving to and discharging from higher elevation area. (Fig. 22.9c)

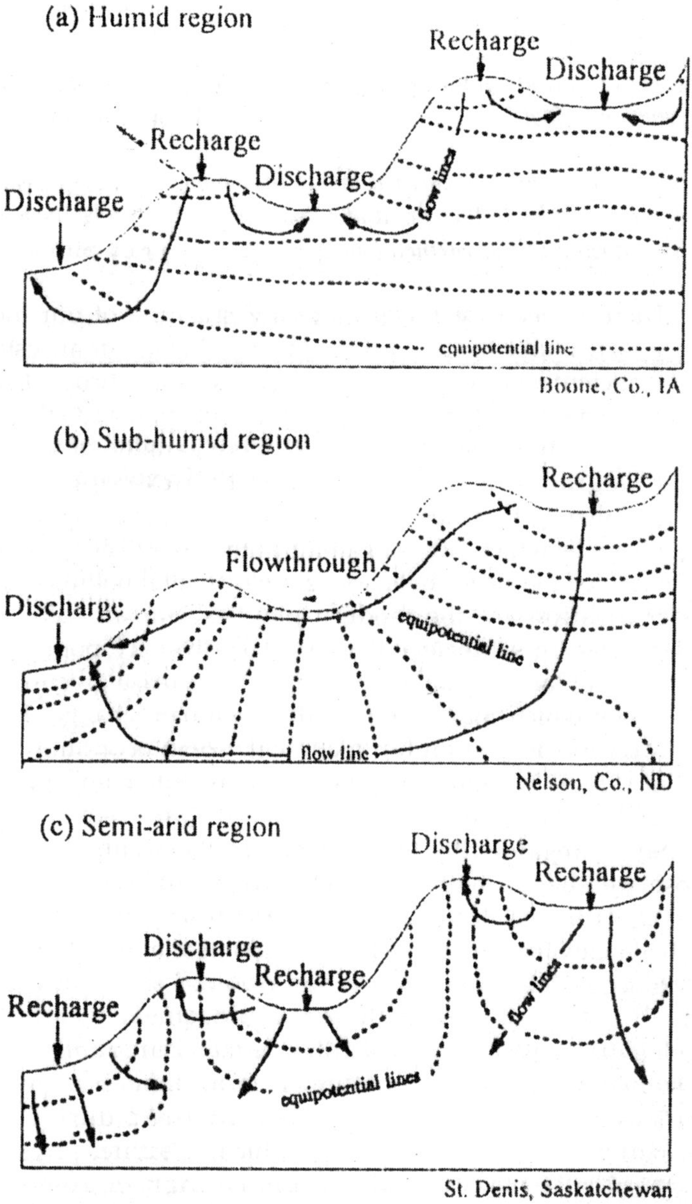

Fig. 22.9a—22.9c: Two-dimensional landscape diagrams with recharge-discharge, flowlines, and equipotential lines in different climatic settings [Modified from Richardson et al., 1991).

Soil

"We know more about the movement of the celestial bodies than about the soil underfoot." —Leonardo da Vinci.

"Soil resources are the earthen looms that shape the living people. The more completely they are understood; the better can be the fabric of life woven on these earthen looms." —Roy Simonson.

There is need for much more investigation of our most precious resource to stave off the fate that befall our ancestors in the ancient empires of Babylon, Egypt, China, Europe, India and the America where soil erosion, salinization and siltation laid waste to their land resources. Modern parallels of soil and land degradation continue in both developing and industrialized countries. In addition to the problems facing the ancients which are still ubiquitous, chemical toxicities irreparable land use conversion, environmental pollution and desertification are the modern plagues. Sustainability of today's culture is threatened by loss of biodiversity, population growth and land degradation leading to increase greenhouse emissions which may result in global climate change. Land degradation is a complex technical, socio-economic and political issue without implementation as public policy, they are likely to be futile.

Soil is regarded by the geologist as the accumulation of loose weathered material which covers much of the land-surface of the earth to a depth ranging from a fraction of an inch to many feet. But geographers and soil scientists owe their existence to the extremely but thin precious skin, called soil, which covers the unweathered and partially weathered geological formations at the Earth's surface with a unique and extremely thin, fragile coating. It is no longer a rock or geological sediments but has been changed during soil formation by geological, topographical, climate, physical, chemical and biological factors to form a living being composed of an association of inorganic or minerals particles inextricably linked with organic matter perfused by the gases. When wetted by fundamental water, the solvent and conveyor of nutrients and major constituents of all living beings this complex becomes the fertile substrate from which all terrestrial life on this planet springs. This is life-sustaining pedosphere zone which is biologically active, porous and structured medium that effectively integrates and dissipates mass fluxes and energy.

Importance and Role of the Soil

Soil is identical to a sponge, regulating and buffering the supply of nutrients and water for the growth of macro- and micro-flora and flora and fauna and determining the partitioning of water into that which flows by surface paths of rivers and lakes and that which percolates into replenish subterranean ground-water reservoirs or aquifers. Not only does it serve to promote and sustain life in its many forms, but it also acts as a living filter for the waste generated by the biota. In this role, it cleanses, purifies and recycles water and detoxifies and renders harmless most toxins and pathogens which otherwise would irreparably pollute and degrade the environment.

Soil is an evolving, living organic/inorganic layer at the Earth's surface in dynamic equilibrium with the atmosphere and biosphere above, and geology below. It acts as an anchor and transmitter of water and nutrients for roots, as a home for vast microorganisms and animals, as a sanitizer of the environment and as a source of raw materials for construction and manufacturing. Soil is the long term capital on which nation builds and grows. It is the basic component of the ecosystem and ecosystem management.

Every soil consists of one to several layers called horizons, a few hundreds of centimetres thick that reflect the physical, chemical and biological processes which have taken place. Horizons are composed of aggregates called peds which are made up of association of minerals and organic particles. Peds and particles are often separated from each other by pores that vary widely in size and shape. In addition, individual peds and particles may be coated by material such as clay, organic matter and precipitated salts.

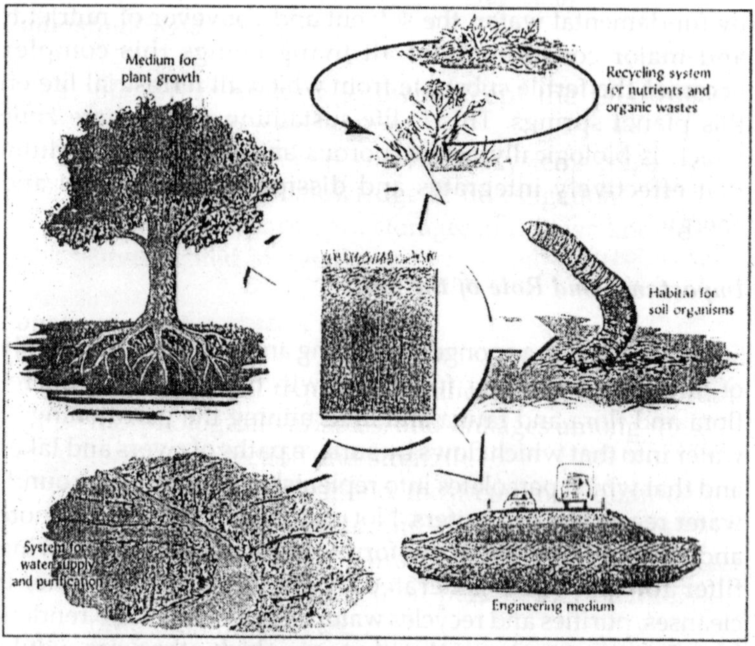

Fig. 22.10: The many functions of soil can be grouped into five crucial ecological roles.

Fig. 22.10 reveals the five key roles played by the soils. First, soil supports the growth of high plants mainly by providing a medium for plants roots and supplying nutrient that are essential to the entire plants. Properties of soil often determine the nature of vegetation present and indirectly, the number and types of animals (including people) that the vegetation can support.

Second, soil properties are principal factor controlling the fate of water in hydraulic cycle. Water loss, utilization, contamination, and purification are all affected by the soil.

Third, the soil functions as nature's recycling system. Within the soil, waste products and dead bodies of plants, animals, and people are assimilated, and their basic elements are made available for reuse by next generation of life.

Fourth, soil provides habitats for a myriad of living organisms, from small mammals and reptiles to tiny insects to microscopic cells of unimaginable numbers and diversity.

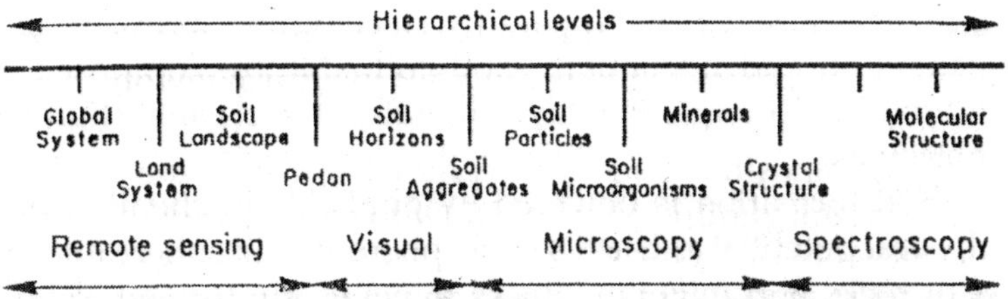

Fig. 22.11: Relationship between hierarchical levels of the soil continuum and the tools used for their study.

Finally, inhuman-built ecosystems, soil plays an important role as an engineering medium. Soil is not only an important building material in the form of earth fill and bricks but provide the foundation for virtually every road, airport, and house we build (Fig. 22.10). Many functions of soil can be grouped into five essential roles.

Pedology

It is the earth science that quantifies the factors and processes of soil formation including the quality, extent, distribution spatial variability of soils from microscopic to megascopic scales. Pedology is an integrative extrapolative science. It provides a vehicle for extrapolation and scaling of spatial variability from components of soils (pedons, toposequence, physiographic entities and pedosphere) to the population of soils within the continuum as a whole illustrated in Fig 22.11 with hierarchical levels in this continuum of soil organization relative to tools used to generalize information content at multiple scales of resolution.

Pedogenic Processes/Soil Formation Processes

Soil is a mixture of geological parent material, living organisms and the residue of their interaction. The residue (deposits) imparts soil with its unique characteristic. Soil is a heterogeneous polyphasic, particulate, disperse and porous system. The colloidal nature of soil exposes a large surface area which gives rise to absorption of water and chemicals, ion exchange, adhesion and capillarity, swelling and shrinking, and dispersion and floccution. Soil colloidal are composed of interbonded inorganic (clay minerals) and organic (humus) molecules.

Soil formation is powered by gradients in chemical and physical potential as Earth's atmosphere and biosphere interact with rocks and minerals. Rocks minerals are formed at high temperature and pressure which makes them susceptible to attack by water and biospherically derived acids existing in soil.

***Four major processes*:** Soil genesis brought about by a series of processes, the most important of which are :

1. Transformations such as weathering and organic matter breakdown by which some soil constituent are modified or destroyed, and others are synthesized.
2. Translocation or movement of inorganic and organic material from one horizon up or down to another, the materials being moved mostly by water but also by soil organisms.
3. Addition of material to soil profile from outside sources such as organic matter from leaves, dust from the atmosphere or soluble salts from the groundwater.
4. Losses of material from the soil profile by leaching to groundwater, erosion of surface material or other forms of removal.

Soil Forming Factors

The concept of soil forming factors is one of the earliest and most important of soil science. It defines soil as a component of ecosystems that must be characterized in terms of both geological substrate and biological inputs (which are called environmental factors). Environmental variables are broadly grouped as geological, climatological, biological and topographical. The geological characteristic constitute a site factor that sets the initial condition for soil formation where the climatologically biological factors represent energy input that drives soil development.

(a) Geologic Factors: Lithology, Mineralogy and Topography

Geological processes ranging from plate tectonic and volcanism to deep ocean carbonate sedimentation to glaciations create conditions that eventually provide deposits for soil formation. Plate tectonics drive global processes that differentiate lighter, silica-rich rocks from mantle into the crust and create relief that produces erosion, accumulation of alluvium and formation of sedimentary rocks. Continental area that are actively uplifted, such as Tibetan plateau and the cordillera of South and North America support thin young soil. In contract those areas on stable rocks such as Central West Africa, Brazil and Australia have deep soils that are superimposed on thick highly weathered saprolite zones.

Volcanism brings locally melted silica-rich rocks in subduction zones to the surface in areas such as Japan, Indonesia and Washington states. Hot spots such as Hawaii, the Canary Islands or Yellowstone provide a direct conduit (medium) from mantle to surface for Iron and Magnesium rich rock.

Globally, much of the variation in soil properties can be attributed directly to differences in parent material mineralogy and dynamic of Earth history. Numerous Pleistocene glaciations have left distinctly younger soils than the deeply weathered soils of stable rock areas in the tropic.

Parent material lithology determines the physical and mineralogical nature of soil. For example, soil formed in marine clay or shale inherits a large quantity of colloids that leads to poor water infiltration, relatively low plant available water and substantial shrink/swell behaviour. In contrast soils formed on quartz sandstone or weathered granite are likely to be sandy with little inherited clay. Commonly, they are excessively drained and often have low fertility and water holding capacity. Parent material mineralogy determines ecological soil properties, such as nutrients supply and retention and water movement. Inorganic parents materials can either be formed in place (sedentary) as residual weathering material or they

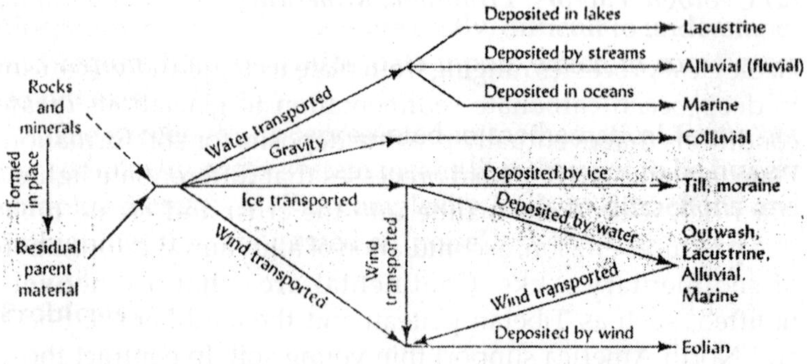

Fig. 22.12: How various kinds of parent material are formed, transported, and deposited,

can be transported from one location and deposited at another as shown in Figure 22.12.

Topography relates to the configuration of the land surface and is described in terms of differences in elevation and slope in other words, the lay of the land. Topography can delay the work of climate forces. In smooth, flat country, excess water is removed less rapidly than in rolling areas. Rolling to hilly topography encourages natural erosion of the surface layers, which reduces the possibility of a deep soil. On the other hand, if water stands for part or all of the year on a given area, the climate is less effective in regulating soil development.

(b) Energy Factors: Water, Temperature, Biology and Time

The developing soil profile is strongly dependent on energy that is ultimately provided by the Sun and Earth's gravitational field. Solar heating is by direct energy absorption at the surface and indirect transfer from the atmosphere. These processes set-up superimposed diurnal and annual waves of energy moving through soil profiles. Soil provides a way station in the hydrological cycle which moves water from ocean to atmosphere.

Temperature and precipitation affect the rate of chemical, physical and biological processes responsible for soil profile development. For every 10°C rise in temperature, the rate of biochemical reaction doubles. Furthermore, biochemical changes by soil organisms are sensitive to temperature as well as moisture. Water from rain and snowfall is primary requisite for weathering of tropical soils. Not only the plant nutrient leached from these soils, but the basic framework of the original soil minerals is destroyed and new minerals are created. Likewise, a deficiency of water is a major factor in determining the characteristics of soil of arid regions. Soluble salts are not leached from these soils, and in some cases they build up to levels that curtail plant growth.

Soil organisms play a major role in profile differentiation, organic matter accumulation, profile mixing, nutrient cycling, and structural stability are all enhanced by the activities of organisms in the soil. Vegetation cover reduces natural soil erosion rates thereby slowing down the rate of mineral surface removal. The mineral element content in the leaves, limbs, and stems of the natural vegetation strongly influences the characteristic of the soils that develop, especially their acidity. Soil acidity is more likely to develop under pine vegetation with its low content of base-forming cations than under oak or maple tree vegetation. Also, the removal of base-forming cations by leaching is more rapid under coniferous vegetation.

The length of time that materials have been subjected to weathering influences soil formation. Time effects can be seen by comparing soils of a glaciated region with those in a comparable unglaciated area nearby. Soil located on alluvial or lacustrine material generally has not had as much time to develop as have the surrounding upland soils.

The Soil Profile

In most places, the rock exposed at the earth's surface has crumbled and decayed to produce a layer of unconsolidated debris overlying the hard, unweathered rock. This unconsolidated layer is called *regolith,* and varies in thickness from non-existent in some places (i.e. exposed bare rock) to ten metres in other places.

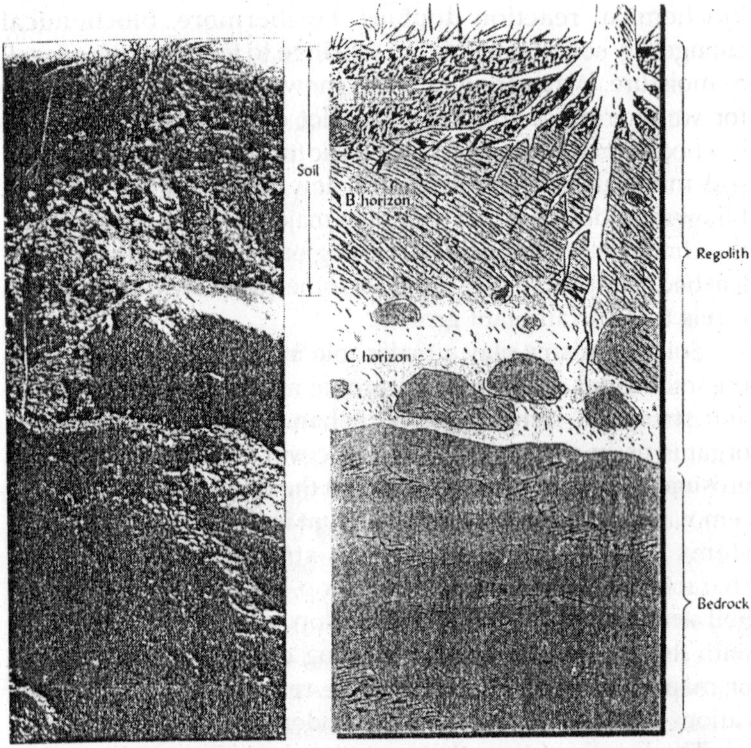

Fig. 22.13: Relative positions of the regolith, its soil, and the underlying bedrock. Note that the soil is a part of the regolith, and that the A and B horizons are part of the **solum** (from the Latin word *solum*, which means soil or land). The C horizon is the part of the regolith that underlies the solum, but may be slowly changing into soil in its upper parts. Sometimes, the regolith is so thin that it has been changed entirely to soil; in such a case, soil rests direcdy on the bedrock. (Photo courtesy of R. Weil)

The upper part of this regolith has been affected by the activities of living organisms, and is quite different from the deeper layers. Here, at the interface between the worlds of rock (the lithosphere), air (atmcsphere) water (the hydrosphere) and living things (the biosphere) soil is born. (Fig. 22.13) The transformation of inorganic rock and debris into a living soil is one of nature's most fascinating display.

Soil Horizons

When you dig a large hole, called a soil pit, usually several metres deep and about a metre wide, to expose soil horizons for study, the vertical section exposing setoff horizons in the wall of such a pit is termed soil profile. Horizons within soil may vary in thickness and have somewhat irregular boundaries but generally parallel to the land surface.

Horizons of the soil: Organic matter from decomposed plant leaves and roots tends to accumulate in the uppermost horizons of a soil profile, giving these horizons darker colours than the lower horizons. The organic matter enriched horizons nearest the soil surface are called the A horizons. In some soils intensely weathered and leached horizons that have not accumulated organic matter occur in the upper part of the profile, usually just below the A horizon. These horizons may be designated E horizon. The layers lying the A horizon contain comparatively less organic matter than the horizons nearer the surface. Varying amounts of silicate clay, iron and aluminium oxide, gypsum, or calcium carbonate may accumulate in the underlying horizon. The accumulated material may have been washed down from the horizons above, or they may have been formed in place through the weathering process. These underlying layers are referred to as B horizons. Below the B horizon only salts accumulated, this layer may be termed as C horizon. (Fig. 22.14)

Soil Texture

Soil texture means different size and shape of the particles in the soil. Soil textural class names are used to convey an idea of the textural make-up of soils and to give an indication of their physical properties. They are classified on the basis of size into gravel, sand, silt and clay. Gravel may range from pebbles down to particles about 1/12" (200 μ micron) in diameter. Sand grains feel gritty and we can see them easily (50 μ micron). Silt is made up of particles which can barely be seen with the naked eye, it looks and feels like flour (5 μ micron).

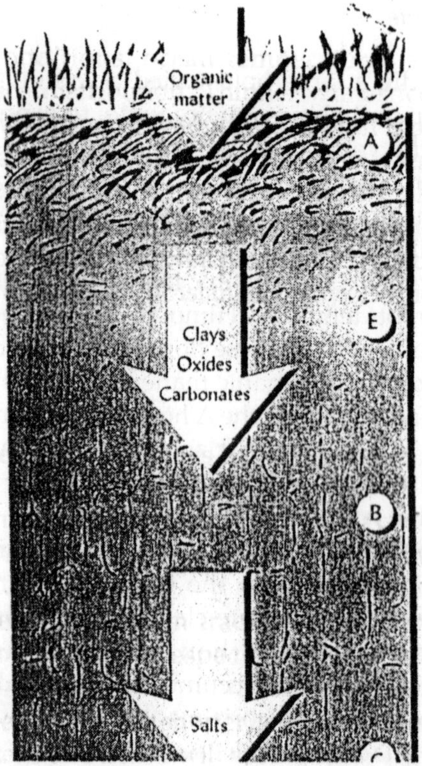

Fig. 22.14: Horizons begin to differentiate as materials are added to the upper part of the profile and other materials are translocated to deeper zones. Under certain conditions, usually associated with forest vegetation and high rainfall, a leached E horizon forms between organic-matter-rich A and the B horizons.

Clay particles are so fine (less than 2μm micron) that one cannot see them with unaided eye and many of them are too small to be seen even under a microscope.

(Size in mm)

	Clay	*Silt*	*Fine Sand*	*Coarse Sand*
USDA	Less than 0.002	0.002-0.05	0.05-0.2	0.2-2
International	0.002	0.002-0.02	0.02-0.2	0.2-2

Most soils are mixtures of the various sized particles except that many contain no gravel and some very little sand. If a soil contains considerable qualities of at least two sizes of particles, it is known as a loam. Texture depends on the size of particle most common in soil. Sand, loamy sand, sandy loam, fine sandy loam, loam, clay loam, sandy clay loam, silt, silt loam, silty clay, silty clay loam, sandy clay and clay textural classes are normally recognized.

Three broad groups of these classes are recognized sandy soil, loamy soils, and clayey soils. Within each group specific textual class names have been devised:

(a) **Sandy Soils:** These soils are dominated by the properties of sand, for the sand separate comprises at least 70 per cent of the material by weight (less than 15 per cent of the material is clay). Two specific textural classes are recognized—sand and loamy sand. Although in practice two subclasses are also used: loamy fine sand and loamy very fine sand.

(b) **Clayey Soil:** Characteristics of the clay separate are distinctly dominant in these soils. Class name in this group and the amount of clay separate required to be classified in each of the classes are clay (at least 40 per cent clay) sandy clay (at least 30% clay and sandy clay (at least 40 per cent clay).

(c) **Loamy Soil:** The loam group contains many subdivisions. An ideal loam may be defined as a mixture of sand, silt and clay particles that exhibits the properties of those separates in about equal proportions. This definition does not mean that in a loam the three separates are present in equal amounts. This anomaly exists because only relatively small amounts of sand and silt have a lesser influence on how a soil behaves. Thus the clay modifier is used in the class name of soil with as little as 20 per cent clay, but to quality for the modifiers sandy or silt, a soil must have at least 40 or 45 per cent of these separates respectively. Most soils are some type of loam. They

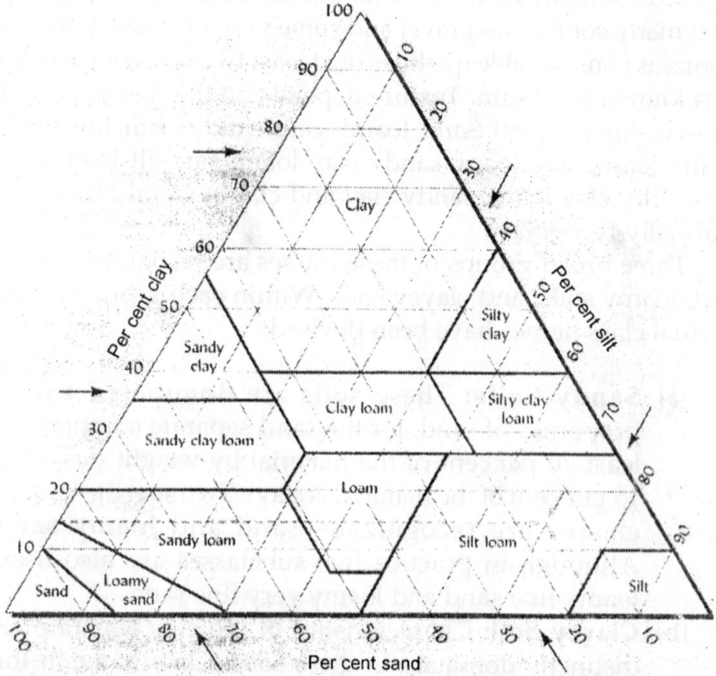

Fig. 22.15: The major soil textural classes are defined by the percentages of sand, silt, and clay according to the heavy boundary lines shown on the textural triangle. If these percentages have been determined for a soil sample by particle size analysis, then the triangle can be used to determine the soil textural class name that applies to that soil sample. To use the graph, first find the appropriate clay percentage along the left side of the triangle, then draw a line from that location across the graph going parallel to the base of the triangle. Next find the sand percentage along the base of the triangle, then draw a line inward going parallel to the triangle side labelled "silt". The small arrows indicate the proper direction in which to draw the lines. The name of the compartment in which these two lines intersect indicates the textural class of the soil sample. Percentages for any two of the three soil separates is all that is required. Because the percentages for sand, silt, and clay add up to 100 per cent, the third percentage can easily be calculated if the other two are known. If all three percentages are used, the three lines will all intersect at the same point. Consider, as an example, a soil which has been determined to contain 15 per cent sand, 15 per cent clay, and 70 per cent silt. This example is indicated by the light dashed lines that intersect in the compartment labelled "silt loam". What is the textural class of another soil sample which has 33 per cent clay, 33 per cent silt, and 33 per cent clay? The lines (not shown) for this second example would intersect in the centre of the "clay loam" compartment.

possess the ideal make-up of equal proportions described above and be classed simply as loam. Usually, however, the varying quantity of sand is dominant and is classified as a sandy loam, in the same way, there may occur silt loams, silty clay loams and clay loams. (Fig. 22.15)

Soil Structure

The term structure relates to the grouping or arrangement of soil particles. It describes the gross, overall combination or arrangement of the primary soil separates into secondary groupings called aggregate or peds.

Types of Soil Structure

The dominant shape of peds or aggregates in a horizon determines its structure type. There are four main types of soil structure:

1. Spheroidal
2. Platy
3. Prismlike
4. Blocklike

1. Spheroidal: Granular and crumb subtype. Rounded peds or aggregates are placed in this category. They usually lie loosely and are separated from each other. Relatively non-porous, aggregate are called granular and their shape granular. However, when the granules are especially porous, the term crumb is applied.

Granular and crumb structures are characteristic of many surface soils, particularly those high in organic matter and are especially prominent in grassland soils and soils that have been worked by earthworms.

2. Platy: In this structural type the aggregates are arranged in relatively thin horizontal plates, leaflets or lenses. The plate structure is found in the surface layer of some virgin soils, but may characterize the lower horizons as well. Although most

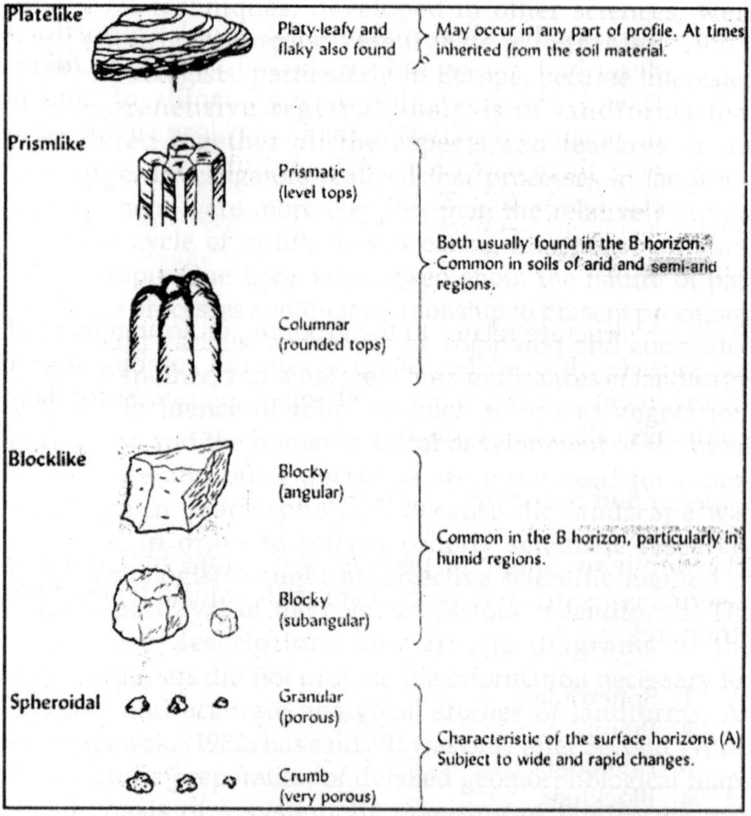

Fig. 22.16: Various structural types found in mineral soils. Their typical location in the profile is suggested. In arable topsoils, a stable granular structure is prized.

structural features are usually a product of soil-forming forces, the platy type is often inherited from soil parent materials, especially those laid down by water.

3. *Prismlike*: Columnar or prismatic subtypes are characterized oriented aggregates or pillars that vary height among different soils and may reach a diameter of 15 cm or more. They commonly occur in subsurface horizons in arid and semi-arid region and, when well developed or a very striking feature of profile. They also occur in some poorly drained soils of humid regions, and they often are associated

with presence of swelling types of clay. When the tops of the prisms are rounded, the term columnar is used. Columnar structure is especially common in subsoils high in sodium. When the top of the prisms are angular and relatively flat horizontally, the structure pattern is called Prismatic.

4.Blocklike: Angular blocky and subangular blocky subtypes. In this case the aggregates have been reduced blocks irregularities six faced with their three dimensions more or less equal. In size these peds range from about 1—10 cm in thickness. In general the shape is so unique that identification is so easy. When edges of the cubes are sharp and rectangular faces distinct, the subtype is designated Anguiar Blocky. When some rounding has occurred the aggregates are referred to as Subangular Blocky. These types are usually confined to subsoil and their stages of development and other characteristics have much to do with soil drainage aeration and root penetration. (Fig. 22.16)

Soil Classification

The sustainable use of land anywhere in the world is largely dependent on soil properties. To make proper land use decision we must know the differences in soil properties from one site to an other. Our ancient civilizations have noticed differences in soils and classified them, if only in term of good or bad.

First scientific study of the soil was carried out by the Russian soil scientist V.V. Dukochaev and his associates in the 1880s, they considered soils as a natural bodies. Russian studies of soils were seriously considered in late 1920s. C.F. Marbut of the US Department of Agriculture was one who grasped the concept of soils as natural bodies and in 1927 he developed a soil classification scheme based on these principles.

The natural body concept of soils assumes the existence of individual entities, each of which we call a **Soil.** A set of groups each having some characteristic in common can occur as one moves up the classification ladder to the **Soil**.

There are no sharp demarcation between the properties of one soil individual and those of another. Rather there is a

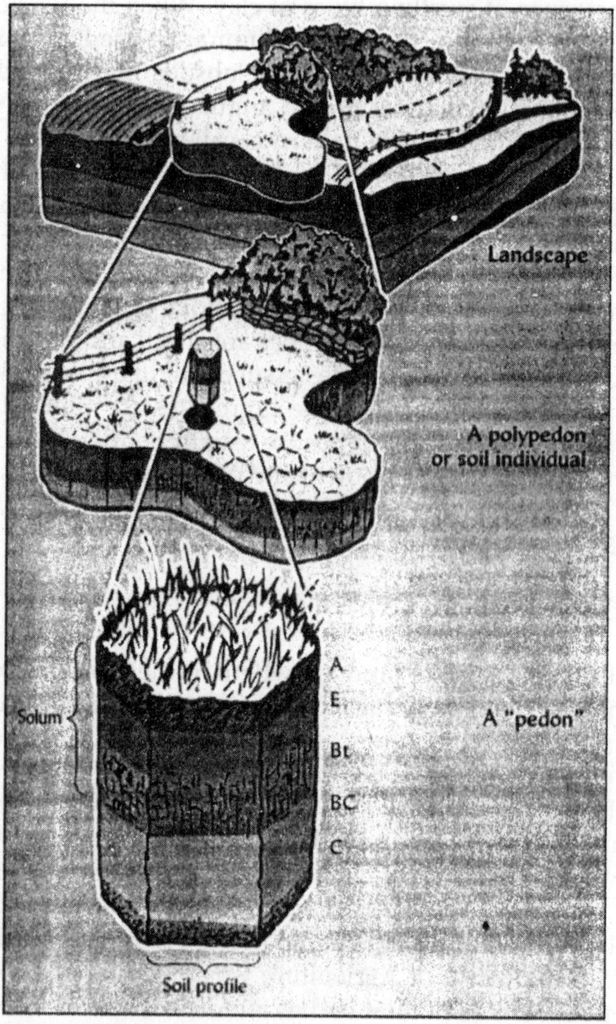

Fig. 22.17: A schematic diagram to illustrate the concept of pedon and of the soil profile that characterizes it. Note that several contiguous pedons with similar characteristics are grouped together in a larger area (outlined by broken lines) called a *polypedon* or *individual soil.* Several individual soils are present in this landscape.

gradation in such properties as one moves from one soil individual to a second. It is necessary to characterize a soil individual in terms of smallest three-dimensional unit that embodies the primary characteristic of a soil. Such three-dimensional unit is termed as *Pedon* (Derived rhymes with "head on". From the Greek pedon, ground). It is the smallest sampling unit that can characterize a soil individual. However, a group of pedons is termed as a *Polypedon*. (Fig. 22.17)

Comprehensive Classification System: Soil Taxonomy

There are a number of soil classification systems in different parts of the world. US soil survey staff of Agriculture in cooperation with soil scientists in United States and other countries developed a new comprehensive system of soil classification based on soil properties. This system has been in use in the US since 1965 and used by more than 45 countries of the world.

The comprehensive soil classification system called SOIL TAXONOMY maintains the natural body with two other major features. First the system is based on soil properties (Physical, Chemical and Biological) that are easily verified. Second important feature of Soil Taxonomy is the unique nomenclature employed, which gives a definite connotation of the major characteristics of the soils in question.

Diagnostic Surface Horizons: The investigative horizon that occur at the soil surface are called **Epipedons** (from Greek *epi*, over and *pedon*, soil). The upper horizon of the soil profiles roughly corresponding to the A horizon. It may include part of B horizon. Seven epipedons are recognized but only five occur naturally over wide area. The other two called anthropic and plaggen are the result of intensive human use of soils. (Fig. 22.18) The following are the major five surface horizons = Epipedons:

1. Mollic Epipedon: (Word mollic derived from latin *mollis* means soft). This mineral surface horizon is noted for its dark colour associated with organic matter content (> 0.6 % organic carbon throughout the thickness) for its thickness (generally >25cm) and for its softness even when dry. It has a high base saturation (greater than 50% base-forming cations). Mollic

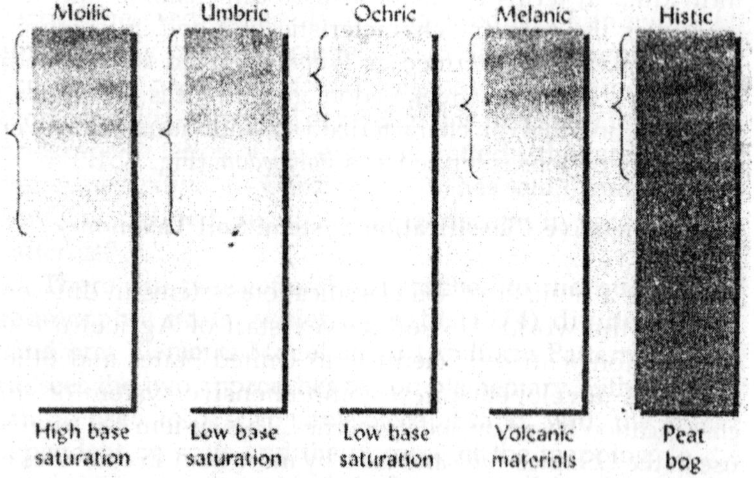

Fig. 22.18: Representative profile characteristics of five surface diagnostic horizons (epipedons). The comparative organic matter levels and distribution are indicated by the darkness of colours. The Mollic and Umbric epipedons have similar organic matter distribution but the percentage base saturation is higher (greater than 50%) in the Umbric epipedon. The Ochric epipedon is much lower in organic matter content; consequently, it is light in colour and sometimes hard when dry. Two other epipedons have very high organic matter contents and are very dark in colour. The Melanic epipedon is formed on recently deposited volcanic materials, usually in cool wet areas. The Histic epipedon is formed from organic deposits laid down in peat bogs. The relative depths of each epipedon is shown by the brackets.

epipedons are moist at least three month a year when the soil temperature is usually 5°C or higher to a depth of 50 cm.

2. Umbric Epipedon: (word umbric derived from latin *umbra* means shade; hence dark) has the same general characteristic as the mollic epipedons expect that the percentage base saturation is less than 50 per cent. This mineral horizon commonly develops in areas with somewhat higher rainfall and where the parent material has lower content of base forming cations compared to the mollic epipedon.

3. Ochric Epipedon: (word ochric derived from Greek *ochros* means pale) A mineral horizon that is either too thin, too light in colour, or too low in organic matter to be either a mollic or umbric epipedon. As a consequence of its low organic matter content it may be hard and massive when dry.

4. Melanic Epipedon: (word melanic derived from Greek *melas, melan,* means black). This is very black in colour due to its high organic matter content (organic carbon >6%). It is characteristic of soils high in such minerals as allopane developed from volcanic ash. It is more than 30 cm thick and is light in weight and fluffy.

5. Histic Epipedon: (word histic derived from Greek *histos* means tissue) consist mainly of organic materials (peat and muck). Formed in wet areas, it too is black in colour. It is very low in density.

Diagnostic Subsurface Horizons

There are eighteen subsurface horizons used to characterize different soils in soil Taxonomy. Figure 22.19 explains that the focus is primarily on soil properties. Among the characteristics emphasized is accumulation of silicate clays, organic matter, iron and aluminium oxides, calcium compounds and soluble salt as well as materials that become cemented or high acidified, thereby containing root growth. The presence or absence of these horizons play a major role in determining in which class a soil falls in Soil Taxonomy. (Fig. 22.19)

Categories of Soil Taxonomy

There are six categories of classification in Soil Taxonomy:

1. Order
2. Suborder
3. Great group
4. Subgroup
5. Family
6. Series

1. *Order*: There are 11 soil orders in Soil Taxonomy. These order category are based on soil forming processes. (Fig. 22.20)

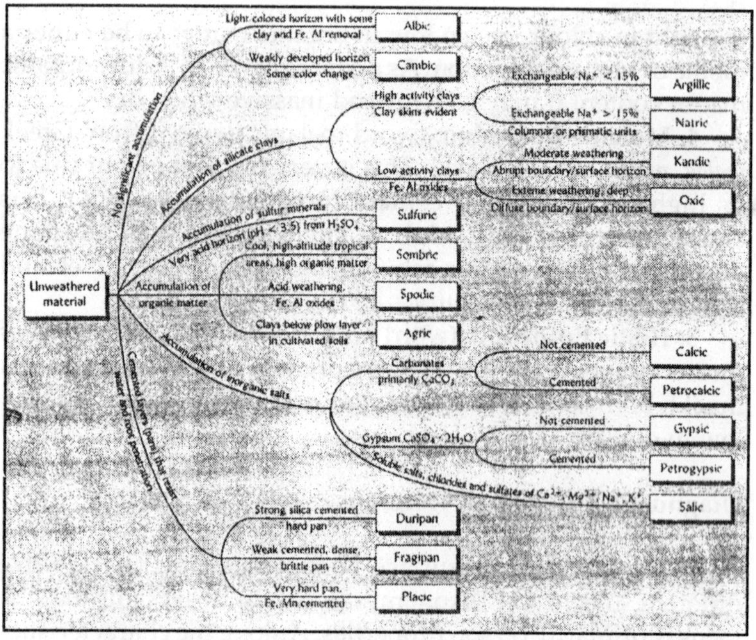

Fig. 22.19: Names and major distinguishing characteristics of subsurface diagnostic horizons. Note that the focus is primarily on soil properties, not on how the properties have presumably evolved. Among the characteristics emphasized is the accumulation of silicate clays, organic matter, Fe and Al oxides, calcium compounds, and soluble salts, as well as materials that become cemented or highly acidified, thereby constraining root growth. The presence or absence of these horizons play a major role in determining in which class a soil falls in *Soil Taxonomy*.

A given order includes soils whose properties suggest that they are not too dissimilar in their gneiss. The names and major characteristics of each soil order are shown in Table 22.1.

2. *Suborder*: The suborders are subdivision of orders that emphasize properties that some common features of soil genesis. Thus the soil moisture and temperature regimes associated natural vegetation, which help determine the nature of the genetic processes, also will help to determine the suborder in which soil is categorized. Some 55 suborders are recognized.

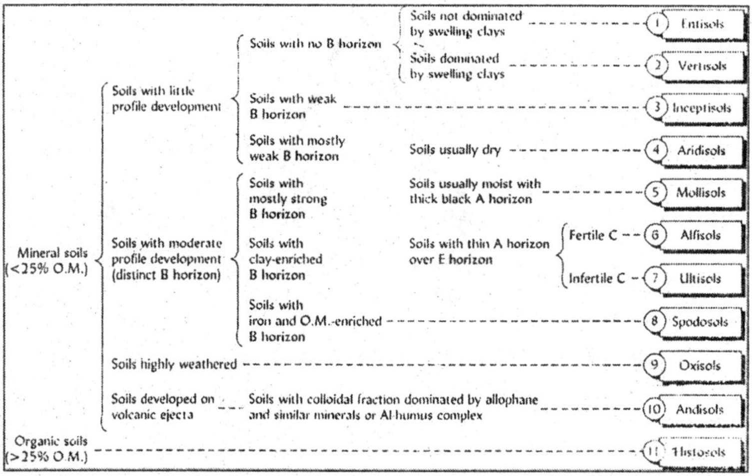

Fig. 22.20: Simplified key to help differentiate among the 11 soil orders of *Soil Taxonomy* (see Appendix A for a more detailed key). (Reprinted by permission from *Soil Science Simplified*, second edition, by M. Harpstead, F.D. Hole, and W.F. Bennett, © 1988 by Iowa State University Press, Ames, IA 50010).

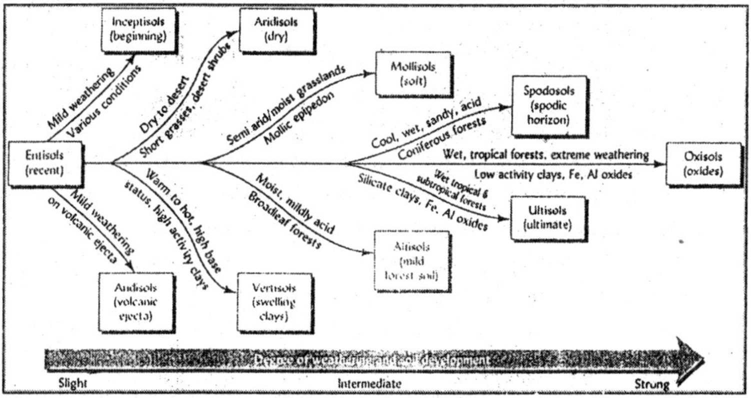

Fig. 22.21: Diagram showing general degree of weathering and soil development in the different orders of mineral soils classified in *Soil Taxonomy*. Also shown are the general climatic and vegetative conditions under which soils in each order are formed.

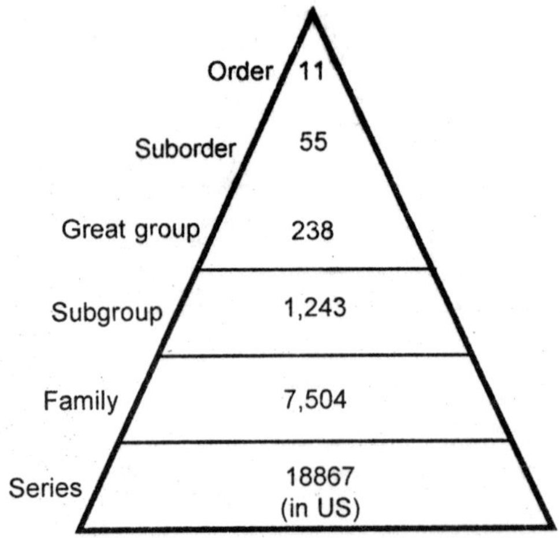

Fig. 22.22: The categories of *Soil Taxonomy* and approximate number of units in each category. (Data from U.S. Department of Agriculture, personal correspondence)

3. *Great Groups*: Diagnostic horizons are primary bases for differentiating the great groups in a given suborder. Soil in the given great group has the same kind and arrangement as these horizons. More than 230 great groups are recognized in the world.

4. *Subgroups*: The subgroups are subdivisions of great groups. The central concept of a great group makes up one subgroup (typic). Other subgroup may have characteristic that are integrates between those of the central concept and soils of other orders, suborders, subgroups or great groups. More than 1240 subgroups are recognized.

5. *Family*: In the family category are found soils with a subgroup having similar physical and chemical properties affecting their response to management and especially to the penetration of plants roots (e.g. soil-water relationship). Difference in texture, mineralogy, temperature and soil depth are primary base for family differentiation. Some 7500 families have been recognized. (Fig. 22.22)

Table 22.1: Names of Soil Orders in *Soil Taxonomy* with their Derivation and Major Characteristics.

	Formative Element		
Names	*Derivation*	*Pronunciation*	*Major Characteristics*
Entisols	Nonsense symbol	Recent	Little profile development, ochric epipedon common
Inceptisols	L. *inceptum*, beginning	Inception	Embryonic soils with few diagnostic features, ochric or umbric epipedon; cambic horizon
Mollisols	L. *mollis*, soft	Mollify	Mollic epipedon, high base saturation, dark soils, some with argillic or nitric horizons
Alfisols	Nonsense symbol	Pedalfer	Argillic or nitric horizon; high to medium base saturation
Ultisols	L. *ultimus*, last	Ultimate	Argillic horizon, low base saturation
Oxisols	Fr. *oxide*, oxide	Oxide	Oxic horizon, no argillic horizon, highly weathered
Vertisols	L. *verto*, turn	Invert	High in swelling clays; deep cracks when soil dry
Aridisols	L. *aridus*, dry	*Arid*	Dry soil, ochric epipedon, sometimes argillic or nitric horizon
Spodosols	Gk. *Spodos*, wood ash	*Pod*zol; odd	Spodic horizon commonly with Fe, Al, and humus accumulation
Histosols	Gk. *Histos*, tissue	*Histology*	Peat or bog; >30% organic matter
Andisols	Modified from Ando	*Andesite*	From volcanic ejecta, dominated by allophane or Al-humic complexes.

Soil orders: Eleven orders are recognized. The names of these orders and their major characteristics are shown in Table 22.1. Note that all names have a common end, sol (from the Latin *solum*, soil). The general conditions that enhance the formation of mineral soils in different orders and the degree of development of soils is shown in Fig. 22.21. Note that soils with essentially no profile layering (Entisols) have the least development, while the deeply weathered soils of humid

Table 22.2: Estimated Losses of Total and Available Nitrogen, Phosphorus, and Potassium in Eroded Sediments.

(In thousands of metric tons)

	Nitrogen		*Phosphorus*		*Potassium*	
Region	*Total*	*Available*	*Total*	*Available*	*Total*	*Available*
Pacific	100	18	29	0.6	1,154	23
Mountain	176	32	64	1.3	2,550	51
Southern Plains	512	94	101	2.0	3,043	61
Northern Plains	2,068	380	293	5.9	11,711	234
Lake states	622	114	107	2.1	3,643	73
Corn belt	4,360	802	624	12.5	24,959	499
Delta states	478	88	141	2.8	4,220	84
South-eastern States	202	37	101	2.0	1,007	20
Appalachian states	676	124	169	3.4	3,381	67
North-eastern States	300	55	75	1.5	2,252	45
Total	9,494	1,744	1,704	34.1	57,920	1,158

tropics (Oxisols and Ultisols) show the greatest soil development. Global Soil Region Map. (Fig. 22.23)

Soil Erosion and its Control

Soil erosion is far less a problem of the old area of the world than of the new ones. It is an abnormal feature of man's expansion over the earth, based primarily on specialized economics directed towards maximum cash returns. Where man has settled with a primary regard for permanent living there has rarely been such trouble. Neither agriculture nor grazing *per se* leads soil erosion. It is principally where men have considered land in terms of quick and large monetary returns that this ill has arisen. Hence, any scientific study of the processes and expressions of soil erosion must regard both physical and cultural factors and forms. Because, the conditioning may lie in nature, the cause is man operating unnaturally, and in the long run unsocial and uneconomical.

No soil phenomena are more destructive world-wide than soil erosion. (Table 22.2)

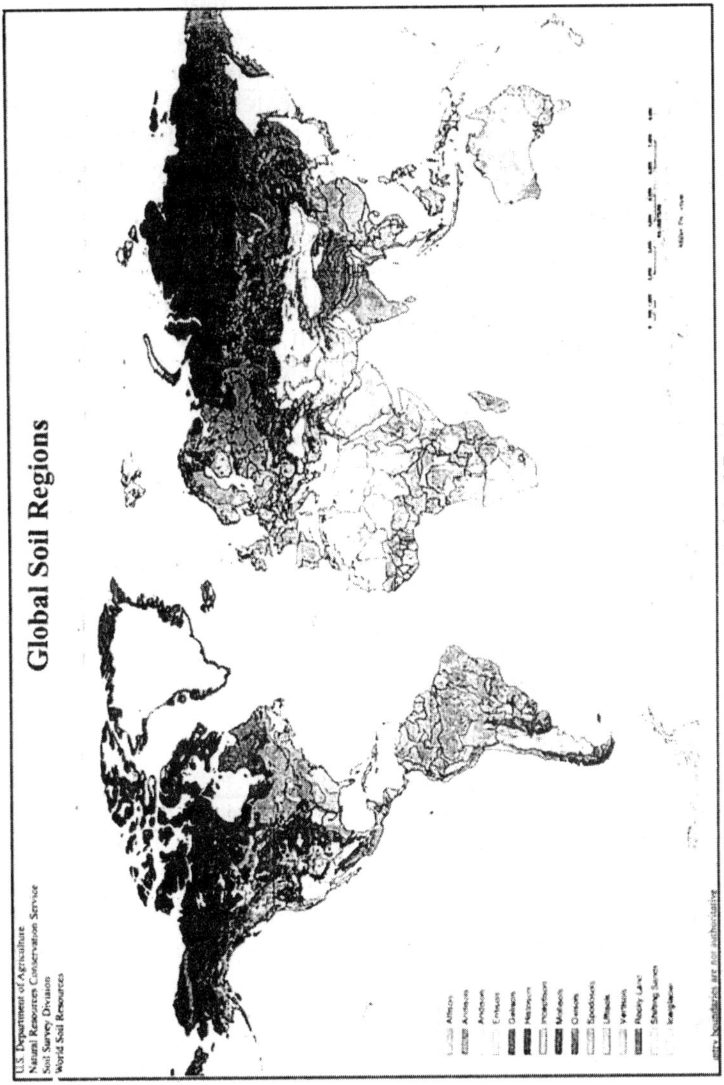

Fig. 22.23: Global Soil Regions

Soil erosion refers to a series of processes leading to soil depletion *in situ* and the export of sediments by exogenetic forces. Soil erosion is a world-wide problem. Preliminary investigations in the tropical area suggest that erosion rates for part of Africa and Latin America may be at least as high as those found for Asia. Table 22.4 indicates the high sediment losses from eight major river systems, including the Mississippi-Missouri. The Yellow river in China and Kosi river in Nepal/India carry very heavy loads of sediment—four to six times that of Mississippi. The threat of soil erosion today is more warning than at any time in history. Subsequent surveys have confirmed the seriousness of the soil erosion problems but also suggested that wise conservation practices can reduce somewhat both water and wind erosion.

1. Wind velocity and turbulence.
2. Soils surface conditions.
3. Soil characteristics.
4. The nature and orientation of the vegetation.

A wind erosion prediction equation has been used. WEQ Comparative Soil Loss by Erosion in the US and other countries of Europe and Asia. (Table 22.3)

Table 22.3

Country	*Total cropland*	*Excessive Soil loss*	
	(million ha)	*Million Mg*	*Mg/ha*
China	99	3628	36.6
India	140	4716	33.7
United States	167	1524	9.1
Others	607	11201	18.5
Total	1013	21069	97.9

Table 22.4: Annual Sediment Load

Rivers	*Countries*	*Annual sediments load (Million Mg)*	*Erosion (Mg/ha drained)*
Yangtze	China	1600	479
Ganges	India	1455	270
Amazon	Brazil, Peru etc.	363	13
Mississippi	US	300	93
Irrawaddy	Burma	299	139
Kosi	India, Nepal	172	555
Mekong	Vietnam, Thailand, etc.	170	43
Red	China, Vietnam	130	217
Nile	Sudan, Egypt, etc.	111	8

Soil erosion can be understood under the following headings:

Nutrient Losses: There is great loss of essential nutrient from the soil by erosion processes. The total quantity of nitrogen, phosphorus and potassium removed by erosion from the soil nation-wide is estimated at more than 38 million Mg. (Table 22.2). Much of nitrogen and phosphorus lost in soil is in the organic matter, fine particles of which are included in the eroded sediments. Total potassium losses shown in Table 22.2 are higher than those for nitrogen or phosphorus because this element is commonly present in larger total quantities in soils.

Water Erosion: Water erosion is one of the most common geological phenomena. The normal erosion amounts annually to about 0.2-0.5 Mg/ha (0.1-0.2 tonnes/acre). It operate slowly, yet unavoidably. Erosion that exceeds this normal rate becomes usually destructive and is referred to as accelerated erosion. This is usually caused by human activities that make the soil more susceptible to the action of water and wind. Water erosion takes place through three processes:

1. Detachment of soil particles, aggregates, clods, and large size volumes from soil mass.
2. Movement of detached material (e.g. by gravity and overflow).
3. Deposition.

Processes are subdivided based on other criteria into:

1. Diffuse surface erosion (e.g. splash and interrill erosion)
2. Linear erosion (rill and gulley)
3. Subsurface erosion (piping erosion)
4. Shallow mass movement.

Splash Erosion: In soil translocation, runoff water plays the major role. On a soil subject to easy detachment, a very heavy rain may splash as much as 225 Mg/ha of soil, some of particles rising as high as 0.7 m and moving horizontally perhaps 1-2 m. On a slope if a wind is blowing, splashing greatly aids and enhance runoff translocation of soil; the two together account for total soil movement that finally occurs. In sheet erosion, soil is removed more or less uniformly from every part of the slope.

Rill Erosion and Gully Erosion: Overland flow tends to rapidly join together into rivulets whose paths are unstable and can quickly change due to local depression or erision episode. The wind around small protuberances until local micro impluvia where they merge with other rivulets forming well-defined flow patterns that can eventually develop into rills. Where the volume of runoff water is further concentrated, the formation of larger channels or gullies occurs by undermining and down cutting. This is gully erosion. (Fig. 22.24)

Piping Erosion: Pipes develop in particular situation. Some are linked to the karst processes, involving soil chemical dissolution and removal by subsurface flow while other or linked to excessive subsurface water that evacuates through cracks or macrospores in which the water flow fast enough to transport sediment. Subsoil, sod city further promotes clay dispersion and transport. These processes result in a pipe or karst like depressions.

Shallow Mass Movement: Shallow mass movements are important in water erosion because (i) They are responsible for the evolution of rill and gulley slopes during and between rainstorm. (ii) The large majority of mass movement in cultivated fields are due to excess water in the soil profile in

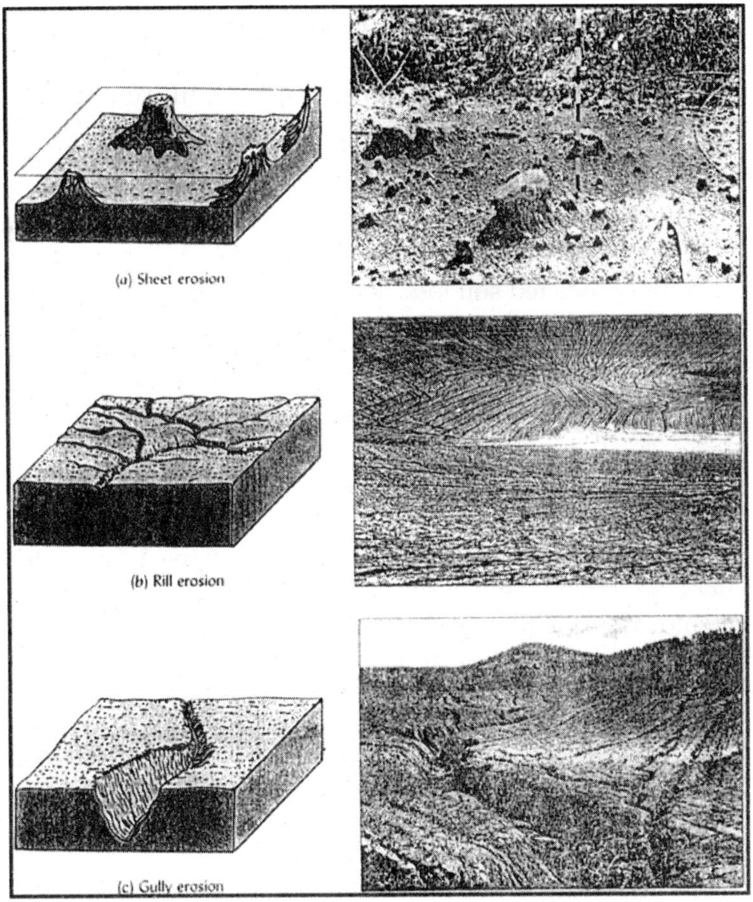

Fig. 22.24: Three major types of soil erosion. Sheet erosion is relatively uniform erosion from the entire soil surface. Note the perched stones and pebbles have protected the soil underneath from sheet erosion. Rill erosion is initiated when the water concentrates in small channels (rills) as it runs off the soil. Subsequent cultivation may erase rills but it does not replace the lost soil. Gully erosion creates deep channels that cannot be erased by cultivation. Although gully erosion looks more catastrophic, far more total soil is lost by the less obvious sheet and rill erosion. [Drawings from FAO (1987); photos courtesy USDA Natural Resources Conservation Service]

the vicinity of more impervious layers. (iii) Mass movement may trigger gullies and pipes and *vice versa.*

Factors Affecting Accelerated Erosion

Decades of agricultural research coupled with centuries of farmers' experience have clearly identified the major factors affecting accelerated erosion. These factors are included in the universal soil-loss equation (USLE).

A = RKLSCP

A, the predicted soil loss, is the product of:

R = climate erosivity (rainfall)
K = Soil erodibility
L = slope length
S = slope gradient or steepness
C = cover and management
P = erosion control practice

Working together, these factors determine how water enters the soil, how much runoff, and the manner and rates of its removal. This equation has been used widely since the 1970s. It now serves as a basis for a modern computerized erosion prediction tool called the Revised Universal Soil-loss Equation (RUSE). RUSE uses the same basic factors of USLE shown above. RUSE is a readily available computer software package. The major differences between USLE and RUSE are given in Table 22.5.

Factors Affecting Wind Erosion

Susceptibility to wind erosion is related to the moisture content of soils. Other factors that influence the wind erosion are:

E = f(ICKLV)

The predicted wind erosion (E) is a function (f) of:

I = soil erodibility factor
C = climate factor
K = soil-ridge roughness factor

Table 22.5: Summary of Major Differences Between USLE and RUSLE

Factor	*Universal soil-loss equation (USLE)*	*Revised universal soil-loss equation (RUSLE)*
R	Based on long-term average rainfall conditions for specific geographic areas in the United States	Generally the same as USLE in the eastern United States. Values for western states (Montana to New Mexico and west) are based on data from more weather stations and thus are more precise for any given location. *RUSLE* computes a correction to R to reflect, for flat land, the effect of raindrop impact on water ponded on the surface.
K	Based on soil texture, organic matter content, permeability, and other factors inherent to soil type.	Same as USLE but adjusted to account for seasonal changes such as freezing and thawing, soil moisture, and soil consolidation.
LS	Based on length and steepness of slope, regardless of land use.	Refines USLE by assigning new equations based on the ratio of rill to interrill erosion, and accommodates complex slopes.
C	Based on cropping sequence, surface residue, surface roughness, and canopy cover, which are weighted by the percentage of erosive rainfall during the six crop stages. Lumps these factors into a table of soil-loss ratios, by crop and tillage scheme.	Uses these subfactors: prior land use, canopy cover, surface cover, surface roughness, and soil moisture. Refines USLE by dividing each year in the rotation into 15-day intervals, calculating the soil loss ratio for each period. Recalculates a new soil-loss every time a tillage operation changes one of the subfactors. RUSLE provides improved estimates of soil-loss changes as they occur throughout the year, especially relating to surface and near-surface residue and the effects of climate on residue decomposition.
P	Based on installation of practices that slow runoff and thus reduce soil movement. *P* factor values change according to slope ranges with some distinction for various ridge heights.	*P* factor values are based on hydrologic soil groups, slope, row grade, ridge height, and the 10-year single storm erosion index value. RUSLE computes the effect of strip-cropping based on the transport capacity of flow in dense strips relative to the amount of sediment reaching the strip. The *P* factor for conservation planning considers the amount and location of deposition.

From Renard *et al.*, 1994.

L = width of field factor
V = vegetative cover factor

The WEQ involves the major factors that determine the severity of erosion, but it also considers how these factors interact with each other.

The erodibility factor (I) relates to the prosperities of the soil and to degree of slope of the site in question. The soil ridge roughness factor (K) takes into consideration the cloddiness of the soil surface vegetation Cover (V) and ridges on the soil surface. The climate factor (C) involves wind velocity, soil temperature and precipitation (which helps control soil moisture). The width of the field factors (L) is the width of the field in the downwind direction.

Control of Wind Erosion

The factors of wind erosion equation clues to the methods of reducing wind erosion. Obviously, if the soil can be kept moist, there is little danger of wind erosion. A vegitative cover also discourages soil blowing especially if the plant roots are well established. Conservation tillage practices should be used for wind erosion. Windbreaks and tenacious grasses and shrubs are especially recommended for wind erosion. (Fig. 22.25)

Soil Conservation is set of management strategies for prevention of soil being eroded from the earth's surface or becoming chemically altered by overuse, salinization, acidification, or other chemical soil contamination. The principal approaches these strategies take are:

- choice of vegetative cover
- erosion prevention
- salinity management
- acidity control
- encouraging health of beneficial soil organisms
- prevention and remediation of soil contamination
- mineralization

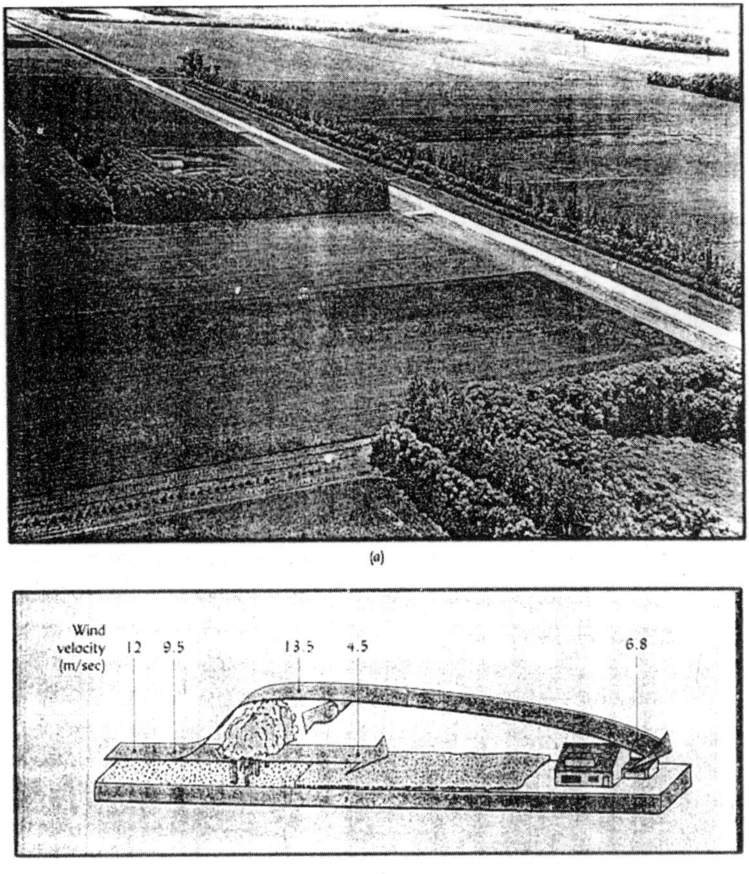

Fig. 22.25: (a) Shrubs and trees make good windbreaks and add beauty to a North Dakota farm homestead. (b) Effect of a windbreak on wind velocity. The wind is deflected upward by the trees and is slowed down even before reaching them. On the leeward side further reduction occurs, the effect being felt as far as 20 times the height of the trees. [(a) Courtesy USDA Natural Resources Conservation Service; (b) from FAO (1987)]

Other ways are:

- no till farming
- contour plowing
- wind rows

Fig. 22.26: Contour ridges must be carefully laid out with sufficient height to hold back water from even heavy rainfall. Here, surface retention is fast becoming surface runoff. (Photo courtesy of R. Weil)

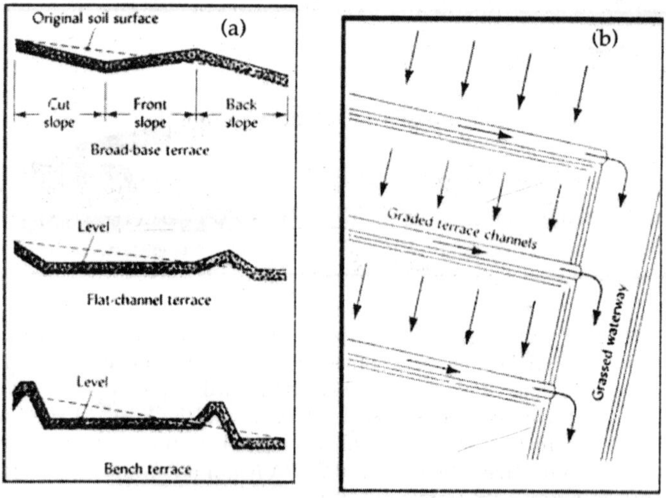

Fig. 22.27: (a) Three types of terraces in use around the world. Broad-based terraces permit the entire surface to be cropped and are widely used in the United States. Flat-channel terraces allow large volumes of water to move off the soil without erosion. Bench terraces can keep water on the land and are used widely in rice production. (b) Diagram of controlled flow of runoff water from a field to terrace channels and onto a grassed waterway from which it can move to a canal or stream channel.

Fig. 22.28: The use of vegetative barriers to create natural terraces. (Photo) A tropical grass (vetiver) has been planted vegetatively on the contour in a cassava field by pushing root sprigs into the soil. (a) The root cuttings are planted perpendicular to the slope direction. In a year or so the grass will be well established and its dense root and shoot growth will serve as a barrier to hold soil particles while permitting some water to pass on through. (b) Note the buildup of soil above the grass, basically forming a terrace wall. Perennial tall wheatgrass is being used experimentally in the Northern Plains area of the United States to serve as a barrier against snow movement and later against wind erosion. (Photo courtesy of Centro Internacional Agricultura Tropical in Cali, Colombia).

- crop rotation
- the use of natural and man-made fertilizer
- resting the land

Many scientific disciplines are involved in these pursuits, including *agronomy, hydrology, soil science, meteorology, microbiology,* and *environmental chemistry.*

Decisions regarding appropriate *crop rotation, cover crops,* and planted *windbreaks* are central to the ability of surface soils to retain their integrity, both with respect to erosive forces and chemical change from nutrient depletion. Crop rotation is

Table 22.6: General Classification of Different Conservation Tillage Systems: All systems maintain at least 30 per cent of the crop residues on the surface

Tillage System	*Operation Involved*
No-till	Soil undisturbed prior to planting, which occurs in narrow seedbed, 2.5-7.5 cm wide. Weed control primarily by herbicides.
Ridge till (till, plant)	Soil undisturbed prior to planting, which is done on ridges 10-15 cm higher than row middles. Residues moved aside or incorporated on about one-third of soil surface. Herbicides and cultivation to control weeds.
Strip till	Soil undisturbed prior to planting. Narrow and shallow tillage in row using rotary tiller, in-row chisel, etc. Up to one-third of soil surface is tilled at planting time. Herbicides and cultivation to control weeds.
Mulch till	Soil surface disturbed by tillage prior to planting, but at least 30 per cent or residue left on or near soil surface. Tools such as chisels, field cultivators, disks, and sweeps are used (e.g., stubble mulch). Herbicides and cultivation to control weeds.
Reduced till	Any other tillage and planting system that keeps at least 30 per cent of residues on the surface.

Note: Definitions used by Conservation Technology Information Center, West Lafayette, Ind.

simply the conventional alteration of crops on a given field, so that nutrient depletion is avoided from repetitive chemical uptake/deposition of single crop growth.

Cover crops serve the function of protecting the soil from erosion, weed establishment or excess *evapotranspiration*; however, they may also serve vital soil *chemistry* functions. For example, *legumes* can be ploughed under to augment soil *nitrates*, and other plants have the ability to metabolize soil contaminants or alter adverse *pH*. The cover crop Mucuna pruriens (*velvet bean*) has been used in Nigeria to increase *phosphorus* availability after application of rock *phosphate*. Some of these same precepts are applicable to urban landscaping, especially with respect to ground cover selection for erosion control and weed suppression.

Erosion barriers on disturbed slope, Marin County, California.

Windbreaks are created by planting sufficiently dense rows or strands of trees at the *windward* exposure of an agricultural field subject to wind erosion. *Evergreen species* are preferred to achieve year round protection; however, as long as *foliage* is present in the seasons of bare soil surfaces, the effect of deciduous trees may also be adequate. Trees, *shrubs* and *groundcovers* are also effective perimeter treatment for soil erosion prevention, by insuring any surface flows are impeded. A special form of this perimeter or inter-row treatment is the use of a "grassway" that both *channels* and dissipates runoff through *surface friction*, impeding *surface runoff*, and encouraging infiltration of the slowed surface water.

Erosion Prevention

Erosion

When surface planting is not feasible, there are a variety of mechanical management tactics to protect surface soils from

Terraced potato farming on Taquile Island, Peru.

water and wind erosion. Need for these tools arises on construction sites and other situations of transition, where bare soils are exposed. The primary tactics applied are mulching of soil surfaces and use of *surface runoff* barriers. From 1990 to 2005 considerable innovation has occurred in manufacture of plastic confined hay-bale products, so that a variety of shapes and sizes of runoff barriers can be delivered to the construction site.

There are also conventional practices that farmers have invoked for centuries. These fall into two main categories: *contour farming* and *terracing*, standard methods recommended by the U.S. *Natural Resources Conservation Service*, whose Code 330 is the common standard. Contour farming was practised by the ancient *Phoenicians*, and is known to be effective for slopes between two and ten per cent. Contour ploughing can increase crop yields from 10 to 50 per cent, partially as a result from greater soil retention.

Keyline design is an enhancement of contour farming, where the total watershed properties are taken into account in forming the *contour lines*. Terracing is the practice of creating benches

Salt deposits on the former bed of the Aral Sea.

or nearly level layers on a hillside setting. Terraced farming is more common on small farms and in underdeveloped countries, since mechanized equipment is difficult to deploy in this setting.

Human *overpopulation* is leading to destruction of *tropical forests* due to widening practices of *slash and burn* and other methods of *subsistence farming* necessitated by famines in lesser developed countries. A sequel to the deforestation is typically large scale erosion, loss of soil nutrients and sometimes total *desertification.*

Salinity Management

Soil Salinity

The ions responsible for salination are: Na^+, K^+, Ca^{2+}, Mg^{2+} and Cl^-. Salinity is estimated to affect about one-third of all the earth's *arable land*. Soil salinity adversely affects the *metabolism* of most crops, and erosion effects usually follow *vegetation* failure. Salinity occurs on *drylands* from *overirrigation* and in areas with shallow saline water tables. In the case of over-irrigation, salts are deposited in upper soil layers as a by product of most soil *infiltration*; excessive irrigation merely increases the rate of salt deposition. The best known case of shallow *saline* water table *capillary action* occurred in *Egypt* after the 1970 construction of the *Aswan Dam*. The change in the groundwater level due to *dam* construction led to high concentration of salts in the water table. After the construction, the continuous high level of the *water table* led to *soil salination* of previously arable land.

Use of *humic acids* may prevent excess salination, especially in locales where excessive irrigation was practised. The mechanism involved is that humic acids can fix both *anions* and *cations* and eliminate them from *root zones*. In some cases, it may be valuable to find plants that can tolerate saline conditions to use as surface cover until salinity can be reduced; there are a number of such saline-tolerant plants, such as *saltbush*, a plant found in much of *North America* and in the *Mediterranean* regions of *Europe*.

Soil pH

Soil pH levels adverse to crop growth can occur naturally in some regions; it can also be induced by *acid rain* or *soil contamination* from *acids* or *bases*. The role of soil pH is to control nutrient availability to vegetation. The principal *macronutrients* (*calcium, phosphorus, nitrogen, potassium, magnesium, sulphur*) prefer neutral to slightly *alkaline* soils. Calcium, magnesium and potassium are usually made available to plants via cation exchange surfaces of *organic material* and clay soil surface particles. While acidification increases the initial availability of these *cations*, the residual soil moisture concentrations of nutrient cations can fall to alarmingly low levels after initial nutrient uptake. Moreover, there is no simple relationship of pH to nutrient availability because of the complex combination of soil types, soil moisture regimes and meteorological factors.

Yellow fungus, a mushroom that assists in organic decay.

The important observation is that pH is the regulatory mechanism to plant nutrient uptake, and that the theoretical concentration of soil nutrients is meaningless until pH levels are in the optimum range for uptake. Soil pH can be raised by amendment by *agricultural lime*. The pH of an alkaline soil is lowered by adding *sulphur*, *iron sulphate* or *aluminium sulphate*, although these tend to provide costly short-term benefits. *Urea*, *urea phosphate*, *ammonium nitrate*, *ammonium phosphates*, *ammonium sulphate* and *monopotassium phosphate* also reduce soil pH.

Soil Organisms

Promoting the viability of beneficial soil organisms is an element of soil conservation; moreover this includes macroscopic *species*, notably the *earthworm*, as well as *microorganisms*. Positive effects of the earthworm are known well, as to *aeration* and promotion of *macronutrient* availability. When worms excrete *egesta* in the form of *casts*, a balanced selection of minerals and plant nutrients is made into a form accessible for *root* uptake. *US* research shows that earthworm casts are five times richer in available *nitrogen*, seven times richer in available *phosphates* and eleven times richer in available *potash* than the surrounding upper 150 mm of soil. The weight of casts produced may be greater than 4.5 kg per worm per year. By burrowing, the earthworm is of value in creating soil *porosity*, creating channels enhancing the processes of aeration and drainage.

Soil *microorganisms* play a vital role in *macronutrient* wildlife. For example, *nitrogen fixation* is carried out by free-living or *symbiotic bacteria*. These bacteria have the *nitrogenase* enzyme that combines *gaseous* nitrogen with hydrogen to produce *ammonia*, which is then further converted by the bacteria to make other *organic compounds*. Some nitrogen fixing bacteria such as *rhizobia* live in the *root nodules* of *legumes*. Here they form a *mutualistic* relationship with the plant, producing ammonia in exchange for *carbohydrates*. In the case of the *carbon cycle*, carbon is transferred within the biosphere as *heterotrophs* feed on other organisms. This process includes the uptake of dead organic material (*detritus*) by *fungi* and *bacteria* in the form of *fermentation* or *decay* phenomena.

Mycorrhizae are symbiotic associations between soil-dwelling fungi and the roots of vascular plants. The mycorrhizal fungi increase the availability of minerals, water, and organic nutrients to the plant, while extracting sugars and amino acids from the plant. There are two main types, endomycorrhizae (which penetrate the roots) and ectomycorrhizae (which resemble 'socks', forming a sheath around the roots). They were discovered when scientists observed that certain seedlings failed to grow or prosper without soil from their native environment.

Some soil microorganisms known as *extremophiles* have remarkable properties of adaptation to extreme environmental conditions including temperature, pH and water deprivation.

The viability of soil organisms can be compromised when *insecticides* and *herbicides* are applied to planting regimes. Often there are unforeseen and *unintended consequences* of such chemical use in the form of death of impaired functioning of soil organisms. Thus any use of pesticides should only be undertaken after thorough understanding of residual toxicities upon soil organisms as well as terrestrial *ecological* components.

Killing soil microorganisms is a deleterious impact of *slash and burn* agricultural methods. With the surface temperatures generated, virtual annilation of soil and vegetative cover organisms are destroyed, and in many environments these effects can be virtually irreversible (at least for generations of mankind). *Shifting cultivation* is also a farming system that often employs *slash and burn* as one of its elements.

Soil Contamination

There are thousands of *anthropogenic chemicals* that enter soil systems, most of which have an adverse effect upon soil quality and plant metabolism. While the role of pH has been discussed above, heavy metals, *solvents*, petroleum *hydrocarbons*, *herbicides* and *pesticides* also contribute soil residues that are of potential concern. Some of these chemicals are totally extraneous to the agricultural landscape, but others (notably herbicides and pesticides) are intentionally introduced to serve a short-term function. Many of these added chemicals have long *half-lives* in soil, and others degrade to produce derivative chemicals that may be either persistent or pernicious.

Typically the expense of soil contamination remediation cannot be justified in an agricultural *economic* analysis, since cleanup costs are generally quite high; often remediation is mandated by state and county *environmental health* agencies based upon *human health* risk issues.

Mineralization

To allow plants full realization of their phytonutrient potential, active mineralization of the soil is sometimes undertaken. This can be in the natural form of adding crushed rock or can take the form of chemical soil supplement. In either case the purpose is to combat mineral depletion of the soil. There are a broad range of minerals that can be added including common substances such as phosphorus and more exotic substances such as zinc and selenium. There is extensive research on the phase transitions of minerals in soil with aqueous contact.

The process of flooding can bring significant bedload sediment to an *alluvial* plain. While this effect may not be desirable if floods endanger life or if the eroded sediment originates from productive land, this process of addition to a *floodplain* is a natural process that can rejuvenate soil chemistry through mineralization and *macronutrient* addition.

REFERENCES

Arthur T. Hubbard, *Encyclopedia of Surface and Colloid Science Vol. 3*, Santa Barbara, California Science Project, Marcel Dekker, New York (2004) *ISBN 0-8247-0759-1.*

Bill Mollison, *Permaculture: A Designer's Manual*, Tagari Press, (1988). Increases in porosity enhance infiltration and thus reduce adverse effects of *surface runoff.*

Dan Yaron, *Salinity in Irrigation and Water Resources*, Marcel Dekker, New York (1981) *ISBN 0-8247-6741-1.*

B.O. Vanlauwe, C. Nwoke, J. Diels, N. Sanginga, R. J. Carsky, J. Deckers, and R. Merckx, Utilization of rock phosphate by crops on a representative topo-sequence in the Northern Guinea savanna zone of Nigeria: response by Mucuna pruriens, Lablab purpureus and maize, *Soil Biology and Biochemistry* 32:2063-2077. (2000).

Perimeter landscaping of Carneros Business Park, Lumina Technologies, Santa Rosa, Ca., prepared for Sonoma County, Ca. (2002).

Predicting soil erosion by water, a guide to conservation planning in the Revised Universal Soil Loss Equation, U.S. *USDA* Agricultural Research Service, Agricultural handbook no. 703 (1997).

Wolfgang Summer, *Modelling Soil Erosion, Sediment Transport and Closely Related Hydrological Processes* entry by Mingyuan Du, Peiming Du, Taichi Maki and Shigeto Kawashima, "Numerical modelling of air flow over complex terrain concerning wind erosion", International Association of Hydrological Sciences publication no. 249 (1998) *ISBN 1-901502-50-3*.

Y.C. Lu, K. B. Watkins, J. R. Teasdale, and A. A. Abdul-Baki. Cover crops in sustainable food production. *Food Reviews International* 16:121-157 (2000).

23

THE ROLE AND NATURE OF TIME IN GEOMORPHOLOGY

- Time is a fundamental concept given the traditional preoccupation with landscape change over time
- Until recently geomorphologists have deferred the study of material properties and process mechanics to engineers
- There is no geomorphic time scale, that is, there are standard time periods besides the divisions of the geologic time scale, which are based on the age of rocks and sediments (i.e. mostly fossil evidence)
- Thus geomorphologists have tended to treat time in a relative way as discussed below
- There is a broad positive correlation between the size of a landscape unit and the duration of change
  - o e.g. ripples change quickly, dunes less quickly; or rills *vs.* gullies *vs.* stream channels
  - o thus the relevant time period can vary with nature of a study (e.g. process *vs.* climatic geomorphology)
- All models of landscape change are of two types:

1. Time-dependent (developmental or relaxation-time models)

- Landforms change in response to an initial disturbance or change in input (climatic change, tectonism), and then gradual progressive sequential change is characterized by the association of morphology with stage of development
- e.g. the Davisian cycle of erosion that dominated geomorphological thinking for decades

- It was a time-decay model because no further inputs were envisioned after the initial uplift of the landscape
- Just gradual, inevitable landscape evolution at an increasingly slower rate as the initial energy is dissipated.

2. Time-independent (characteristic-form models)

- The alternative and subsequent view of landscape evolution
- Rather than subject to ever-decreasing energy, landscapes are open systems, constantly adjusting to changing inputs and levels of internal energy
- Geomorphic systems adjust to approach an equilibrium with the inputs acting on them
- Because the inputs constantly vary, equilibrium can probably never be attained and thus characteristic is preferred to equilibrium to describe forms which reflect the landscape after a period of adjustment (relaxation) to a disturbance or major change in inputs.

With the emergence of equilibrium concepts, the historical significance of a time-dependent view, and the development of isotope techniques for establishing the absolute age of sediments and thereby surfaces, time is a subject of ongoing interest and research with four recurring themes:

1. Classification and Sampling of Time

Schumm, S.A. and R.W. Lichty (1965): Time, space and causality in geomorphology. *American Journal of Science,* 263: 110-119.

The subdivision of time according to whether variables of landscape evolution are independent, semi-dependent or dependent:

- *Cyclic (geologic, 106 yrs)*
  - o longest timespans of landscape evolution,
  - o time is an independent variable and the most important one, because there are no specific or constant temporal

relationships between independent and dependent variables (cause and effect),
  - climate, initial relief and geology are the other independent variables, all others are dependent.
- *Graded (modern, 102 yrs)*
  - a short segment of cyclic time during which a graded condition or dynamic equilibrium exists, that is, fluctuations about (approaches to) a steady state and tendency for negative feedback,
  - applies to some but not all components of a geomorphic system because relief is still being reduced throughout the entire system,
  - time and initial relief are irrelevant because parts of a geomorphic system are adjusted to prevailing inputs,
  - the only dependent variables are responses (i.e. morphology and output from the system).
- *Steady (present, 10-2 yrs or 3-4 days)*
  - a brief period during which some part of a system is unchanging and thus truly time-independent,
  - once again time and initial relief are irrelevant,
  - the only dependent variable is outputs of energy and mass since morphology is constant,
  - more recently referred to as engineering time.

Sampling of Time

- According to technique,
  - (a) direct observation and measurement (< 10 years),
  - (b) historical (archival) data, usually generated for other purposes, e.g., stream flow records,
  - (c) absolute dating of materials and surfaces, limit of conventional 14C dating is about 50,000 years,
  - (d) stratigraphic methods employing both absolute and relative dating techniques, in particular, the age correlation of surfaces and strata.
- According to duration of observation,
  - (a) *continuous*: these observations are rarely reported rather nearly always reduced before analysis, that is,

(b) *quantized*: division of continuous time into useful or familiar segments (e.g. hour, day, month, year), when data are summarized by segment some are always lost depending on the duration of the segment and the manner of summarization (weekly means from daily means or from average of max and min); e.g. weather data,

(c) *discrete*: one observation per interval (day, month, etc.); also a means of expressing frequency,

(d) *sampled*: a continuous process observed intermittently (e.g. one day per month).

2. Magnitude and Frequency

- How often an event of a certain magnitude occurs expressed as
  - o the number of occurrences relative to a period of time or number of observations,
  - o the probability of such an event, or
  - o the recurrence interval (return period).
- Magnitude and frequency are inversely related,
- Lognormal frequency distributions are common, because the lowest magnitude is fixed at just above zero but the upper magnitude (rare events) is limited only by earth and atmospheric physics,
- Which magnitude is most important geomorphologically?

Wolman, M.G. and W.P Miller (1960): Magnitude and frequency of forces in geomorphic processes. *Journal of Geology*, 68: 54-74.

Wolman and Miller determined that at least 50 per cent of sediment transport was by flows of a magnitude that occurred at least once per year, that is, that events of intermediate magnitude and frequency generally dominate fluvial sediment transport. This benchmark analysis sparked much further consideration of magnitude and frequency including elaboration of the concepts and recognition of its limitations and complexities:

1. The common calculation of recurrence intervals for meteorological and hydrological events is useful but the geomorphic significance of these events also depends on other controls (i.e. vegetation, soil, morphology and rock).
2. Frequency and magnitude assume stationarity of the series, that the mean and variance do not vary over time (basically equivalent to substantive uniformitarianism), however trend, persistence and intermittency in the behaviour of geomorphic systems makes the frequency distribution an incomplete description of a sequence of geomorphic events:
   - trend, the direction of change, can include cyclic patterns
   - persistence (autocorrelation) occurs when a value influences an adjoining value in a series (e.g. tree ring widths) and implies that the true variability of a series may not be evident in a short record
   - intermittency, the nonperiodic clustering of similar values over long periods, is perhaps the most problematic because cumulative effects can have large geomorphic impact.
3. Extremely rare events (e.g., once in half a million years or an annual probability of 10^{-5}) may have tremendous geomorphic consequences as postulated by neocatastrophists (e.g. draining of glacial Lake Missoula creating the eastern Washington scablands).
4. There is a fundamental difference between the processes of quasi-continuous sediment transport and the processes that create landforms in terms of magnitude/frequency characteristics,
   - therefore, the effectiveness of geomorphic events can be expressed in terms of timing and magnitude relative to other events and the recovery (relaxation) of the geomorphic system,
   - if two low frequency events occur in consecutive years, they will differ in effectiveness as the landscape will be rendered more (e.g., lag deposit) or less resistant (e.g., lack of vegetation) by the first event.
5. Just as magnitude and frequency relationships differ between transport and formative processes, they also differ

for various parts of a landscape, particularly between stream channels and interfluves, and over larger areas between regional climates (note: the original magnitude/frequency work was based in humid temperate climates)

6. For the above reasons, comparisons of the effectiveness of various geomorphic processes are increasingly based, not on magnitude and frequency, but on more uniform measures of erosion such as mass of sediment moved or amounts of geomorphic work.

3. Change through Time

Brunsden, D. and Thornes, J.B. (1979). Landscape sensitivity and change. *Transactions, Institute of British Geographers,* 4: 463-484.

- Disturbances or changes in controls, that exceed geomorphic thresholds, can be classified as either
  - o *pulsed*: short duration compared to time period under consideration (e.g., tectonism), or
  - o *ramped*: sustained at a new level once initiated (e.g., climatic change).
- The response of a landscape in quasi-equilibrium is to absorb the input without changes.
- Adjustment with or without perceptible change are described by the following concepts:

Reaction and Relaxation

- Absorbing and adjusting to change or disturbance, respectively.
- The response of landscapes and geomorphic systems often occurs over longer time periods than the duration of the disturbance or change in controls (inputs).

Reaction time

The interval between a disturbance or change in controls and observable morphological change.

Relaxation time

The ensuing period of adjustment up to a new equilibrium or another geomorphic threshold adjust.

Hysteresis

- Incomplete reversibility (Gr. a deficiency, same root as hysteria).
- When a condition changes from A to B and back to A, the pathways may differ and thus condition A may not be regained.
- For example, soil moisture or suspended sediment load before and after a storm.
- This usually results from a change in boundary conditions caused by the process (e.g., residual stress from loading and unloading of clay).

4. Entropy and Equilibrium

- The concept of entropy, the degree of disorder in an isolated system, and the second law of thermodynamics dictate that, irrespective of the initial level of energy, energy becomes more evenly distributed over time (i.e., energy gradients are reduced) and processes proceed at an ever decreasing rate.
- Ultimately, energy is evenly distributed, entropy is maximum and free energy (available for conversion to work) is zero, this is, the state of equilibrium where the most probable energy distribution is uniform with least work (least landscape change).
- These concepts explain characteristic (least-work, graded) landforms like longitudinal stream profiles, stream channel meanders, dendritic drainage networks, and concave-convex slope profiles.
- Entropy is maximized when all possible energy states have equal probability (i.e., random distribution).

- Thus there is rapid change away from any highly abnormal configuration of a system, corresponding to the notion of initially rapid change in response to an input or disturbance and then a decreasing rate of change over time.

However geomorphic systems are also subject to order imposed by structural (geologic) control and to repeated and variable inputs, suggesting various kinds of equilibrium:

1. ***Static***: Some system properties are unchanging, absolutely and relatively, over a period of time.
2. ***Stable***: Tendency to revert to a previous condition after a limited disturbance.
3. ***Unstable***: A small disturbance results in continuous movement away from an old equilibrium and towards a new stable one.
4. ***Metastable***: An incremental change (trigger mechanism) pushes a system across a threshold from stable equilibrium to a new equilibrium.
5. ***Steady State***: Numerous small-scale about a mean that has no trend.
6. ***Thermodynamic***: A tendency towards maximum equilibrium, second law of thermodynamics.
7. ***Dynamic***: Balanced fluctuations about a trending, non-repetitive mean value; often called quasi-equilibrium because of the tendency towards a steady state with a trending mean (i.e., equilibrium is never attained).
8. ***Dynamic Metastable***: Both dynamic and metastable, fluctuations about a trending mean interspersed by large jumps as thresholds are crossed; used by many geomorphologists to explain landscape behaviour.

Related Equilibrium Concepts

1. Feedback

- accounts for movement away from (positive) and back to negative) a mean value,

- that is, the effect of a disturbance or change in controls is either magnified or dampened, respectively.

2. *Threshold*

- a transition in behaviour, operation or state,
- a large response to a small or incremental change,
- extrinsic and intrinsic, i.e., response to change in an external (e.g., climate) or internal (e.g., shear strength) variable, respectively.

3. *A tendency toward equilibrium*

- favoured by a variety of reaction and relaxation times, and a constant stream of inputs which all but preclude equilibrium,
- landform change is precluded only by static equilibrium, which is highly unlikely in the natural world,
- equilibrium is relative to the timescale of interest, since
- a landform can be in equilibrium with respect to one timescale (graded or steady) and in disequilibrium with respect to another (cyclic).

4. *Complex Response*

- parts of a geomorphic system approach equilibrium at different rates and a single part will exhibit different tendencies at different times,
- e.g. lack of correlation between cross-sections in the response of Douglas Creek, Colorado to overgrazing in the "cowboy era",
- difficult to recognize from short-term process studies,
- serious implications for interpretations of earth history, by precluding the interpretation of single causes for geomorphic responses identified in the quaternary record,
- e.g., the incision of One Tree Creek, Alberta, and formation of terraces, was the response to both base level change as the Red Deer River shifted across its floodplain

and to Holocene climatic changes; the first factor is imposed from the mouth of stream, the second controls the supply of water and sediment to the channel from upstream,

- thus, geomorphologists need to be much more rigorous and precise in defining their temporal and spatial scales.

SPACE IN GEOMORPHOLOGY

General Themes

- There are no natural measures of space,
- Unlike time where we use days, seasons and years according to the rotation and orbit of the earth,
- That is, other than the space within which we conduct our daily affairs and the increasing larger spaces with which we are increasingly less familiar,
- The arbitrary spatial measures are linear (m, km) or areal (acre, ha),
- Thus there are no natural measures or standard scales for the study of the dimensions and distribution of landforms,
- The obvious spatial units, the slope profile (linear) and drainage basin (areal) have large size variation,
- Spatial patterns often reflect past events,
  - o generally slow rates of landscape change represent a constraint on the study of process and form,
  - o inherited patterns are the response to processes not currently operating (paleoprocess),
  - o thus just as time scale depends on the spatial perspective (scale), time intrudes on any consideration of space in geomorphology,
  - o landscapes that bear the imprint of both former and current geomorphic processes have been described as a palimpsest,

    palimpsest: a piece of parchment that had been written upon more than once ,
  - o models of landscape evolution vary in the extent to which they describe these residual effects (inherited forms).

- In addition to the historical tension in landscape studies, there is dimensional tension, i.e., interaction between point, line and areal features (e.g., rock outcrop, stream and drainage basin).

Fundamental Spatial Properties

1. Orientation

- Direction or aspect of linear and areal features,
- In geomorphology, this property is particularly significant relative to the direction of winds, insolation and geologic structures.

Examples

  - o North *versus* south-facing slopes with respect to the solar energy balance.
  - o East *versus* west-facing slopes with respect to prevailing winds and latent heat flux: valley asymmetry and maximum geomorphic activity on north- to east-facing mountain rockwalls.
  - o Orientation of shorelines with respect to wave fetch, winds, longshore processes.
  - o Direction and distance from a point source: glacial erratics, tephra thickness and texture.
  - o Orientation of linear features (e.g., valleys) relative to geologic structures.
- These are all examples of orientation in the horizontal plane.
- Orientation in the vertical plane is measured as dip of bedded rocks and sediments (strata), imbrication of clasts (fabric) in alluvium and till, and the gradient of slope profiles.
- Orientation can be a strong and complex characteristic of active landscapes with short reaction and relaxation times (e.g., aeolian and coastal landscapes).
- Orientation is measured on a circular scale and thus most summary statistics are inappropriate (the arithmetic mean of 1° and 359° is 180°); there is a separate set of statistical

techniques for analyzing circular data (e.g., using sines and cosines which range from –1 to +1).
- The location of any feature can be determined from orientation (horizontal and/or vertical) and distance.

2. *Distance*

- The shortest path between two objects measured in time or length units.
- Often asymmetrical (scalar *versus* vector) because paths represents a flux (e.g., flow in a stream channel) or cause at one location results in an effect at the other.
- Distance is an important variable in most relationships among geomorphic parameters

Examples

1. Decay with increasing distance
   - o changes in the thickness and texture (sorting) of deposits in aeolian, fluvial, coastal and lacustrine environments, and even with mass wasting (i.e., fall sorting).
   - o however, sediments are easily reworked (redistributed) such that original relationships between distance, thickness and texture are quickly masked.
2. Constant up to a boundary
   - o spatial uniformity never exists, but is assumed or portrayed as a common means of storing, processing and presenting geographic information.
   - o all thematic mapping (stratification) and spatial correlation of areal features is based on identification of uniform areas or consistent patterns with respect to specific criteria and scale.
3. No effect except at a certain distance
   - o best examples are coastal where waves that develop in open sea represent the transfer of energy that eventually causes geomorphic work on a distant shoreline.
   - o the extreme example is a tsunami, energy derived from the tectonic disturbance of an ocean basin is transferred

as waves with low amplitude and very long wavelength.

4 Distances in the vertical plane tend to be much shorter but just as important
   - o exponential decay of rates of weathering and soil creep.
   - o concentration of saltation load near the ground.
   - o depth to water table and permafrost as a control on many earth surface processes.

3. *Connectiveness*

- Relative location, contiguity, adjacency,
- Well developed concept in human geography: nearest neighbour analysis, central place theory.
- Most easily understood with respect to distribution of point features (everything at a small enough scale): clustered, random and regular distributions.
- Three basic and independent characteristics of connectiveness:

Pattern: areal or geometric arrangement irrespective of the size of a study area.

Density: frequency of occurrence expressed as number of occurrences per unit area (i.e., relative to the size of the study area).

Dispersion: spread of phenomena relative to the size of the study area; a clustered distribution has the least dispersion.

These concepts provide an objective theoretical basis for the measurement and comparisons of spatial distributions, but this type of analysis is of little value without some reason to suspect (hypothesize) that a particular process would produce a definite distribution.

- Linear patterns
   - o e.g., stream networks.
   - o original work by Horton and Strahler and subsequent stochastic modelling as random networks.

- Areal patterns
  - o two-dimensional shape of landforms (morphology).
  - o the distribution of groups of areas usually corresponding to groups of landforms.
  - o the most compact (least work) shape that provides total coverage of areas is the hexagon; this geometric principle is a theoretical basis for the study of the arrangement and genesis of drainage basins, patterned ground, desiccation cracks, and columnar joint blocks.
- Boundaries
  - o abstract, a concept to satisfy the "intellectual need for simplification that mandates division of the indivisible".
  - o thus there will never be a truly satisfactory theoretical basis for boundary delimitation.
  - o regional boundaries
    - various criteria, not formally defined, vague.
    - e.g. physiographic regions.
  - o drainage divide
    - topographic boundary and thus easily approximated.
    - assumes no transfer of water across boundary, but this does not necessarily imply that geomorphic systems are integrated within a basin.
    - there is often little interaction between major valleys and interfluve landscapes; much of the sediment transported by streams is reworked floodplain alluvium of derived from valley sides.
  - o boundary delimiting a discrete landform
    - these boundaries require definition of landforms according to mappable criteria, which is problematic for incomplete and gradational forms (polygenesis).

Scale Issues

1. Sampling (spatial resolution)
   - usually purposive, especially when a limited number of occurrences are under investigation.
   - use of inferential statistical methods to estimate population parameters and establish relationships is

constrained by lack of random sampling and by spatial autocorrelation, that is, neighbouring observations often influence one another, although the degree of later can be used to characterize landscapes.
- stratification is usually a necessity since there are large spatial variations in resistance and stability of surface materials (e.g., vegetation cover) and thus effectiveness of geomorphic processes.

2. Linkages between scales
   - scale itself is a significant variable and the results of research at one scale do not necessarily apply to other scales.
   - for practical reasons, research is often done at a different scale than the scale of the phenomenon (process or feature) under study.
   - three strategies developed to overcome this problem are:

     (a) using ratio data or dimensionless numbers
     - o e.g., bifurcation ratio, standardized data (relative to the mean), relative relief,
     - o eliminates scale as a variable, size is not implied by the units used.

     (b) smoothing local observations
     - o to extract the dominate trends over an area and detect significant deviations (residuals),
     - o e.g., splines and trend surface analysis.

     (c) synthetic landscapes
     - o placing isolated observations in a sequence (e.g., Caine, 1979) to depict variation along a transect or over an area,
     - o enables extrapolation from detailed observations to a broader scale while retaining statistical treatment and a physical model.

3. Standardization
   - standard scale for studies of a similar nature has not been recognized in geomorphology.
   - thus, for example, there is no standard size of research plots for the study of soil erosion or standard length of slope facet for the study of hillslopes.

- although there has been little conscious attempt to adopt standard scales, and particularly to investigate the impact of scale variation on research results, some standardization is a function of confining research to specific landforms or to drainage basins of a specific order.
- there are also logistical constraints on the size of area that can be studied in detail in the field or on the scale of a laboratory experiment.
- the apparent disparity among the results of individual studies can be an artefact of scale divergence.

REFERENCES

Astinov, V.I., B.N. Mozaev, V.G. Mozaeva and N.V. Skublova, Remote sensing and origin of large relief forms, in *Geomorphology and Paleogeography*, pp. 255-258, XXIII International Geographical Congress, Moscow, 1976.

Barsch, D. and H. Liedtke, Principles, Scientific Value and Practical Applicability of the Geomorphological Map of the Federal Republic of Germany at the Scale 1:25000 (GMK25) and 1:100000 (GMK100), pp. 296-313, Zeitschrift fÿr Geomorphologie Supplementband 36, Berlin, 1980.

Bashenina, N.V., Principles of Geomorphological Classification, in *Guide to Medium Scale Geomorphological Mapping*, edited by J. Demek and C. Embleton, pp. 20-24, Schweizerbarts, Stuttgart, 1978.

Bashenina, N.V. *et al.*, Legend to the International Geomorphological Map of Europe 1:2500000 IGU Commission on Geomorphic Survey and Mapping, Czechoslovak Academy of Sciences, Institute of Geography, Brno, 1971.

Bashenina, N.V. and N.N. TalÖSkaya, Space imagery analysis for a geomorphological map of the world, *Soviet J. Remote Sensing*, No. 6, 861-871, 1981.

Bedi, N., Applied geomorphic studies on Landsat imagery in *Perspectives in Geomorphology, Vol. III* ÑApplied Geomorphology, edited by H.S. Sharma, pp. 251-269, Concept Publishing Company Pvt. Ltd., New Delhi, India, 1982.

Blackinton, J.G. and D.M. Hussong, First results from a combination side-scan sonar and seafloor mapping system (Sea MARC II), paper presented at 15th Annual Offshore Technology Conference, Houston, Texas, May 2-5, 1983.

Canoba, C.A., Geomorphological mapping using Landsat imagery: A case study in Argentina, *ITC J.*, 1982-03, 324-329, 1982.

Christian, C.S., Concept of land units and Land systems, in *Proc. Ninth Pacific Science Conf. of Pacific Science Assn.*, **20**, 74-81, 1958.

Christian, C.S. and G.A. Stewart, Aerial surveys and integrated surveys, in *Proc. Toulouse Conf. on Natural Resources Research*, pp. 223-280, UNESCO, Paris, France, 1968.

Cochran, G.R. and G.H. Browne, Geomorphic mapping from Landsat-3 Return Beam Vidicon (RBV) imagery, *Photogramm. Eng. Remote Sens.*, **47(8)**, 1205-1213, 1981.

Cook, R.U. and J.C. Doornkamp, Geomorphology in *Environmental Management: An Introduction*, 413 p., Clarendon Press, Oxford, 1974.

Crofts, R.S., Detailed geomorphological mapping and land evaluation in highland Scotland, in *Progress in Geomorphology*, edited by E.H. Brown and R.S. Waters, pp. 231-249, Institute of British Geographers, Special Publ. 7, 1974.

Damuth, J.F., V. Kolla, B.D. Flood, B.O. Kowsmann, M.C. Monteiro, M.A. Gorini, J.J.C. Palma and R.H. Belderson, Distributary channel meandering and bifurcation patterns on the Amazon deep-sea fan as revealed by long-range side-scan sonar (GLORIA), *Geology*, **11**, 94-98, 1983.

Davis, W.M., The geographical cycle, *Geographical J.*, **14**, 481-503, 1899.

Demek, J., (Ed.), *Manual for Detailed Geomorphological Mapping*, 320 pp., IGU Commission on Geomorphic Survey and Mapping, Academia, Prague, 1972.

Demek, J., Geomorphological maping: progress and problems in *Perspectives in Geomorphology, Vol. IIIÑApplied Geomorphology*, edited by H.S. Sharma, pp. 221-235, Concept Publishing Company Pvt. Ltd., New Delhi, India, 1982.

Demek, J. and C. Embleton, (Ed.), *Guide to Medium Scale Geomorphological Mapping*, 348 p., Schweizerbarts, Stuttgart, 1978.

Embleton, C. (Ed.), *Geomorphology of Europe*, 429 p., Macmillan, London 1984.

Embleton, C., Techniques, problems and uses of mega-geomorphological mapping, in *Global Mega-Geomorphology*, edited by R.S. Hayden, pp. 84-88, NASA CP-2312, 1985.

Fairbridge, R.W., Kara-Bogaz-Gulf, in *Encyclopedia of Geomorphology*, edited by R.W. Fairbridge, pp. 579-581, Reinhold Book Corp., New York, 1968.

Fenneman, N.H., (1931-1938) *Physiography of Eastern United States and Physiography of Western United States*, McGraw-Hill, New York, 1931-1938.

Gardiner, V., Land evaluation and the numerical delimitation of natural regions, *Geografia Polonica*, **34**, 11-30, 1976.

Garrison, L.E., N.H. Kenyon and A.H. Bouma, Channel systems and lobe construction in the Mississippi Fan, *Geo-Marine Letters*, pp. 31-39, March-June 1982.

Gastil, R.G., R.P. Phillips and E.C. Allison, Reconnaissance Geologic Map of the State of Baja California, *Geol. Soc. Amer. Mem.*, **140,** Plate 1A, 1975.

Gellert, J.F., Principles of geomorphological regionalization and encoding to geomorphological characteristics for computer processing of data, in *Prospectives in Geomorphology, Vol. IIIÑApplied Geomorphology,* edited by H.S. Sharma, pp. 237-242, Concept Publishing Company Pvt. Ltd., New Delhi, India, 1972.

Hammond, E.H., Small scale continental landform maps, *Annals AAG,* **44,** 34-42, 1954.

Howard, J.A. and C.W. Mitchell, Phyto-geomorphic classification of landscape, *Geoforum,* **11,** 85-106, 1980.

Hussong, D.M. and P. Fryer, Back-arc seamounts and the SeaMARC II Seafloor Mapping System, *EOS,* **64,** 627-632, 1983.

Kayan, I. and V. Klemas, Application of Landsat imagery to studies of structural geology and geomorphology of Mentese region of southwestern Turkey, *Remote Sens. Environ.*, **7,** 51-60, 1978.

King, L., Geomorphology, in *Encyclopedia of Geomorphology,* edited by R.W. Fairbridge, p. 403, Reinhold Book Corp., 1968.

Klimaszewski, M., Detailed geomorphological maps, *ITC J.*, **1982-83,** 265-271, 1982.

Laughton, A.S., The first decade of GLORIA, *Annual Report Institute of Oceanographic SciencesÑ 1980,* pp. 60-85, Surrey, England, 1980.

Lowman, P.D., *Space Panorama,* Weltflugbild, Zurich, Switzerland, 1968.

Mabbutt, J.H. and G.A. Stewart, Application of geomorphology in resources surveys in Australia and New Guinea, *Revue de Géomorphiologie Dynamique,* **14,** pp. 97-109, 1963.

Mahrer, K.O. and C. Bradley, Magnetic terrain anomalies from arroyos in an alluvial fan, Amer, Geophys. Union, 1983 Fall Meeting, *Eos,* **64,** 684, November 8, 1983.

Merifield, P.M. and D.L. Lamar, Faults on Skylab Imagery of the Salton Trough Area, Southern California, *Technical Report 75-1,* NASA Lyndon B. Johnson Space Center, Houston, Texas, 23 p., 1975.

Miyogi, D., S.K. Sarkar and S. Mallick, Geomorphic mapping in the plains of West Bengal, India, *Selected Papers, Vol. lÑPhysical Geography,* pp. 89-94, XXI International Geographical Congress, India 1968, National Committee for Geography, Calcutta, 1970.

Mukerjee, S., Remote sensing in geomorphological mapping, in *Perspectives in Geomorphology, Vol. IIIÑApplied Geomorphology,* edited by H.S. Sharma, pp. 271-277, Concept Publishing Company Pvt. Ltd., New Delhi, India, 1982.

Nicholayev, V.A., Principles of a landscape classification, *Soviet Geography: Review and Translation*, **15**, 654-661, 1974.
Ollier, C.D., The great escarpment of eastern Australia: Tectonic and geomorphic significance, *Geol. Soc. Australia J.*, **29**, 13-23, 1982.
Pain, C.F., Mapping of landforms from Landsat imagery; An example from New South Wales, Australia, *Remote Sens. Environ.*, **17**, 55-65, 1985.
Palabekinoglu, S. ERTS-1 imagery: The valuable tool of geoscientists, in *Proc. Symp. Remote Sensing and Photointerpretation*, International Society for Photogrammetry Commission, **VII**, 597-609, 1974.
Panizza, M., Analysis and mapping of geomorphological processes in environmental management, *Geoforum*, **9**, 1-15, 1978.
Rao, K.R. and N. Bedi, Remote sensing for geomorphology, in *Perspectives in Geomorphology, Vol. IÑRecent Trends*, pp. 401-417, Concept Publishing Company Pvt. Ltd., New Delhi, India, 1982.
Rosenfield, C.L., Terrain sensitivity mapping at Resolute Bay, N.W.T., in *Association of Pacific Geographers Yearbook*, **39**, 77-87, 1976.
Savigear, R.A.G., A Technique of Morphological Mapping, *Annals AAG*, **55(3)**, 514-538, 1965.
Searle, R.C., Side-scan sonar studies of North Atlantic fracture zones, *J. Geo. Soc. London*, **136**, 283-292, 1979.
Searle, R.C. and A.S. Laughton, Fine scale sonar study of tectonics and volcanism on the Reykjanes Ridge, *Oceanologica Acta*, Actes 26e Congres International de Geologie, Colloque Geologie des Oceans, Paris, 7-17, Juil, 1980, pp. 5-13, 1981.
Slaney, V.R., Satellite imagery applied to earth science, in *Proc. Symp. Remote Sensing and Photointerpretation*, International Society for Photogrammetry Commission, **VII**, 555-57 1, 1974.
Speight, J.G., A parametric approach to landform regions, *Progress in Geomorphology*, pp. 213-229, Institute of British Geographers Special Publ. 7, 1974.
St-Onge, D., Geomorphic Maps, in *Encyclopedia of Geomorphology*, edited by R.W. Fairbridge, pp. 388-403, Reinhold Book Corp., 1968.
Strahler, A.N., Quantitative Analysis of Watershed Geomorphology, *Amer. Geophys. Union Trans.*, **38**(6), 913-920, 1957.
Thompson, R.W., Tidal flat sedimentation on the Colorado River Delta, Northwestern Gulf of California, *Geo. Socl. Amer, Mem.*, **107**, 1968.
Townshend, J.R.G., Regionalization of terrain and remotely sensed data in *Terrain Analysis and Remote Sensing*, edited by J.R.G. Townshend, pp. 109-132, George Allen and Unwin, London, 1981.
Tricart, J., Cartographic aspects of geomorphological surveys in relation to development programs, *World Cartography*, **9**, pp. 75-83, U.N. Department of Economic and Social Affairs, New York, 1969.

Ulaby, F.T. and J. McNaughton, Classification of physiography from ERTS imagery, *Photogramm. Eng. Remote Sens.*, **41,** 1019-1027, 1975.

Verstappen, H.T., *Remote Sensing in Geomorphology,* Elsevier Scientific, New York, 1977.

Verstappen, H.T. and R.A. vanZuidam, ITC system of geomorphological survey, in *ITC Textbook of Photo-Interpretation, Vol. VII: Use of Aerial Photographs,* International Institute for Aerial Survey and Earth Sciences, Delft, The Netherlands, 1968.

Verstappen, H.T. and R.A. vanZuidam, Orbital photography and the geosciences: A geomorphological example from the Central Sahara, *Geoforum,* **2,** 33-47, 1970.

Wright, R.L., Principles in a geomorphological approach to land classification, *Zeitschrift für Geomorphologie,* **16**(4), 351-373, 1972.

24

DATING ENVIRONMENTAL CHANGE

Why do We need Dating Techniques ?

The conventional lithostratigraphic dating of Geology is not sufficient for the Quaternary. Lithological dating of Quaternary events is only possible at the local scale, as deposits are not sufficiently widespread. Lithological dating also suffers from being imprecise. Comparison of faunal and floral horizons may be more precise, but inaccurate if an incorrect interpretation of a faunal assemblage has been made. Therefore, it is necessary to use other means of dating past sediments and events. The techniques used to date events and estimate rates of change are constantly being refined as greater demands are made of them. Hitherto it has been usual to obtain dates simply in order to establish relative chronologies. Now that this is largely understood Quaternary Science is becoming more concerned with the precise timing of events and the determination of rates of change. A greater level of precision and accuracy is required.

The difference between precision and accuracy is important. The simplest analogy is to compare two kinds of wristwatch. On one hand imagine you have a child's 'Mickey Mouse' watch. It has no second hand but it is set to the right time. On the other you have an expensive 'Roland Cartier' watch, with a second hand and an alarm but is set a few (unknown) minutes fast. Now imagine that at 12 O'clock you need to take your pulse. The watch you would rely upon to tell you when it was midday would be the Mickey Mouse watch. It could accurately tell you when it was midday (i.e. correctly), but not precisely (to the second). But the watch you would rely upon for taking your pulse would be the Cartier, because the second hand would allow you to precisely monitor your pulse over 15 seconds. (Nurses do not wear Mickey Mouse watches.) Thus accuracy and precision are not the same thing. It is above all important to be accurate, and then to be as precise as possible. The search for accurate dating techniques still goes on, as does the refinement of existing techniques to improve their precision.

Age Estimating Techniques

Age estimates give a real point in time BP. They are often referred to as Absolute Dates. This gives an entirely false impression of the accuracy of the dates.

Incremental Dating

This relies upon the gradual and regular additions of organic or sedimentary material to a sequence. In sedimentology such regular additions are known as **Rhythmites,** and if these can be proved to be annual, they are known as *Varves.* If a varve sequence can be found for which one of the varves can be dated, then it is possible to count back individual years, as recognised by Swedish geologist, De Geer. If a pollen sequence can also be extracted, then it will be possible to reconstruct year on year pollen diagrams. (This has not yet been done!)

Lichens grow incrementally, and a certain species, *Rhizocarpon geographicum,* grows in a circular fashion. Measurement of the diameter of the largest lichens on dated surfaces can enabie the construction of a local growth curve. One can then analyse surfaces of unknown age and measure the size of lichens in order to calculate the age of that surface. The tecnique of dating surfaces using lichens is known as *Lichenometry.*

The most important incremental method is that of *dendrochronology*. Trees produce annual growth rings, and in a stressed environment they will produce different widths of rings every year to form distinct marker bands. If we have a pattern of rings from a tree, and we know the date that the tree was felled, it is possible to correlate tree ring growth patterns back through time. In Europe analysis of contemporary and subfossil oak trees has enabled a year on year pattern of ring growth to be established that goes back nearly 9,000 years. In North America where 4,000 years old living species of the Bristlecone Pine can be found, a similar 9,000 year old record exists. The matching of 'free floating' records from Europe has enabled a tentative curve going back 13,000 calendar years to be established, but it needs refining.

Radiometric Dating

All radiometric methods rely upon the same basic principles. On formation of a particular mineral a quantity of radioactive elements will often be incorporated within it. These elements decay over time. If the proportion of radioactive element incorporated within the mineral at formation is known then the difference between the expected observation of the fresh mineral and the actual observation must be due to radioactive decay. Radioactive elements decay at an average constant rate that can be determined (the half life) and thus an estimation of the number of elapsed years can be made.

There are a number of radiometric methods, including Uranium series dating, Potassium/Argon and Argon/Argon dating, Fission track dating and Thermoluminesence dating. A new development is the use of a cosmogenic Chlorine isotope, ^{36}Cl. This technique relies upon the formation of ^{36}Cl upon exposed rock surfaces due to cosmic ray bombardment. Once a rock surface has been exposed, it will accumulate ^{36}Cl, (which has a half life of 3,00,000 years), until it is either buried or the surface is removed (by weathering or erosion). Stable surfaces potentially offer the possibility to date events throughout the Quaternary. The most important and widespread radiometric dating technique, Radiocarbon dating, is discussed in more detail below.

Radiocarbon Dating

Radiocarbon dating (^{14}C) is another radiometric dating technique similar to those described above. It was in fact the first radiometric technique to be developed. As such it has the longest history of use and is the most intensely analysed technique. It is also the most widely used due to the ubiquity of carbon within the environment.

Principles

^{14}C is an unstable heavy isotope of carbon (^{12}C). It is formed in the upper atmosphere by a radioactive np reaction, whereby ^{14}N is transformed into ^{14}C by the addition of one neutron and loss of one proton. Over time ^{14}C will decay back to ^{14}N by beta transformation. The ^{14}C produced readily combines with oxygen to form carbon dioxide and is spread throughout the atmosphere, and thence into the biosphere, geosphere and the oceans.

As ^{14}C is constantly produced, replacing that which decays, then a dynamic equilibrium of production and loss of ^{14}C in the atmosphere. As atmospheric carbon dioxide is in equilibrium with carbon dioxide in the biosphere, it follows that all living things will contain the same proportion of ^{14}C as the atmosphere, and that this will gradually decay away upon death.

During the 1950s, Professor W.F. Libby developed the technique of ^{14}C dating based upon the theory outlined above. He determined the half life of ^{14}C (557030 years, later corrected to 573040 years) and calculated the decay curve. Radiocarbon dating was developed based upon the following principles:

1. The amount of radiocarbon in an organism when alive is known.
2. The decay rate of ^{14}C is known.

3. The amount of radiocarbon remaining can be determined.

Pilcher (1993) recognised three phases in the evolution of ^{14}C dating, namely the pioneer phase, the routine phase and the questioning phase. As a result of the large number of dating exercises undertaken during phase 2, discrepancies began to appear. The present questioning phase is engaged in assessing the validity of the principles outlined above. In many cases the principles have proven to be invalid, although not beyond redemption.

The main problem is the assumption that living organisms have a ^{14}C content in equilibrium with the atmosphere. This has been shown to be false. In some cases, the disequilibrium is due to the presence of contaminants (sources of 'young' or 'old' carbon), in others it is due to isotopic fractionation, as organisms preferably absorb ^{12}C rather than ^{14}C. Fortunately, this can be monitored through the analysis of another heavy (stable) isotope, ^{13}C. Another big problem is the assumption of the constant rate of production of ^{14}C. It is now known that ^{14}C production has varied over time, and that far from a smooth decay curve, there is in fact a ^{14}C 'wiggle' which can seriously hamper precise absolute dating (dating in terms of calendar years). Techniques have been developed at Belfast, a very accurate facility and one of only two or three reliably accurate laboratories in the world, which enable 'wiggle' matching, a means by which calibration of radiocarbon dates with calendrical dates is feasible.

The lecture will consider the problems associated with ^{14}C dating, and use a case study from Llanilid, South Wales, to exemplify how careful analysis can produce remarkably precise dates.

Age Equivalent Methods

Age equivalence establishes the age of something with relation to other events that were contemporaneous, and can be widespread (describing a building as Georgian in age is an equivalence method). Relative age (such as lithostratigraphy) puts events in the correct order, and may establish rates of change relative to that scale, but needs other techniques to establish real dates. Oxygen isotope analysis is such a method. Age equivalence and relative age dating are, therefore, comparative dating methods. Other dating techniques are required in association with them in order that they provide true age estimates.

The most important age equivalence method is undoubtedly *oxygen isotope dating*. With this method of dating environmental events can be dated with respect to the relevant oxygen isotope stage, and as the curve is independently dated it provides a method of direct dating. Many of the other dating methods referred to in today's lecture are correlated with the oxygen isotope curve. Another important age equivalence method is *tephrochronology*.

Palaeomagnetism provides us with a further method of age equivalence dating. The earth's magnetic field is not constant, but fluctuating, and indeed occasionally reverses. Magnetised sediments and volcanic lavas will align themselves according to the contemporary magnetic field, and become 'fossilised', recording that past magnetic pattern. The pattern of change over time is known, and therefore, one can determine a magnetostratigraphy. At present this is a somewhat imprecise method of dating but it is a useful means of separating long sequences of sediment. It has the further added advantage that it is a global technique, independent of local factors.

Relative Age Methods

There are a number of relative age methods, but by far the most important is *amino-acid dating*. Amino-acids are the building blocks of proteins. All but one amino-acid exist as two non-superimposable mirror images. These *optical isomers* are referred to as D- and L-amino-acids. The L-isomers are only found in living organisms, and upon death they interconvert with D-isomers (*racemisation*) until equilibrium is reached. This process is very rapid in exposed locations, but if it is in a closed system (such as a mollusc shell) that undergoes rapid diagenesis, then the racemisation process can be very slow. Analysis of the ratio of D-isomer to L-isomer can enable relative dates to be obtained.

The most commonly used amino-acid is *L-isoleucine*. This interconverts with its *diastereoisomer—D-alloisoleucine*. This is done by a process known as *epimerisation*. Amino-acid dating is extremely useful as it is a method that can date the whole of the Quaternary, from 2.4Ma to (and including) the Holocene. The only drawback is that it requires an independent date to calibrate it. This has been done successfully in Britain (Bowen *et al.* 1985, Bowen *et al.* 1989, Miller *et al.* 1979).

A note of caution has to be mentioned, however, as the rate of epimerisation is temperature controlled, and therefore, ages from one climatic region to another are not directly comparable. Indeed a recent study (Sejrup and Haugen, 1994) has shown that intra shell epimerisation rates can differ, and therefore, great care is required during the sampling stage.

There are other, relatively uncommon methods of relative dating. Analysis of the Fluorine or Uranium content of fossil bones (Fluorine and Uranium can be absorbed from ground-water by bones over time) can help distinguish between contemporary and non-synchronous deposits, whilst analysis of the nitrogen content of bones can lead to the same result, albeit that nitrogen concentrations decrease with time as collagen decays.

Obsidian hydration is a technique that is widely used in the Americas. The technique establishes a curve for the rate of weathering of fresh obsidian surfaces. The creation of fresh surfaces was very common, for obsidian was the most important source lithology for the manufacture of stone tools in many native meso-American cultures. Other weathering techniques have also been developed. In the U.S.A., relative dates for glaciations in Colorado have been established by analysis of the degree of weathering of boulders in tills and through analysis of soil development, although this is questionable.

25

CAUSES OF CLIMATIC CHANGE

Introduction

There is no single accepted cause of climatic change, at whatever timescale is chosen. For any theory proposed, there is an argument contrary to the idea. A number of theories carry a lot of support and seem to have the weight of evidence behind them, but no single theory seems to be able to explain the increasingly well known phenomena associated with climatic change. The chapter will outline and consider the main theories of climatic change that carry the greatest support in contemporary science, but that does not mean that the theories are correct, nor uncontested.

What is Climatic Change?

This is not as easy a question to answer as it might first appear. Climate is defined as the average weather for a given area over a certain time period. Therefore, any change in certain climatic variables that is statistically abnormal for a given area can be viewed as a change in climate. There are a number of problems with this. Firstly, how long does the perceived change in certain variables have to persist before the change is considered to be real? How representative of the true 'average' weather of an area is the statistical base from which we are working? For example, most weather data are only 300 years old at best. If we are looking for signals of Holocene climate change, how representative of the Holocene (10,000 years) are our 300-year records? It may be that an apparent climate change, which is perceived as unusual, is in fact a return to normal conditions.

Most of the foregoing is just the hypothetical meanderings and ramblings of my mind. But they do illustrate part of the point. A further point about climate change is that it is often discussed on a global scale. To discuss world climate is to suggest 'average' weather for the globe. This is patently silly, but we do not have a better alternative. Consequently, discussions of global climatic change can only focus upon one or two climatic variables, and that at a rather imprecise level. The one advantage we have is that examination of past climatic change relies upon proxy data, which cannot provide a greater level of precision than analytical processes can allow, and the proxy data is restricted to relatively few climatic variables. Therefore, our data and theories can be analysed at the same level, which is nice.

What are the Climatic Changes that have Occurred in Recent Earth History?

Lectures and seminars have illustrated that global climates have changed enormously during the Quaternary. The methods by which the patterns of change are discerned have been outlined to you. The simple pattern of climatic change is one of a gradually cooling Earth since the early Tertiary (Late Cretaceous?), and a rapid and massive cooling during the Quaternary period. The Quaternary is synonymous with the Ice Age in common understanding, yet the reality of the Quaternary is much more complex.

The simple picture of the Quaternary is one of Ice Ages (glacial episodes) separated by interglacial episodes. Recent discoveries have shown that the picture of climate change is much more complex than this. Cold periods (glacials) have not been uniform in character throughout the Quaternary, and we know that climate alters massively during cold periods. Interglacial periods may be equally variable in terms of climate, although the evidence suggests that the degree of variability may itself vary with each successive interglacial episode. For example, the last interglacial period in Western Europe (the 'Ipswichian' in Britain, 'Eemian' in Europe, alternatively Oxygen Isotope stage 5) seems to have been punctuated by regular rapid periods of cooling and subsequent warming; whereas the current interglacial (the Holocene) has been (relatively) stable, in terms of temperature at least.

Therefore, climatic change has to be viewed at a number of different scales. For example, the explanation for general global climatic cooling cannot also be the explanation for the appearance and disappearance of major ice sheets. It is likely that a number of different factors have come together to create a number of different effects, and the importance of each factor varies with time and place. Consequently, the factors believed to be responsible for general cooling, the factors believed responsible for 'Ice Ages' themselves and the factors responsible for shorter term climatic fluctuations will have to be considered separately.

The Causes of Post-Cretaceous Climate Cooling

Undoubtedly, a major factor in post-Cretaceous climatic change has been the pole-ward drift of large landmasses. The proximity of both the Eurasian and North American landmasses to the North Pole has contributed to global cooling. The net radiation loss from these large landmasses has contributed towards a cooling Earth, but cannot alone explain global cooling, since the land mass of Antarctica has been located at the south pole since the early Cretaceous, yet the east Antarctic ice sheet only began to accumulate during the middle to late Eocene, and was not fully formed until the middle Miocene (Hansom and Gordon, 1998). The west Antarctic ice sheet did not fully form until the Pliocene.

Two other important factors are believed to have contributed towards global cooling, and both are related to tectonic and orogenic effects. The closure of the Tethys Ocean and the straits of Panama radically altered global oceanic circulation. The separation of the Atlantic and Pacific oceans had a significant impact upon the redistribution of heat around the earth. At the same time, Antarctica became separated from South America. This led to the cooling of Antarctica and the build up of ice sheets. In the northern hemisphere, continental collision led to the uplift of landmasses into the critical mid-latitude circulation regions. The uplift was to such an elevation that these areas became zones of net radiation loss and contributed towards global cooling. The most significant area of uplift was the Tibetan plateau.

Ice Ages: The Milankovitch Theory

Milutin Milankovitch was a Serbian astronomer who calculated the periodicity of certain of the Earth's orbital tracks around the sun, and the effect of these changing orbits upon the amount of solar radiation received in the northern hemisphere. The consequence of his studies, which took many years, was a recognition that three critical cycles of insolation minima occurred whose combined periodical effect would be to lower the latitude of glaciation at sea level. In essence, the southern limit of glaciation in Europe would move from central Norway to Scotland. The three critical cycles were 100k (orbital eccentricity), 41k (obliquity of ecliptic) and 19-21k (precession of equinoxes) cycles.

At the time that the theory was first developed, it was generally accepted that there had been four glacial episodes during the Quaternary. The Milankovitch theory suggested that there could have been at least 24 periods of ice expansion during the 2.4Ma of the Quaternary, if not more. Therefore, at the time, the theory was not taken very seriously. But as evidence emerged, particularly as a result of the search for North Sea oil and gas, for many glaciations during the Quaternary, the theory became widely accepted during the 1970s. It is now believed that Milankovitch orbital forcing of glaciations is a major cause of the rapid and extreme changes from relatively-warm to cool or cold conditions on planet Earth. There are a number of issues that cannot be explained by Milankovitch forcing alone, and thus there must be other factors that affect the timing and number of glaciations during the Quaternary period.

The first problem derives from the theory itself. The Milankovitch theory recognises that changes in northern hemisphere insolation will effectively depress the snowline across the northern hemisphere, and permit the growth, for example, of glaciers in Scotland. What the theory does not suggest is why a whole ice sheet should develop across the British Isles. In essence, the Earth cools a lot more than the theory would suggest. If reduced solar input cannot explain the degree of cooling, it suggests that the Earth's climate system has some internal amplifier, probably through complex positive feedback mechanisms, which exaggerates the effect of reduced solar insolation in the northern hemisphere.

The second problem concerns the mid-Pleistocene change from glaciations every 41,000 years or so to glaciations every 1,00,000 years. The periodicities of glaciations are still consistent with Milankovitch, but the frequency of glaciation has apparently reduced. It seems likely that the intensity of cooling has become greater. The last four glacial episodes seem to have been much more intense, and of greater longevity, than the previous glacial episodes. In order to address the problem, we need to suggest why ice sheets collapse, and why the change from 41,000 to 1,00,000 year cycles might occur.

Although Milankovitch forcing undoubtedly accounts for part of the reason why northern hemisphere ice sheets collapse; and other positive feedback mechanisms associated with the release of greenhouse gases from a warming planet may also partially explain ice sheet collapse; we have evidence that during any glacial cycle the northern hemisphere ice sheets expand and contract many times (see next section). Why, therefore, should ice sheets ultimately collapse after a 1,00,000 years? Furthermore, why did ice sheets collapse after only 41,000 years before the mid-Pleistocene? The answer may lie within the sub-glacial dynamics of glaciated areas.

There is evidence that the extent of glaciation prior to the mid-Pleistocene was no less than after, but that after the mid-

Pleistocene ice sheets were much thicker. It has been suggested that early Pleistocene ice sheets grew upon, and moved over, a thick deformable layer of Tertiary weathering regolith. The deforming bed material allowed ice sheets to spread faster, but were consequently thinner. After 41,000 years orbital forcing would lead to the rapid decay of these thin ice sheets. Later ice sheets grew upon solid, non-deformable beds that had been stripped of their surface cover by the preceding glaciations. These ice sheets were not able to spread as quickly to the limits of glaciation, and could only do so once a sufficient thickness of overlying ice created sufficient pressure for movement by deformation or sliding over a solid, perhaps frozen, bed. The thickness of the more recent ice sheets also permitted them to survive the 41,000-year cause of ice sheet collapse, but may ultimately have led to the 1,00,000-year collapse. It is thought that at their maxima, the northern hemisphere ice sheets would cause such great crustal subsidence that the ice sheets, particularly at their margins, would become depressed well below regional snow lines. The resulting rapid ablation would exceed regional crustal rebound rates and the ice sheets would enter a negative feedback phase of warm local temperatures and encroaching sea levels coincident with the 1,00,000-year orbital forcing parameter.

Sub-Milankovitch Cycles

The explanation of the change in ice sheet behaviour following the mid-Pleistocene demonstrates that Milankovitch style orbital factors are not the only factors governing the behaviour of ice sheets. Analysis of ice and ocean sediment cores spanning the last glacial episode have revealed that there are significant, cyclical changes in both climate and ice sheet behaviour that are unrelated to Milankovitch cycles. Evidence from ice cores shows that, even during the last interglacial period, there were periods of slow cooling followed by relatively short periods of rapid warming. The warm periods tend to be relatively short-lived, before another period of slow cooling ensues. A period of slow cooling followed by rapid warming is known as a *Dansgaard-Oeschger* event. These events characterise the end of the last interglacial period in the northern hemisphere, and become accentuated during the last glacial period. The duration of such events is on the millennial scale, and is therefore not related to the earth's orbital variations around the sun.

During the last cold period repeated cycles of Dansgaard-Oeschger events are evident from oxygen-isotope records. Each successive event results in warming that does not return to the greatest warmth of the previous cycle, and subsequently an even greater level of cooling than the previous event. After a number of cooling/warming episodes there is a terminal dramatic cooling, before a major warming event. Such a cycle of warming and cooling is known as a *Bond Cycle.* The termination of a Bond cycle in the oxygen isotope record is mirrored in North Atlantic marine sediment cores. Bond cycle end points are noted by the interruption of normal microfossil rich sedimentation with clastic sedimentation. The layers of clastic sediment that mark the terminus of a Bond cycle are believed to consist of Ice Rafted Debris (IRD) and are known as Heinrich layers. The terminal phase of a Bond cycle is, therefore, known as a *Heinrich* event.

The recognition of Heinrich events in the late 1980s and 1990s was an important breakthrough in our understanding of the Earth's climate system. Heinrich events are relatively short-lived events, but they have a profound impact upon the climate of Europe, in particular. The causes of Heinrich events are not known for sure, but it seems that northern hemisphere ice sheets expand to a critical size and then their oceanic margins break up rapidly in an iceberg super-discharge episode. The 'armadas' of icebergs, which enter the North Atlantic, have a cooling effect upon the sea, and also a diluting effect. The input of large volumes of freshwater from the icebergs themselves, and later from the rapidly retreating ice shelves (sheets) causes the dilution of salt water in the North Atlantic that has a further dramatic consequence. The diluted waters can no longer sink as density currents and form southward flowing North Atlantic Deep Water (NADW). NADW is important on a global scale, as it is the driving force behind the Atlantic conveyor, the current that brings warm tropical water into the north-east Atlantic and keeps Europe unusually warm. Heinrich events act as a 'switch' to the Atlantic conveyor, and cause rapid climate changes of the order of 5-10°C within perhaps a decade. Both cooling and warming are equally rapid.

When they were first recognised it was at first believed that IRDs were derived solely from the enormous Laurentide ice sheet (I emphasise the size of this ice sheet, it was as big as the current Antarctic ice sheet before it decayed. Think about it.) Later analysis revealed that elements of IRDs were also derived from the Greenland, British and Scandinavian ice sheets and that the growth and retreat of these ice-sheets was synchronous. More recent research suggests that IRDs derived from the different Atlantic ice sheets may not be truly synchronous, but they are closely related in time. In any event, the short time span of a Heinrich event makes the behavioural synchronicity of circum-Atlantic ice sheets quite remarkable. This says something about global climatic forcing at a sub-Milankovitch scale, but we are not quite sure what that is, yet. It is possible that the Laurentide ice sheet exerts a strong controlling influence over the other Atlantic ice sheets, but as yet this is unknown. Suffice to say that iceberg events have continued, at a much-reduced scale, throughout the Holocene on something like a 1500-year cycle.

The fact that Heinrich-type events appear to continue may provide a clue as to the origin of Bond cycles. During a cold period, in which the overall trend of cooling on a 1,00,000-year cycle is dominant, occasional warming episodes probably occur on a short 1500-year cycle (or thereabouts). The cooling episode is probably accompanied by ice sheet growth, and the effect of the Atlantic conveyor is weak. (Due to cold winds blowing from the expanding ice sheets). A sudden warming marks the end of a cooling episode, as the conveyor is switched back on.

We can consider gradual cooling as a time period when the conveyor slowly recovers its potential. Once full potential is achieved and the Atlantic conveyor is strong, a brief warm episode ensues. Since the overall (Milankovitch forced) climatic trend is towards cooling, the return to warmth is insufficient to melt ice-sheets, and, through increased precipitation from a warmer Atlantic, possibly results in even greater ice accumulation. With greater ice accumulation, input of more freshwater into the north Atlantic is likely and the conveyor is once more diminished.

This cycle repeats itself until the circum-Atlantic ice sheets reach a critical threshold, and break up, discharging such vast quantities of frozen fresh water that the Atlantic conveyor is completely switched off, plunging the northern hemisphere into its most intense cold periods, and the globe into great aridity. At this point ice sheets may retreat, sometimes considerably if the retreat is coincident with a 41,000-year Milankovitch cycle, but will re-advance should astronomical conditions allow. The ultimate disappearance of the northern hemisphere ice sheets (all bar Greenland) is a final submission to orbital forcing.

A critical question is this: what about Dansgaard-Oeschger events during the last interglacial period? Could they happen again? It may be that the current interglacial period is affected by an even longer pattern of solar forcing which has not been seen for 8,00,000 years, and that the climatic swings apparent during the last interglacial are unlikely this time around. What we do know is that maximum temperatures during the last interglacial period were higher than the current interglacial, and that certain Greenhouse gases, principally CO_2, were more abundant. Quite what the effect of the anthropogenic rise in greenhouse gases and potential global warming will have upon the delicate balance of fresh and salt water in the North Atlantic is not known, and maybe we would not want to find out!

We will consider future climatic change and predictions that arise from the models Global Climatic Change. One thing you might bear in mind, is the extent to which studies of the past provide us with realistic models of global climate change. The study of past climate and environment change has enormously enlightened our knowledge of the Earth's environment systems, particularly their sensitivity to climate change.

REFERENCES

Hansom, J.D. and Gordon, J.E. 1998. *Antarctic Environments and Resources.*

Lowe, J.J. and Walker, M.J.C. 1997 (2nd edn). *Reconstructing Quaternary Environments.* Longman, London.

Mannion, A.M. 1999. *Natural Environmental Change.* Routledge, London.

Slaymaker, O. and Spencer, T. 1998. *Physical Geography and Global Environmental Change.* Longman, Harlow.

26

FUTURE CLIMATE CHANGE AND GLOBAL CLIMATE MODELS

Understanding Climate Change

Predicting future climate change is a minefield. Correlating observed effects with postulated causes is the only means we have at our disposal yet this leads to a qualitative understanding that is scientifically flawed. Phrases such as 'statistically abnormal' are constantly used—what timescale do they refer to? For some cycles of 'natural' variation we simply do not know (a) what is normal and (b) whether there is anything that we can call 'normal'? The problem is that the past is all that we have to go on—we don't know many things. These will become apparent.

Simulation and Sensitivity Models

Most climate models are ambitious. They are designed to simulate past climates and to predict likely future climatic scenarios. Most models require quantitative definitions of components of the climate system, e.g. parameters such as: solar radiation, solar insolation, extent of ice sheets, deep and surface ocean temperatures, land albedo, cloud cover, CO_2 and water vapour content of atmosphere. The processes involved are represented in mathematical terms but the complexity of the system means that the processes get to be hugely complex themselves.

Practicalities

In order to represent processes operating, simplifying assumptions have to be made. Physical qualities that vary continuously in 3D are represented by their values at a finite number of points arranged in a 3D grid. The spacing between grid is referred to as the spatial resolution—the finer the resolution the more points that exist and the more calculations that have to be done. Models are limited by the computing resources available. The formulation of the effect of small scale process on a large scale is adopted—parameterisation—all climate models use this. Typically 100s of km in the horizontal are used. More important elements are at smaller scales than this (e.g. clouds, land, surface variations). Detailed models at high resolutions are available for the processes themselves but they are too complex.

The Models

(i) One-Dimensional Radiative-Connective Atmosphere

Globally (horizontally) averaged, but contain many layers in the atmosphere. Detailed in the transfer of solar and infra-red radiation. Useful in modelling the radiative perturbations associated with the changes in atmosphere's composition.

(ii) One-Dimensional Upwelling-Diffusion Ocean Models

Here the atmosphere is modelled as a single well-mixed box that exchanges heat with underlying ocean and land surface. Deep sea ocean water mixing is treated separately. Sinking and compensating upwelling represent global scale advective overturning (thermohaline circulation).

(iii) One-Dimensional Energy Balance Models

The only dimension represented is variation with latitude; atmosphere is averaged vertically and E-W. Multiple processes of N-S heat transport by atmosphere and oceans are usually represented as diffusion. Such models are useful for modelling the interaction of heat transport feedbacks.

(iv) Two-Dimensional Atmosphere and Ocean Models

2D models that permit more physically-based computation of horizontal heat transport than 1D.

(v) Three-Dimensional Atmosphere and Ocean General Circulation Models

These models are the most complex. The models divide atmosphere or ocean into a horizontal grid with a typical resolution of 2-4° latitude by 2-4° longitude and 10-20 layers in the vertical. They directly simulate winds, ocean currents and many other processes. Feedback processes are simulated in the coupled atmosphere and ocean GCMs—water vapour, clouds, seasonal snow and ice.

There are also:

- Models of the oceanic part of the carbon cycle.
- Models of atmospheric chemistry and aerosols.
- Models of ice sheets—high resolution (20 × 20 grid) 2D and 3D models of polar ice sheets—to assess the impact on global mean sea level of various idealised scenarios for temperature and precipitation changes over ice sheets.

Utilisation of Simple and Complex Models

Simple models allow the investigation of basic relationships between components of the climate system, the factors driving climate change and the overall response of the system. They are inexpensive to run. However, they represent only the most critical processes. 3D models are the only prospect for reliably simulating key processes that determine climate sensitivity and longer term feedback involving the terrestrial and marine biosphere. But these models are costly, difficult to understand, and require high-resolution data inputs which in some cases do not exist.

The processes are represented in mathematical terms but the complexity of the system means that they get to be hugely complex themselves. The complexity leads to ditractions and ambiguities that are termed as 'noise'.

Problems

Apart from the glacial maximum 18,000 BP there is not high enough resolution palaeoclimatic data available on which to base simulations of past global climate systems. Many models, therefore, use simplified versions of the climate system to conduct sensitivity experiments rather than try to produce the minutest detail of past global climate episodes. In other words, the models are used to explore the sensitivity of the climate system to changes in specific forcing factors (e.g changes in the proportion of GHGs or distribution of land and sea). Sensitivity results from the feedback processes that amplify the effect of the original forcing mechanism. Difficulties arise in that feedback processes operate at different time-scales—they are dependent on the rate of response to change. The more that nature is simplified, the faster the models can run but there is a fine line between simplification and models being so unrepresentative that they are of no use.

The Current Climate Picture

Once there are models that satisfactorily simulate observations of the behaviour of the atmosphere, oceans and cryosphere, the next step is coupling them together to refine simulations of possible past climatic conditions and to predict more confidently future climate change. New computer models are for ever being invented. If the programmes can successfully match the waxing and waning of temperatures over the last 1000 years, then there is a greater chance of predicting the future accurately.

Earleir in the course TH pointed out some of the proxy data sources: tree rings, ocean cores, pollen, beetles, ice cores etc.

They help us to model by proxy what the climate has been doing.

The Record: The Last 1000 Years

- First 5 centuries of last millennium were mild—medieval warm period extended certainly from 900-1200 AD.
- From 1550-1850—LIA (increased vulcanism?)
- 1601—coldest summer with frosts in June (historic records)
- Coolest—16th, 17th, 18th centuries (1500s, 1600s, 1700s)

All of the mild centuries prior to the LIA were cooler than the 20th. People have used this as an argument for global warming—

- The global mean surface air temperature record has been broken 4 times since 1990—1990, 1995, 1997, 1998.
- The hottest years in the modern global temperature record have all been in the 1990s—1990, 1995, 1997, 1998, 1999.
- 1998 hottest of them all, notably hotter than 1106 which was a record summer—What then might the future hold?

We have looked at climate change over the Quaternary and its impact on landscape development in the last 7 weeks. Dating gives us the power to put data into models to predict. What are the likely results?

Predictions for the Future

Background

The Intergovernmental Panel on Climate Change (IPCC) has estimated that over the next century human activities will do much to alter climate, especially increasing CO_2. By AD 2050, average surface temperatures may have increased by 1.5–4.5°C.

During the last 1000 years average global temperatures have not varied by more than 1°C. Concentrations of CO_2 are rising faster in the Arctic than S latitudes. Thinning of ozone over Arctic regions promotes more solar radiation to enter the atmosphere. Northern latitudes are a major source of natural GHGs world's largest area of wetland generating and releasing methane.

Issues

Ice Sheets, Glaciers and Snow Cover: Greenland

Little change in C20th. Response time of ice sheets is 1000s of years. Likely that the Greenland ice sheet (GIS) is not in equilibrium with present climate. Complete melting of Greenland ice sheet could be triggered with as little as x 2 CO_2 cannot tell whether the mass balance has increased or decreased.

Greenland ice sheet is unusual as it cannot advance much because of a deep continental shelf which destabilises it and causes melting and calving. So even with advance likely from short-term increased snowfall, it is unlikely to grow hugely. During the last interglacial (Ipswichian) 1,30,000 years BP it may have almost disappeared.

Complete melting would increase sea level by 7m over a period of 1000-5000 years.

Likely to get increased icebergs from fast-moving tidewater glaciers (increased melt rates at ice margins and increased snowfall).

Antarctica

The Antarctic Ice Sheet (AIS) is 7 x the size of the Greenland Ice Sheet and and has 10 x the volume of ice.

3 uncertainties are associated with the AIS:

(a) Surging from the West Antarctic Ice Sheet (WAIS)—new evidence seems to indicate that surging is not affected by global warming.

(b) Ice sheet mass balance may become more positive with a temperature rise, on the grounds that their temperature rise will still be inadequate to generate ice melt.

(c) The stability of the WAIS. This ice sheet is grounded below sea level and held back by ice shelves that could rapidly disintegrate because of global warming.

EAIS melt would need a regional warming of over 20°C. Unlikely except in case of very high GHG emissions.

Sea Level

There is likely to be a world-wide increase of 0.5m in global sea levels (IPCC) by the year 2100. Estimates vary from 20-86cm.

N. Canada, Alaska and Greenland have a relative sea level which is still comparatively mobile, associated with isostatic rebound after glaciation, and coastal changes associated with the melting of ground ice. Coastal erosion, flooding and increase in marine sedimentation would be prevalent.

N.B. Sea level response to climate change involves very long time scales and entails commitments that will endure for centuries after consumption of fossil fuels has ended. SLR by end of 21st century will only form a small per cent of final sea level rise, even if GHGs are stabilised by early part of 21st century.

Permafrost

The Northward migration of the southern limits of permafrost could have enormous consequences; increased erosion, subsidence and slope instability being the main problems. Melting could release enormous pockets of methane—a potent GHG. When ground ice melts it will not be replaced for millennia even if the climate cools again. The response of permafrost landscape to warming will be dependent on the spatial distribution of ground ice.

Sea Ice

Atmospheric conditions, ocean temperatures and salinity determine sea ice distribution. Thinner winter ice and a 20-day thaw advance are conservative estimates. It is possible that the Arctic Ocean's ice cover (1.5m thick on average) could disappear. The resultant change in net global albido could increase the source of atmospheric moisture from the free ocean and increase cloudiness and precipitation. This could alter the environment of the high Arctic islands, many of which are polar deserts.

Impacts

Modelled future climate changes have predicted:

- Autumn and winter temperatures may rise as much as 10°C, summer temperatures showing a more modest 2 or 3°C.
- Displacement of storm tracks to the N will increase Arctic precipitation especially in the autumn and winter.
- Snow seasons will be shorter, snow accumulations will be greater but then so will spring runoff and floods.
- Ice seasons will be shorter. Ice thicknesses will decrease possibly resulting in ice-free Arctic summers. In winter, the S limits of sea ice will retreat N.
- Slow permafrost decay will result in increased land instability in some areas. Limits of permafrost will retreat N.
- N hemisphere temperatures will rise. Longer growing seasons will result with accompanying increased summer heat.
- Warmer winters will aid marine transport and offshore exploration. The absence of sea ice will result in more severe storms; consequently there will be increased snow in accumulation areas and possibly increased glacial surging and the production of icebergs in N Arctic.

- Water resources—average annual precipitation increase by 20-30 per cent in the N.
- Spring runoff would be increased along with average river flow.
- Water management programmes and dam construction/reservoir development are designed to last for next 50-100 years (they use past flow and climate data).
- Arctic heating costs will decrease.
- Treelines will migrate N by 100km for each °C rise in temperature.
- Changes in landscape.

Despite all of the above, it is worth bearing in mind that climate modellers sometimes use statistics like a drunk uses a lamp-post—for support, rather than illumination.

Changes in landscape? Many and varied...

REFERENCES

Alley, R.B. and Whillans, I.M. 1991. 'Changes in the West Antarctic ice sheet'. *Science* **254,** 959-963.

Barnola, J.M., Raynaud, D., Korotkevich, Y.S. and Lorius, C. 1987. 'Vostok ice core provides 160,000 year record of atmospheric CO_2'. *Nature* **329,** 408-414.

Battle, M., Bender, M., Sowers, T., Tans, P.P., Butler, J.H., Elkins, J.W., Ellis, J.T., Conway, T., Zhang, N., Lang, P. and Clark, A.D. 1996. 'Atmospheric gas concentrations over the past century measured in air from firn at the South Pole.' *Nature,* **383,** 231-235.

Bentley, C.R. 1997. 'West Antarctic Ice Sheet collapse'. *Science,* **276,** 663-664.

Bentley, C.R. 1998. 'Rapid sea-level rise from a west Antarctic ice-sheet collapse: a short-term perspective'. *Journal of Glaciology* **44,** 157-163.

Briffa, K.R., Jones, P.D., Schweingruber, F.H., Shiyator, S.G. and Cook, E.R. 1995. 'Unusual twentieth-century summer warmth in a 1,000-year temperature record from Siberia'. *Nature* **376,** 156-159.

Broecker, W. 1997. 'Thermohaline circulation, the Achilles of our climate system: Will man-made CO_2 upset?

27

MAPPING

The study of landforms, their structure and development, includes the need to illustrate both the findings of an investigation and the character of the landforms investigated. Geomorphologists have used a variety of methods of illustration, including sketches, block diagrams, and various types of photography and other imagery, both from the ground and from the air, to show features of the Earth's land surface. The geomorphological map, in its various forms, represents recent efforts of many geomorphologists, including most European practitioners, to develop a method of graphical display of the features of the Earth's physical surface.

Throughout the 19th century and into the early years of the 20th century, the study of landforms was dominated by a static descriptive physiography. Some investigators in Europe and the United States recognized the influence of dynamic forces on the landscape. In Switzerland, Agassiz saw the erosive power of ice in the Alps. John Wesley Powell recognized the force of water in the carving of the Grand Canyon. Ferdinand Hayden noted the presence of volcanism and the landscape-shaping power of endogenic force beneath the geysers and boiling mud pots of the Yellowstone volcanic region. G.K. Gilbert posed the idea of a landscape in dynamic equilibrium in his studies in the Henry Mountains in Utah. In 1899, with the publication of the Geographical Cycle, William Morris Davis introduced into the stagnant physiography of his time what King (1968) calls "a semblance of dynamism." Davis' concept of the "cycle of erosion" produced a radical change in geomorphology. It suggested forcefully that the landscape was dyamic and constantly evolving through a cycle of change due to the external forces of water and the atmosphere. Although his ideas have been extensively criticized and to some extent discredited in their original form, Davis' dynamic approach to the landscape has been indelibly stamped on geomorphology. A similar dynamism was introduced into French geomorphology at about the same time through the development of the "Geologie Dynamique." However, in practice, most geomorphological research continued to be dominated by a static descriptive physiography in which the landscape was described in writing, generally accompanied by artistic block diagrams drawn to illustrate the author's conclusions. Although these diagrams were often excellent illustrations of geomorphological processes, they tended to be idealized qualitative portraits rather than quantitatively verifiable graphic analyses of the landscape.

By the 1920s, photography had developed to a fairly high level of sophistication and should have been useful to geomorphologists. It was generally ignored in favour of block diagrams or given a very low rating as a research tool. There were a few early users of photography in landform study. As early as 1820, the Frenchman, J. Arago, suggested that photography could be useful in topographic mapping. The early photographers and balloonists, Nadar and Triboulet, experimented with aerial photography and the view it provided of the landscape. In 1899, Albert Heim published his photographs and observations made during a balloon flight over the Alps; he is probably the first person to use aerial photography in geomorphological research. During the First World War, aerial photographs were used to investigate terrain and map battlefield positions.

Second World War represented a fundamental divide for geomorphology both theoretically and technically. Great advances were made, technologically, in the use of aerial photographs and airphoto interpretation for the terrain analysis necessary for the highly mobile style of instrumentation for interpretation became available, increasing the ability to study landform features by means of aerial photographs. Quantitative analytical techniques, developed in other sciences, were applied to geomorphic research (e.g., Strahler, 1957). Geomorphologists, particularly in Europe, became interested in comprehensive regional analysis of landforms that considered together all the aspects and features of the landscape. Investigators realized that processes in landform development were more complex than the relatively simple Davisian cycle of uplift, downwearing to a peneplain, and rejuvenation. Questions were asked about the nature of past landform processes and their relationship to present processes. How could various landscapes be compared and contrasted with one another, and what were the significances of landforms and the influence of relief on such things as vegetation, hydrology, and the human cultural development of the area? From these and other questions arose the need for a new paradigm in geomorphology. Because the landscape was complex, in order to pursue orderly scientific research, geomorphologists sought an objective scientific method of graphic portrayal of the complex factors of landforms. The qualitative descriptions and artistic diagrams of the physiographers did not provide the information necessary for detailed and accurate analytical studies of landforms. As Klimaszewski (1982) has said, "It was only after Second World War that the preparation of detailed geomorphological maps on the basis of a systematic mapping of landforms was advanced as a need or even a necessity." In the 1950s and 1960s, the science of geomorphology became, in the words of Fairbridge (1968), "the analytic physiography of the Earth's surface", and the detailed geomorphological map became, in many countries, "the main research method in geomorphology" (Demek, 1982). Research in modern analytic geomorphology has developed around five fundamental landform concepts:

1. *Morphology*: the appearance, shape, etc. of the landscape. (In Europe, the term "morphography" is often used).
2. *Morphometry*: the measurements, dimensions, and slope values of landforms.

3. *Morphogenesis*: the origin of each landform.
4. *Morphochronology*: the age of each landform.
5. *Morphodynamics*: the land-forming processes presently active on the landscape or those that may become active in the future.

Some of the earliest detailed geomorphological maps were published by Siegfried Passarge in 1914 in his Morphological Atlas. From that time to the end of Second World War, a few detailed maps of local regions were published by some European geomorphologists and detailed geomorphological surveys were occasionally made, but it was not until that 18th Congress of the International Geographical Union (IGU) in Rio de Janeiro in 1956 that the importance of detailed geomorphological maps received international acknowledgement. Two years later, at the IGU Congress in Stockholm, the Subcommission on Geomorphological Mapping was created and was given three tasks:

1. To introduce and develop the methodology of geomorphological mapping.
2. To adopt a uniform system for geomorphological mapping to ensure compatibility.
3. To demonstrate the applications of geomorphological mapping in local and regional economic planning in order to facilitate a rational utilization of the Earth's surface (Klimaszewski, 1982).

The Subcommission on Geomorphological Mapping met regularly through the 1960s. In 1968, at the IGU Congress in New Delhi, India, it was upgraded to The Commission of Geomorphic Survey and Mapping and was charged with the responsibility of developing a Manual of Detailed Geomorphological Mapping and devising the legend for an International Geomorphological Map of Europe on a scale of 1:2500000. The legend, produced in collaboration with many geomorphologists from European countries, was published in 1971 (Bashenina *et al.*, 1971). The Manual, a compilation of articles by 20 geomorphologists, was published in 1972 (Demek, 1972). The first of the 16 projected maps and accompanying texts was published in 1984 (Embleton, 1984).

Geomorphological Mapping Theory: Development and Diversity

In different countries, the development of geomorphological mapping has followed different courses, partially due to the particular interests and emphases of geomorphologists and partially due to real or perceived differences in the landforms found in various regional settings. To date, most detailed geomorphological mapping has been done in Europe by European geomorphologists. In general, American practitioners have been primarily interested in special feature studies or the mapping of special factors of the landscape rather than the development of comprehensive studies. This low degree of interest in comprehensive studies may be due, in part, to a reduced interest in geography and its integrative emphasis in the United States. Some European geomorphologists, including French, Czechoslovakian, and Hungarian workers, have selected the lithological-structural unit as the basic element in landform analysis. Others, such as Polish, Russian, Rumanian, and German geomorphologists, consider the form as the basic unit. The Manual of Detailed Geomorphological Mapping and the Legend to the International Map of Europe on the scale 1:2500000 generally follow the Polish, Russian, and German convention. In Great Britain, geomorphologists developed an "empirical system" based on division of the landscape into "slopes and flats." Similar systems have been adopted by some Belgian and Canadian geomorphologists because they make it possible to give quantitative values to all landforms, generally at the expense of genesis and chronology. Some geomorphologists have attempted to combine approaches. In Australia, the Division of Land Research and Regional Survey of The Commonwealth Scientific and Industrial Research Organization (CSIRO) developed a system of geomorphological mapping for resource survey based on a concept of land units and land systems. Information derived from aerial photographs, has been used to map large areas of Australia and New Guinea (Mabbutt and Stewart, 1963).

The development of the international legend, or "Unified Key", and its application to a projected international geomorphological map of Europe, of which the first sheet has now been published, is considered a step forward by most European geomorphologists, but much remains to be done. There is appreciation for, but currently no universal acceptance of, the Unified Key as the one legend to use. Demek (1982) points out the desirability of a single legend for geomorphological maps on various scales, but, he says, the legends proposed so far "do not meet all the requirements." In spite of the many differences, there is an underlying unity to terrain classification. All geomorphologists are concerned with identifying the simplest land unit with low variability, and all are attempting to combine these into a hierarchy of increasingly more complex regions (Gardiner, 1976).

Geomorphic Units

Geomorphological mapping begins with the identification of the fundamental units that compose the landscape. Establishment of the nature and character of these units is critical to the success of any geomorphological research. However, there is no single agreed-upon unit that satisfies all types and all scales of research mapping. Miyogi *et al.* (1970) have described a geomorphic unit in general terms as "an individual, genetically homogeneous landform produced by a definite constructional or destructional geomorphic process." Most geomorphologists would agree with this definition as far as it goes, but would differ widely as to the descriptive characteristics of the genetically homogeneous landform. In addition, although most landforms may be considered as "genetically homogeneous" in terms of present processes, most landforms owe their characteristics, in part, to past processes of a different sort. Wright (1972) has pointed out that "...each part of the land surface is the end product of an evolution governed by parent geological material, geomorphological processes, past and present climate, and time." Most geomorphologists would agree that it is necessary to view the landscape in terms of recognizable patterns that are, to some degree, repetitive. As Gardiner (1976) has said, "units of land, however defined, recur in the landscape in recognizable patterns."

There are two general approaches to the question of geomorphic units, which Speight (1974) divides into a Landform Elements Model and a Landform Patterns Model. He sees the two approaches as complementary. Either model can be used to describe any piece of land, with the choice dependent on scale and the purpose of the mapping. In the Landform Elements Model, the units of landscape are compared to "a simply curved geometric surface without inflections", focused on slope and slope measurements. In the Landform Patterns Model, the land surface is seen as a "three-dimensional cyclic or repetitive phenomenon in which simpler elements recur at quasi-regular intervals in a definable pattern." Here the units are those elements that form the patterns. The Russian, Polish, and Czechoslovakian systems of geomorphological mapping are among those that follow the landform pattern model in which the basic units are morphostructures generally fitted into a nested classification of forms (Bashenina, 1978). British geomorphologists and some others use a landform elements model that classifies the basic units of the landscape in geometric terms as facets and segments defined by slope and area measurements (Savigear, 1965). Other national groups have developed systems that follow some variation of the two models.

Differences in the identification of geomorphic units are significantly related to matters of regionalization and scale. At different scales, covering regions of different size, different features can be identified as basic homogeneous units. The choice of unit is dependent on the scale of analysis. Howard and Mitchell (1980) have said that the clearest and simplex basis of classification is "classification of landscapes into homogeneous units suitable to the mapping scale required for the particular purpose." Geomorphology is a complex subject with multiple approaches to geomorphological analysis and multiple scale of geomorphological mapping. The nature of the geomorphic unit is controlled by the model of analysis chosen and the scale of mapping required. The basic geomorphic unit is not a single feature or entity, but must be carefully selected to be essentially homogeneous and indivisible at the scale chosen. Although all generally agree that the basic geomorphic unit should be a homogeneous entity, it can be defined in terms of genetic or structural pattern, the approach followed by the IGU and most European geomorphologists, or in terms of location and dimensions of geometric elements, along the lines of the British system.

Applications of Geomorphological Mapping

The modern detailed geomorphological map provides a unique means of displaying all of the various factors and features of the physical landscape in an orderly scientific fashion. Such a map is the one analytical research tool developed so far by which it is possible to approximate a portrayal of the Earth's complex surface and its dynamics. It has direct scientific value for research in theoretical geomorphology and likewise serves as a basis from which applied maps may be drawn, focused on special aspects of the landscape, to aid it a variety of different types of applied geomorphological research.

Detailed geomorphological mapping leads to a number of applications to basic scientific research essential to the continued growth and development of geomorphology. As Klimaszewski (1982) pointed out, mapping can:

1. Provide a precise picture of the dynamics of relief that will, in turn:
   (a) Enable the reconstruction of the development of relief.
   (b) Provide the way to evaluate factors and processes of origin and transformation.

2. Facilitate the search for connections between landforms.
3. Enable regionalization.
4. Facilitate the development of comparative studies.
5. Provide a means of comparison between developed and developing landforms in areas of differing and of similar geologic structure and under varying climatic conditions.
6. Serve as an aid in studying the role of climate in shaping the Earth's surface by distinguishing climatic types of relief.

The complex nature of detailed geomorphological maps tends to limit their usefulness beyond the area of technical geomorphology. As Barsch and Liedtke (1980) point out, these maps are "made by experts for experts." Tricart (1969) notes that "a genuine geomorphological map is an intricate document which can be read only by persons with adequate specialized training." All of these tend to render their information inaccessible to persons outside geomorphology. In spite of this, Tricart maintains that "Geomorphological surveys should constitute one of the basic elements in the preparation of development projects." According to Embleton (1985), geomorphological maps serve several valuable functions. He states that they are primarily "a visual information store." They can also serve to coordinate geomorphic research among workers with diverse concepts and approaches to the subject. In viewing geomorphological phenomena over a wide spatial context, as was required when preparing the IGU Geomorphological Map of Europe on the scale of 2.5 million, all researchers are subject to the discipline of working within an agreed international framework. In education, geomorphological maps can be used with other physical maps to "show the complex integration of the natural environment." Finally, in relation to remote sensing data, Embleton points out that an experienced geomorphologist, mapping the landscape, can more easily and reliably appreciate terrain types depicted on remote sensing images than the analyst without such a background.

Methods of Landform Analysis

Mapping, the essential tool of the geomorphologist mapping techniques have become more important with time as changing technology allows more complex mapping. There are a variety of maps that can be produced, topographic, land use, land type (soil, vegetation), but the most important is the geomorphological map.

Geomorphological Mapping

Geomorphological maps contain information on the morphology, genesis and age of landforms. They mark the presence of key features in the landscape, and depict certain attributes. Geomorphological maps are therefore more 'interpretative' (Lowe and Walker, 1998) than other kinds of map, which is a fancy way of saying that they are 'subjective' (Harris: this publication) or perhaps even a work of fiction, if I may be so cynical.

Procedure

The production of geomorphological maps starts with aerial photography and ground survey: photography (including satellite imagery) can enable the production of a base map which is then 'ground truthed' in the field ground truthing is an important part of the process: small scale features can often be missed, and larger scale features can be misinterpreted. The accuracy and speed of mapping is continuously being improved by more the availability of more sophisticated equipment, in particular GPS is a new and potentially powerful ally, but remote sensing will always be the basic foundation. As the world thaws out from the cold war ever more sophisticated data is becoming available, in particular Radar and GPR extend mapping capabilities. All those spy satellites need to be put to some use!

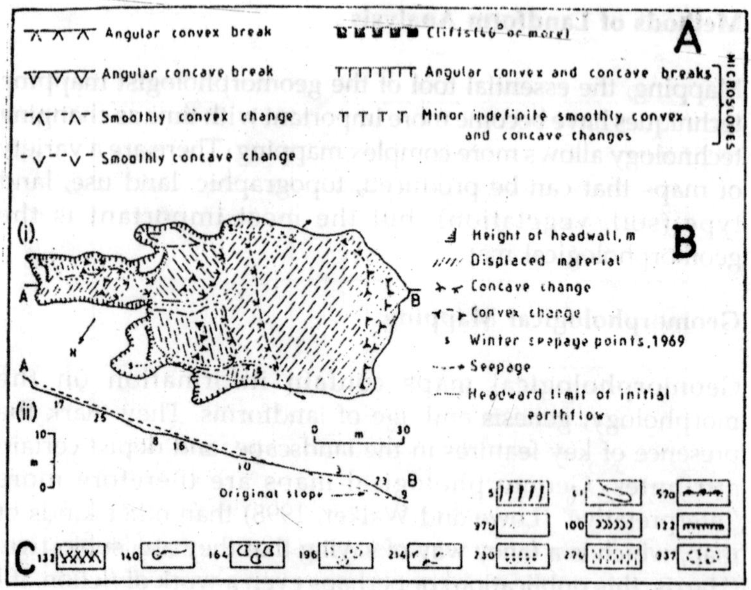

Fig. 27.1: A selection of landform mapping symbols: (A) Some slope-form mapping symbols. (B) Example of a morphological map, featuring an earthflow in the Chittering Valley on the Darling Scarp, north-east of Perth, Western Australia. (i) Map showing form of displaced material, much of which remains within the flow track. Seepage characteristics are recorded since throughflow above an impermeable subsoil is a critical factor in materials in which shear strength decreases rapidly with increased moisture content. (ii) Important supplementary information on a cross-section (A–B on the map), demonstrating the threshold slope of 1·5°, essential for earthflow in this area. (C) Diversity of symbols in the unified key for detailed geomorphological mapping of the world. The example illustrated are: 5 thrust scarp, 41 lava flow, 57a edge of mesa (sandstone); 97b alluvial fan (sand), 100 delta levees, 122 ice-contact slopes, 132 limestone pavament, 140 swallow hole, 164 roche moutonnees, 196 nivation hollow, 214 turf-banked terraces, 230 rock glacier, 274 loess-cover, 337 mining subsidence basins

Sources: Adapted from A. Savigear, 1965; B. Pilgrim and Conacher, 1974; C Demek, 1972.

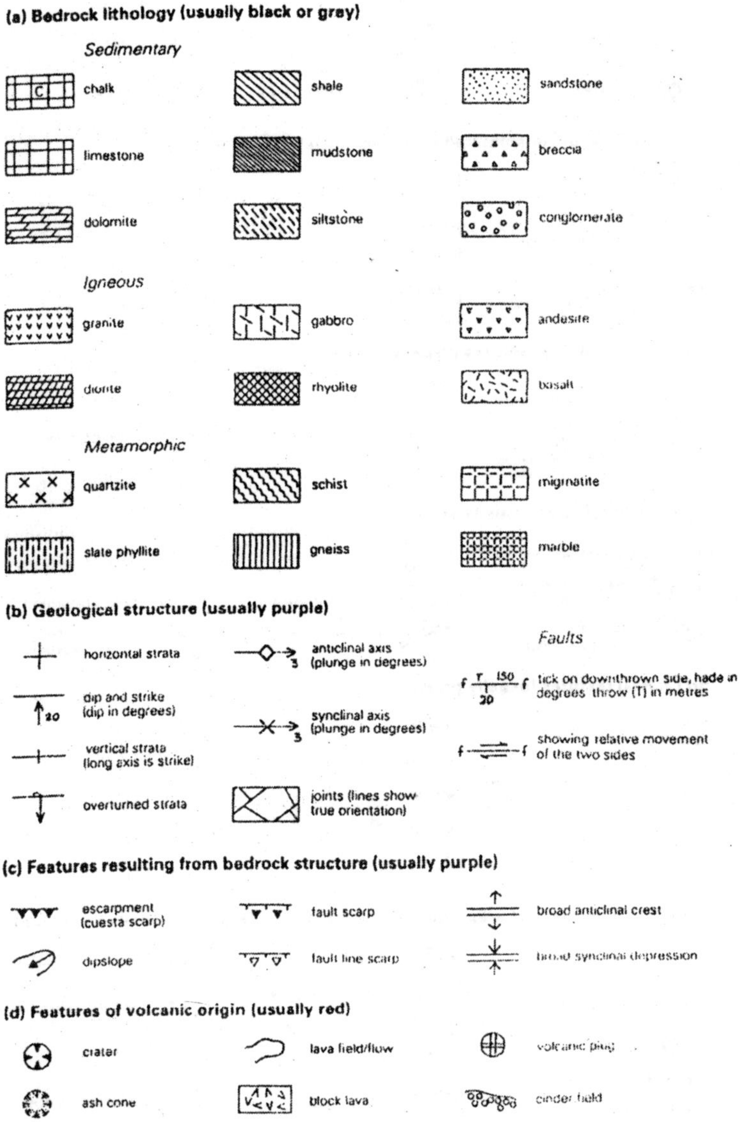

(Fig. 27.2 *contd.*)

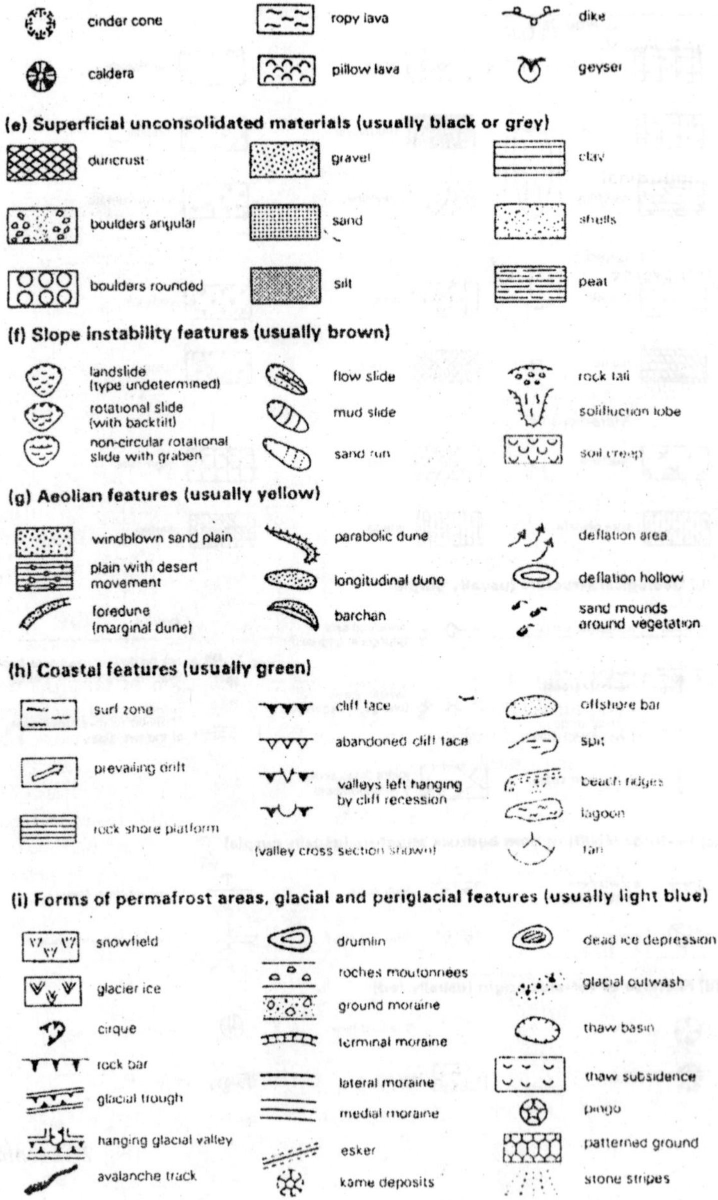

Fig. 27.2: Geomorphological mapping symbols: in (a) and (d), the shading patterns have been chosen from commercially available systems (from Cooke and Doornkamp, 1974.

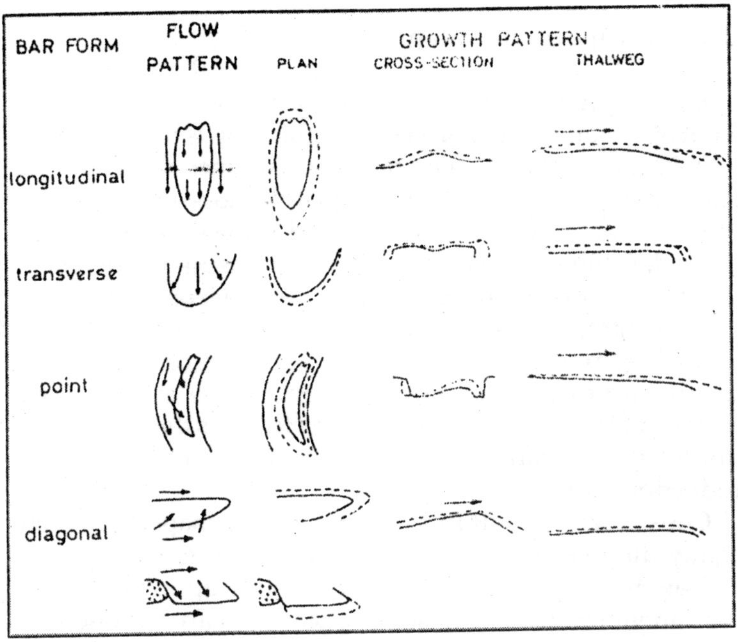

Fig. 27.3: Example of a process-orientated classification of landforms—generalized flow and growth patterns for stream-channel bar forms. Dashed lines indicate accretion. Longitudinal bars, favoured by wide, shallow channels where banks do not affect flow. Transverse bars, with flow lines diverging by channel widening, abrupt depth increase, or at junctions of two channels of upequal flow strength. Point bars, on curved channels with erodible outer banks. Diagonal bars, where channel cross-sectional flow is asymmetrically distributed.

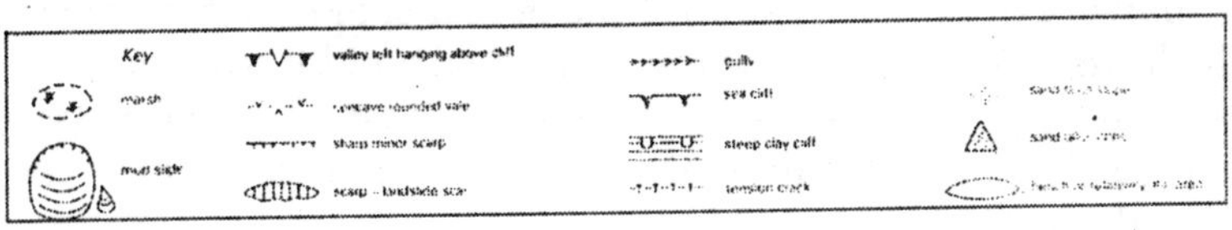

Fig. 27.4

What is a Topographic Map?

A map is a representation of the Earth, or part of it. The distinctive characteristic of a topographic map is that the shape of the Earth's surface is shown by contour lines. Contours are imaginary lines that join points of equal elevation on the surface of the land above or below a reference surface, such as mean sea level. Contours make it possible to measure the height of mountains, depths of the ocean bottom, and steepness of slopes.

A topographic map shows more than contours. The map includes symbols that represent such features as streets, buildings, streams, and vegetation. These symbols are constantly refined to better relate to the features they represent, improve the appearance or readability of the map, or reduce production cost.

Consequently, within the same series, maps may have slightly different symbols for the same feature. Examples of symbols that have changed include built-up areas, roads, intermittent drainage, and some lettering styles. On one type of large-scale topographic map, called provisional, some symbols and lettering are hand drawn.

Reading Topographic Maps

Interpreting the coloured lines, areas, and other symbols is the first step in using topographic maps. Features are shown as points, lines, or areas, depending on their size and extent. For example, individual houses may be shown as small black squares. For larger buildings, the actual shapes are mapped. In densely built-up areas, most individual buildings are omitted and an area tint is shown. On some maps, post offices, churches, city halls, and other landmark buildings are shown within the tinted area.

The first features usually noticed on a topographic map are the area features, such as vegetation (green), water (blue), and densely built-up areas (gray or red).

Many features are shown by lines that may be straight, curved, solid, dashed, dotted, or in any combination. The colours of the lines usually indicate similar classes of information: topographic contours (brown); lakes, streams, irrigation ditches, and other hydrographic features (blue); land grids and important roads (red); and other roads and trails, railroads, boundaries, and other cultural features (black). At one time, purple was used as a revision colour to show all feature changes. Currently, purple is not used in our revision programme, but purple features are still present on many existing maps.

Various point symbols are used to depict features such as buildings, campgrounds, springs, water tanks, mines, survey control points, and wells. Names of places and features are shown in a colour corresponding to the type of feature. Many features are identified by labels, such as "Substation" or "Golf Course", brown by lines of different widths. Each contour is a line of equal elevation; therefore, contours never cross. They show the general shape of the terrain. To help the user determine elevations, index contours are wider. Elevation values are printed in several places along these lines. The narrower intermediate and supplementary contours found between the index contours help to show more details of the land surface shape. Contours that are very close together represent steep slopes. Widely space contours or an absence of contours means that the ground slope is relatively level. The elevation difference between adjacent contour lines, called the contour interval, is selected to best show the general shape of the terrain. A map of a relatively flat area may have a contour interval of 10 feet or less. Maps in mountainous areas may have contour intervals of 100 feet or more. The contour interval is printed in the margin of each U.S. Geological Survey (USGS) map.

Bathymetric contours are shown in blue or black, depending on their location. They show the shape and slope of the ocean bottom surface. The bathymetric contour interval may vary on each map and is explained in the map margin.

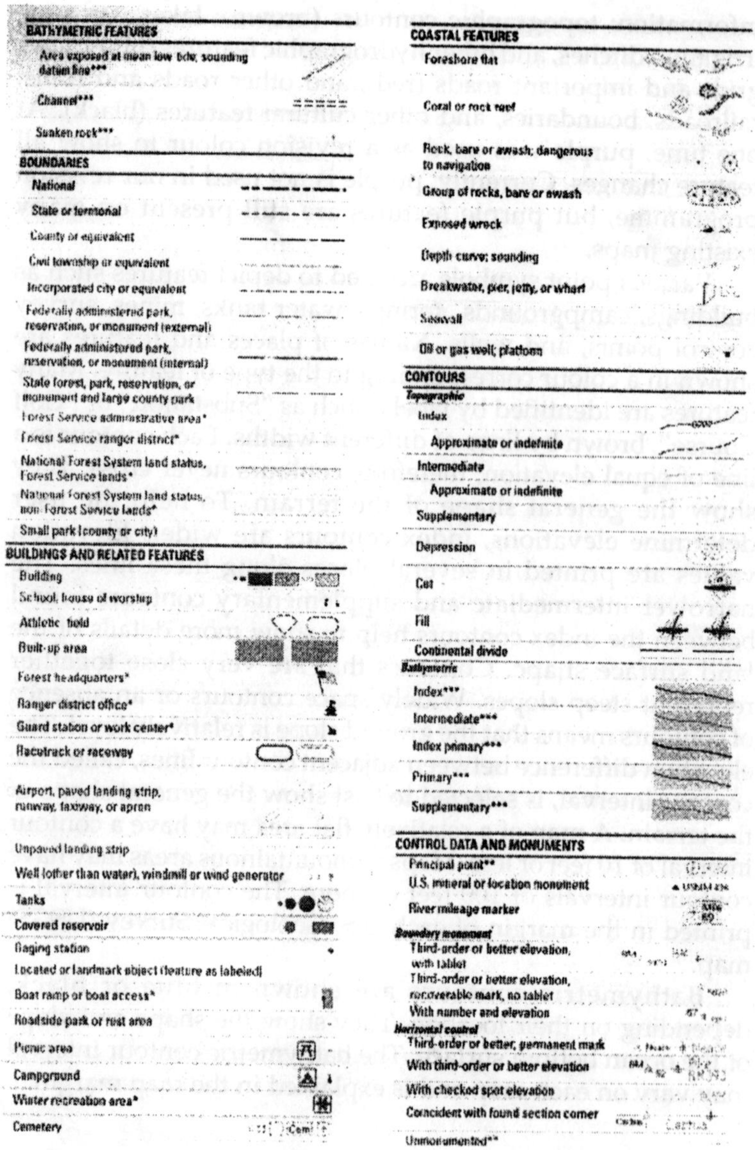
BATHYMETRIC FEATURES
Area exposed at mean low tide; sounding datum line***
Channel***
Sunken rock***
BOUNDARIES
National
State or territorial
County or equivalent
Civil township or equivalent
Incorporated city or equivalent
Federally administered park, reservation, or monument (external)
Federally administered park, reservation, or monument (internal)
State forest, park, reservation, or monument and large county park
Forest Service administrative area*
Forest Service ranger district*
National Forest System land status, Forest Service lands*
National Forest System land status, non-Forest Service lands*
Small park (county or city)
BUILDINGS AND RELATED FEATURES
Building
School; house of worship
Athletic field
Built-up area
Forest headquarters*
Ranger district office*
Guard station or work center*
Racetrack or raceway
Airport, paved landing strip, runway, taxiway, or apron
Unpaved landing strip
Well (other than water), windmill or wind generator
Tanks
Covered reservoir
Gaging station
Located or landmark object (feature as labeled)
Boat ramp or boat access*
Roadside park or rest area
Picnic area
Campground
Winter recreation area*
Cemetery
COASTAL FEATURES
Foreshore flat
Coral or rock reef
Rock, bare or awash; dangerous to navigation
Group of rocks, bare or awash
Exposed wreck
Depth curve; sounding
Breakwater, pier, jetty, or wharf
Seawall
Oil or gas well; platform
CONTOURS
Topographic
Index
Approximate or indefinite
Intermediate
Approximate or indefinite
Supplementary
Depression
Cut
Fill
Continental divide
Bathymetric
Index***
Intermediate***
Index primary***
Primary***
Supplementary***
CONTROL DATA AND MONUMENTS
Principal point**
U.S. mineral or location monument
River mileage marker
Boundary monument
Third-order or better elevation, with tablet
Third-order or better elevation, recoverable mark, no tablet
With number and elevation
Horizontal control
Third-order or better, permanent mark
With third-order or better elevation
With checked spot elevation
Coincident with found section corner
Unmonumented**

CONTROL DATA AND MONUMENTS – *continued*

Vertical control

- Third-order or better elevation, with tablet
- Third-order or better elevation, recoverable mark, no tablet
- Bench mark coincident with found section corner
- Spot elevation

GLACIERS AND PERMANENT SNOWFIELDS

- Contours and limits
- Formlines
- Glacial advance
- Glacial retreat

LAND SURVEYS

Public land survey system

- Range or Township line
  - Location approximate
  - Location doubtful
  - Protracted
  - Protracted (AK 1:63,360-scale)
  - Range or Township labels — R1E T2N
- Section line
  - Location approximate
  - Location doubtful
  - Protracted
  - Protracted (AK 1:63,360 scale)
  - Section numbers — 1 – 36
- Found section corner
- Found closing corner
- Witness corner
- Meander corner
- Weak corner*

Other land surveys

- Range or Township line
- Section line
- Land grant, mining claim, donation land claim, or tract
- Land grant, homestead, mineral, or other special survey monument
- Fence or field lines

MARINE SHORELINES

- Shoreline
- Apparent (edge of vegetation)***
- Indefinite or unsurveyed

MINES AND CAVES

- Quarry or open pit mine
- Gravel, sand, clay, or borrow pit
- Mine tunnel or cave entrance
- Mine shaft
- Prospect
- Tailings
- Mine dump
- Former disposal site or mine

PROJECTION AND GRIDS

- Neatline — 39°15′ 90°37′30″
- Graticule tick — 55′
- Graticule intersection
- Datum shift tick

State plane coordinate systems

- Primary zone tick — 640 000 FEET
- Secondary zone tick — 247 500 METERS
- Tertiary zone tick — 260 000 FEET
- Quaternary zone tick — 98 500 METERS
- Quintary zone tick — 320 000 FEET

Universal transverse mercator grid

- UTM grid (full grid) — 273
- UTM grid ticks* — 269

RAILROADS AND RELATED FEATURES

- Standard guage railroad, single track
- Standard guage railroad, multiple track
- Narrow guage railroad, single track
- Narrow guage railroad, multiple track
- Railroad siding
- Railroad in highway
- Railroad in road
- Railroad in light duty road*
- Railroad underpass; overpass
- Railroad bridge; drawbridge
- Railroad tunnel
- Railroad yard
- Railroad turntable; roundhouse

RIVERS, LAKES, AND CANALS

- Perennial stream
- Perennial river
- Intermittent stream
- Intermittent river
- Disappearing stream
- Falls, small
- Falls, large
- Rapids, small
- Rapids, large
- Masonry dam
- Dam with lock
- Dam carrying road

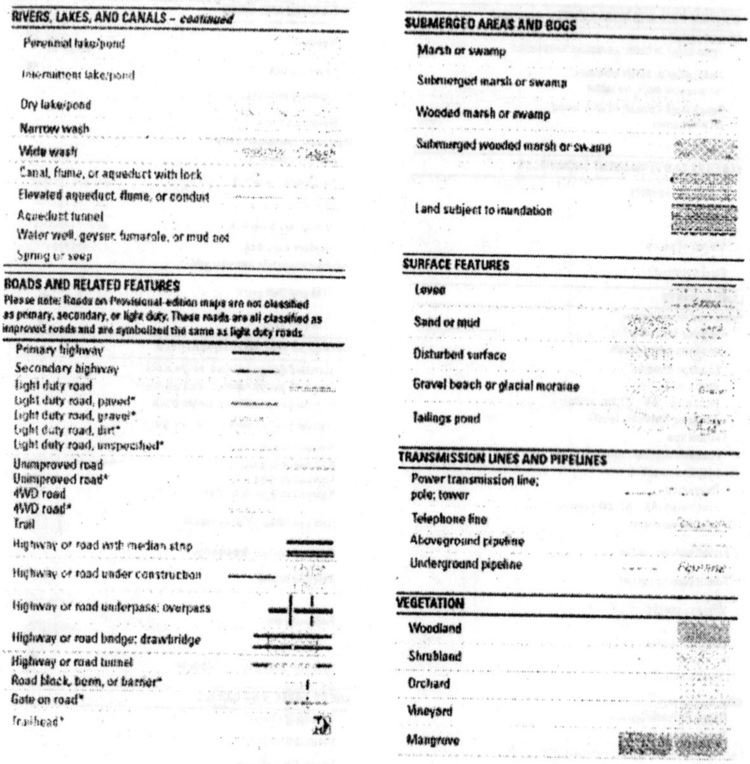

REFERENCES

Bennett, M.R. 1994. 'Morphological evidence as a guide to deglaciation following the Loch Lomond Re-Advance: A review of research approaches and models.' *Scottish Geographical Magazine*, 110, pp. 24-32.

Bennett, M.R. and Boulton, G.S. 1993(a). 'A reinterpretation of Scottish "hummocky moraine" and its significance for the deglaciation of the Scottish Highlands during the Younger Dryas or Loch Lomond Stadial.' *Geological Magazine*, **130,** pp 301-318.

Bennett, M.R. and Boulton, G.S. 1993 (b) 'The deglaciation of the Younger Dryas or Loch Lomond Stadial ice-field in the Northern Highlands, Scotland.' *Journal of Quaternary Science*, **8,** pp. 133-145.

Bennett, M.R. *et al.* 1998. 'Glacial Thrusting and Moraine Mound Formation in Svalbard and Britain: the example of Coire a Cheud—chnoic (valley of 100 hills), Torridon, Scotland'. In: Owen, L.A. (ed) *Mountain Glaciation.* Quaternary Proceedings No. 6, Wiley and Sons, Chichester.

Bromley, C. 1998. An investigation of the river terraces of the Biddulph Valley. Unpublished undergraduate dissertation, Staffordshire University.

Lowe, J.J. and Walker, M.J.C. 1998 (2nd edn.) *Reconstructing Quaternary Environments.* Longman, Harlow.

Slaymaker, O. and Spencer, T. 1998. *Physical Geography and Global Environmental Change.* Routledge, London.

BIBLIOGRAPHY

Ahrens, L.H., *et al.*, (Ed.), *Physics and Chemistry of the Earth 1*, New York, 1956, cl. 317 pages, 3, $ 24 (Contents include: Origin of the Solar System: Temperatures within Earth; Radioactive Methods for Determining Geological Age; Seismology and Hydrodynamics of the Earth's Core; more...).

Ahrens, L.H., *et al.*, (Ed.), *Physics and Chemistry of the Earth 2*, New York, 1957, cl. 259 pages, 3, $ 24 (Contents include: Problems in Oceanography, Data of geochemistry, Solar System, Ocean floor, Geochemistry of Gallium, Indium, Thallium, a review, more...).

Ahrens, T.J., (Ed.), *AGU References Shelf 1: Global Earth Physics—A Handbook of Physical Constants*, Washington, DC., 1995, cl. AGU, 376, pages, 1, $ 99.

Ahrens, T.J., (Ed.), *AGU References Shelf 1: Global Earth Physics—A Handbook of Physical Constants*, Washington, DC., 1995, cl. AGU, 376 pages, 3, $ 50.

American Geological Institute, 1980. *Glossary of Geology*. American Geological Institute, Falls Church, Virginia, 751 p.

American Geological Institute, 1976. *Dictionary of Geological Terms*. Anchor Books, Doubleday, New York, 472 p.

Ashwal, L.D., (Ed.), *Workshop on Early Crustal Genesis: The World's Oldest Rocks*, LPI Technical Report #86-04, Houston, 1985, pb, 185 pages, 3, $ 45 (Precambrian Crustal Evolution; Archean Craton in Greenland and Labrador; Precambrian Crustal Evolution in North and South America; Discussion of Field Excursion to Ameralik Fjord).

Aubouin, J., *et al.*, *Geologie Des Chaines Issues De La Tethys* (*Geology of the Alpine Chains Born of the Tethys*), Paris, 1980, pb, 354 pages, 5, $ 45 (Some chapters written in English, others in French).

Aubouin, J., *Geosynclines, Developments in Geotectonics I,* Amsterdam, 1965, cl, 335 pages, 3, $ 45.

Badgley, P.C., *Structural and Tectonic Principles,* New York, 1965, cl, 521 pages, 3, $ 30.

Bally, A.W., *Geology of North America: An Overview, GSA DNAG Volume A,* Boulder, 1989, cl, 619 pages, 12 plates (in separate slipcase), 1, $ 150.

Bally, A.W., *et al., Dynamics of Plate Interiors,* Geodynamics Series, Volume 1, Washington, DC., 1980, cl, 162 pages, 3, $ 45.

Bartholomew, M.J., Ed., *The Grenville Event in the Appalacians and related Topics, GSA SP 194,* Boulder, 1984, pb, 287 pages, 1, $ 45.

Bartholomew, M.J., (Ed.), *The Grenville Event in the Appalacians and related Topics, GSA SP 194,* Boulder, 1984, pb, 287 pages, 3, $ 30.

Beloussov, V.V., *Structural Geology,* Moscow, 1968, cl, 179 pages, revised English edition of the 1961 Russian edition, 3, $ 50.

Birkeland, P.W., *Pedology, Weathering, and Geomorphological Research,* New York, 1974, cl, 285 pages, 3, $ 45.

British Columbia Ministry of Environment, Assessment and Planning Division, 1981. *Catalogue.* British Columbia Ministry of Environment, 94 p.

Brown, R.J.E. and W.O. Kupsch, 1974. *Permafrost Terminology,* National Research Council of Canada, Technical Memorandum No. 11, Canada Department of Agriculture, 1972.

Canada Soil Survey Committee, 1978. *The Canadian System of Soil Classification.* Canada Department of Agriculture Publication 1646. Supply and Services Canada, Ottawa, 164 p.

Cohen, S.C. and Vanicek, P., (Ed.), *Slow Deformation and Transmission of Stress in the Earth,* Washington, DC., 1989, cl, 138 pages, 2, $ 45 (Geophysical Monograph 49, IUGG Volume 4).

Cohen, S.C. and Vanicek, P., (Ed.), *Slow Deformation and Transmission of Stress in the Earth,* Washington, DC., 1989, cl, 138 pages, 3, $ 40 (Geophysical Monograph 49, IUGG Volume 4).

Condie, K.C., (Ed.), *Proterozoic Crustal Evolution,* Developments in Precambrian Geology 10, Amsterdam, 1992, cl, 537 pages, 3, $ 165.

Condie, K.C., *Plate Tectonics and Crustal Evolution,* 1st Edition, New York, 1976, cl, 288 pages, 2, Several Small Nicks on Bottom of Boards, ovg, $ 30.

Cooke, R.U. and Doornkamp, J.C., *Geomorphology in Environmental Management: A New Introduction,* 2nd Edition, Oxford, 1990, pb, 410 pages, 3, $ 30.

Corry, C.E., *Laccoliths; Mechanics of Emplacement and Growth,* Boulder, 1988, pb, 110 pages, 3 plates in pocket, 1, $ 46.

Costa, J.E. and Fleisher, P.J., (Ed.), *Developments and Applications of Geomorphology,* Berlin, 1984, cl, 372 pages, 5, $ 75.

Cotton, C.A., *Geomorphology: An Introduction to the Study of Landforms,* 5th Edition, Revised, New York, 1949, cl, 505 pages, 3, $ 20.

Cox, A. and Hart, R.B., *Plate Tectonics, How It Works,* Palo Alto, 1986, pb, 392 pages, 3, $ 20.

Cox, A. and Doell, R.R., Review of Paleomagnetism, Boulder, 1960, pb, pages 645-768, in *Bulletin of the GSA* Volume 71, #6, 3, $ 20 (Other articles include River Meanders, Ash Flows, Paleotectonic Evolution of the Central and Western Alps).

Cox, A., *Plate Tectonics and Geomagnetic Reversals: Readings, selected, edited, and with introductions,* San Francisco, 1973, pb, 702 pages, 3, $ 24.

Crowley, T.J. and Burke, K.C., *Tectonic Boundary Conditions for Climate Reconstructions,* Oxford Monographs on Geology and Geophysics # 39, New York, 1998, cl, 285 pages, 2, $ 55.

Dallmeyer, R.D., *Terranes in the Circum—Atlantic Paleozoic Orogens,* GSA SP 230, Boulder, 1989, pb, 277 pages, 2, $ 49.

Dallmeyer, R.D., *Terranes in the Circum-Atlantic Paleozoic Orogens,* GSA SP 230, Boulder, 1989, pb, 277 pages, 3, $ 25.

Dallmeyer, R.D. and Lécorché, J.P., Ed., *The West African Orogens and Circum—Atlantic Correlatives,* Berlin, 1991, cl, 405 pages, 196 figs., 1, $ 159.

Daly, R.A., *Strength and Structure of the Earth,* New York, 1940, cl, 434 pages, 5, $ 15.

Davis, W.M., *Les Côtes Et Les Récifs Coralliens De La Nouvelle=Calédonie,* Paris, 1926, afb, 120 pages, 64 figs., 2, $ 40.

Debiche, M.G., Cox, A. and Engebretson, D., *The Motion of Allochthonous Terranes Across the North Pacific Basin,* GSA SP 207, Boulder, 1987, pb, 49 pages, 3, $ 25.

De Jong, K.A. and Scholten, R. (Ed.), *Gravity and Tectonics,* Van Bremmelen Volume, New York, 1973, cl, 502 pages, 3, $ 25.

Dennis, J.G., *Orogeny,* Stoudsburg, 1982, cl, 379 pages, Benchmark Paper Volume 62, 2, $ 75.

Dennis, J.G., *Structural Geology,* New York, 1972, cl, 532 pages, 3, $ 20.

Dennison, J.M., *Sedimentary Tectonics of the Appalachian Basin,* Dallas, 1983, pb, 129 pages, foxing spots on covers and end plates, contents fine, 4, $ 25.

De Sitter, L.U., *Structural Geology,* 2nd Edition, New York, 1964, cl, 51 pages, 3, $ 20.

D'Lemos, R.S., Strachan, R.A. and Topley, C.G., *The Cadomian Orogeny,* Geological Society Special Paper 51, London, 1990, cl, 423 pages, 1, $ 159.

Doe, B.R., *Studies in Mineralogy and Precambrian Geology,* GSA M 135, Boulder, 1972, cl, 348 pages, 3, $ 38, (Contents include: Investigations of the Pc; Pc of Minnesota-Ontario region; Minerals and Mineralization associated with Pc).

Doornkamp, J.C. and King, C.A.M., *Numerical Analysis in Geomorphology: An Introduction,* New York, 1971, cl, 372 pages, 3, $ 30.

Eardley, A.J., *Structural Geology of North America,* New York, 1951, cl, 624 pages, 3, $ 30.

Edelman, S.H., *et al., Structure Across A Mesozoic Ocean-Continent Suture Zone in the Northern Sierra Nevada, California,* GSA SP 224, Boulder, 1989, pb, 56 pages, 1, $ 30.

Energlyn, W.D., *Through the Crust of the Earth*, New York, 1973, cl, 128 pages, some ink underlining in text, 4, $ 5.

Engelder, T., *Stress Regimes in the Lithosphere*, Princeton, 1993, cl, 457 pages, 1, $ 75.

Environment and Land Use Committee Secretariat (E.L.U.C.S.), 1976. Terrain Classification System. British Columbia Ministry of Environment, Resource Analysis Branch, 56 p.

Ernst, W.G., (Ed.), *Serpentine and Serpentinites: Mineralogy, Petrology, Geochemistry, Ecology, Geophysics and Tectonics*, Columbia, 2004, pb, 606 pages, 1, $ 100.

Fairbridge, R.W. (Ed.), 1968. *Encyclopedia of Geomorphology*. Reinhold Books, New York, 1107 p.

Fairbridge, R.W. (Ed.), *The Encyclopedia of World Regional Geology, Part I: Western Hemisphere* (including Antarctica and Australia), Stroudsburg, 1975, cl, 704 pages, 3, $ 85.

Farb, P., *Face of North America: The Natural Hisory of a Continent*, 1963, New York, cl, 316 pages, 5, $ 20.

Ferb, P., *Face of North America: The Natural History of a Continent*, 1963, New York, cl, 316 pages, 3, $ 25.

Fenneman, N.M., *Physiography of Western United States*, New York, 1931, cl, 534 pages, 3, $ 55.

Fenneman, N.M., *Physiography of Eastern United States*, New York, 1938, cl, 714 pages, 3, $ 5.

Fisher, D.J., *Symposium on Layered Intrusions and Mineralogy of the Sulfides*, International Mineralogical Association Special Paper # 1, Washington, DC., 1963, cl, 332, pages, 3, $ 75.

Flint, R.F., 1971. *Glacial and "U" Quaternary "U" Geology*. John Wiley and Sons, New York, 892 p.

Fowler, C.M.R., *The Solid Earth: An Introduction to Global Geophysics*, Cambridge, 1990, pb, 472 pages, 3, lower corner of several back pages and back cover with slight creases, ovg, $ 20.

Frost, E.G. and Martin, D.L., *Mesozoic-Cenozoic Tectonic Evolution of the Colorado River Region, California, Arizona, and Nevada*, Anderson-Hamilton Volume, San Diego, 1982, pb, 608 pages, 5, $ 35.

Fuchs, K. and Froidevaux, C., (Ed.), *Composition, Structure and Dynamics of the Lithosphere-Asthenosphere System*,

Geodynamics Series Volume # 16, Washington, DC., 1987, cl, 327 pages, 3, $ 60.

Fulton, R.J. (Ed.), *Quaternary Geology of Canada and Greenland, GSA DNAG Volume K1*, Boulder, 1989, cl, 839 pages, five plates in separate slipcase, 2, $ 175.

Fulton, R.J., A.N. Boydell, D.M. Barnett, D.A. Hudgson, V.A. Rampton, 1979. Terrain Mapping in Northern Environments. In: Proceedings of a Technical Workshop—To Develop an Integrated Approach to Base Data Inventories for Canada's Northlands. Compiled and edited by M.J. Romaine and G.R. Ironside, Ecological Land Classification Series, No. O, Supply and Services Canada, Ottawa: 3-21.

Gabelman, J.W., *Metallotectonic Evidence for Continental Drift*, AEC Division of Production and Materials Management, Washington, DC., 1971, pb, 41 pages, 3, $ 23.

Gabrielse, H. and Yorath, C.J., *Geology of the Cordilleran Orogen in Canada*, Ottawa, 1992, cl, 844 pages, 12 plates (in separate slipcase), 2, $ 475.

Gans, P.B. *et al.*, *Synextentional Magmatism in the Basin and Range Province: A Case Study from the Eastern Great Basin*, GSA SP 233, Boulder, 1989, pb, 53 pages, 1, $ 30.

Garland, G.D., *The Earth's Shape and Gravity*, Oxford, 1965, pb, 183 pages, 3, $ 15.

Garland, G.D., Ed., *Continental Drift*, Royal Society of Canada Special Publication # 9, Toronto, 1966, cl, 140 pages, 3, $ 20.

Garland, G.D., *Introduction to Geophysics: Mantle, Core, and Crust*, Philadelphia, 1971, cl, 420 pages, 3, $ 50.

Geological Survey of Canada, 1979. *Guide to Authors*. Miscellaneous Report 29 compiled by: R.G. Blackadar, H. Dumych and P.J. Griffin, Ottawa, Canada, 66 p.

Gilluly, J., *Volcanism, Tectonism, and Plutonism in the Western United States*, GSA SP 80, New York, 1965, pb, 67 pages, 7 figs., 3, $ 25.

Goddard, E.N., *et al.*, *Rock-Color Chart*, Boulder, 1970, pb, 12 pages, 3, $ 10.

Goddard, E.N., *et al.*, *Rock-color Chart*, Boulder, 1948, pb, 12 pages, 3, $ 5.

Goguel, J., *Tectonics*, San Francisco, 1962, cl, 384 pages, 3, Translated into English in 1952 by H.E. Thalmann, $ 25.

Gordon, R.B, *Physics of the Earth*, 1st Edition, New York, 1972, pb, 207 pages, 3, $ 25.

Graf, W.L. (Ed.), *Geomorphic Systems of North America*, GSA Centennial Special Volume 2, Boulder, 1987, cl, 643 pages, 3, $ 125.

Grant, F.S. and West, G.F., *Interpretation Theory in Applied Geophysics*, New York, 1965, cl, 583 pages, 3, $ 25.

Gutenberg, B., *Physics of the Earth's Interior*, 1st Edition, Volume 1 of the Physics of the Earth Series, New York, 1959, cl, 240 pages, 3, $ 10.

Gutenberg, B. (Ed.), *Internal Constitution of the Earth*, 1st Edition, Volume VII of the Physics of the Earth Series, New York, 1951, pb, 439 pages, 3, $ 15.

Gutenberg, B. (Ed.), *Internal Constitution of the Earth*, 1st Edition, Volume VII of the Physics of the Earth Series, New York, 1939, cl, 413 pages, 3, $ 30.

Habeck, J.R. and E. Hartley, 1968. *A Glossary of Alpine Terminology*, University of Montana.

Hamelin, L. and F.A. Cook, 1967. *Illustrated Glossary of Periglacial Phenomena.* University of Laval, Quebec, 237 p.

Hansen, E., *Strain Facies*, New York, 1971, cl, 207 pages, 3 plates in pocket, 2, $ 60.

Hansen, E., *Strain Facies*, New York, 1971, cl 207 pages, 3 plates in pocket, 5, $ 45.

Hapgood, C.H., *The Path of the Pole*, Philadelphia, 1970, cl, 413 pages, 3, $ 100.

Hardy, H.R., Jr. and Stefanko, R., *New Horizons in Rock Mechanics*, Proceedings of the 14th Symposium on Rock Mechanics, New York, 1973, cl, 785 pages, 3, $ 45.

Harmon, R.S. and Barriero, B.A. (Ed.), *Andean Magmatism, Chemical and Isotopic Constraints*, Shiva Geology Series, Cheshire, 1984, cl, 250 pages, 3, $ 139.

Harrison, C.G.A. (Ed.), Marine Geophysics, Plate Tectonics, and the Earth's Core: The Bullard Volume, *Journal of Geophysical Research Reprint*, Washington, DC., 1981, 189 pages, 3, $ 50.

Hart, M.G., *Geomorphology: Pure and Applied*, London, cl, 228 pages, 4, $ 28.

Hart, P.J. (Ed.), *The Earth's Crust and Upper Mantle*, AGI Geophysical Monograph 13, Washington, DC., 1969, pb, 735 pages, 3, $ 25.

Hart, P.J. (Ed.), *The Earth's Crust and Upper Mantle*, AGI Geophysical Monograph 13, Washington, DC., 1969, pb, 735 pages, 3, $ 15 (This copy has slightly more wear than those above).

Hatcher, R.D., Jr., Thomas, W.A. and Viele, G.W. (Ed.), *The Appalachian-Ouachita Orogen in the United States*, GSA DNAG Volume F-2, Boulder, 1989, cl, 767 pages, 12 plates in separate slipcase, 2, $ 400.

Hatcher, R.D., Jr., Williams, H. and Zietz, I. (Eds.), *Contributions to the Tectonics and Geophysics of Mountain Chains*, Boulder, 1983, cl, 223 pages, 1, $ 55.

Hatcher, R.D., Jr., Williams, H. and Zietz, I. (Eds.), *Contributions to the Tectonics and Geophysics of Mountain Chains*, Boulder, 1983, cl, 223 pages, 3, $ 40.

Hatcher, R.D., Jr., Williams, H. and Zietz, I. (Eds.), *Contributions to the Tectonics and Geophysics of Mountain Chains*, Boulder, 1983, cl, 223 pages, 3, shows a little more wear, slight foxing, $ 20.

Heacock, J.G. (Ed.), *The Structure and Physical Properties of the Earth's Crust*, AGU Geophysical Monograph 14, Washington, DC, 1971, cl, 348 pages, 2, $ 55.

Heacock, J.G. (Ed.), *The Earth's Crust*, American Geophysical Union Monograph 20, Washington, DC., 1977, cl, 754 pages, 3, $ 60.

Hereford, R. *et al.*, *Late Holocene Alluvial Geomorphology of the Virgin River in the Zion National Park Area, Southwest Utah*, GSA SP 310, Boulder, 1996, pb, 41 pages, 1, $ 30.

Hills, E.S., *Elements of Structural Geology*, London, 1963, cl, 483 pages, 3, $ 25.

Hinds, N.E.A., *Geomorphology: The Evolution of Landscape*, New York, 1943, cl, 894 pages, 3, $20.

Hobbs, B.E., *An Outline of Structural Geology*, New York, 1976, cl, 571 pages, 3, $ 25.

Hobbs, G.W. (Ed.), *Field Guide to the Geology of the Paleozoic, Mesozoic, and Tertiary Rocks of New Jersey and the Central Hudson Valley*, Atlantic Margin Energy Symposium, Atlantic City, 1981, spiral bound, 232 pages, 3, $ 30.

Hodgson, R.A., Gay, S.P., Jr. and Benjamins, J.Y., *Proceedings of the First International Conference on the New Basement Tectonics*, Salt Lake City, 1976, pb, 636 pages, 3, $ 50.

Hoek, E. and Bray, J.W., *Rock Slope Engineering*, London, 1974, cl, 309 pages, 3, $ 45.

Holmes, A., *Principles of Physical Geology*, 2nd Edition Completely revised, London, 1969, cl, 1288 pages, 3, $ 35.

Horton, J.W., Jr. and Rast, N., *Melanges and Olistostromes of the U.S. Appalachians*, GSA SP 228, Boulder, 1989, pb, 276 pages, 2, $ 45.

Howell, D.G., *Tectonostratigraphic Terranes of the Circum-Pacific Region*, Houston, 1985, cl, 581 pages, plate in pocket, 1, $ 95.

Howes, D.E., 1987. *A Terrain Evaluation Method for Predicting Terrain Susceptible to Post-Logging Landslide Activity*. Ministry of Environment and Parks Technical Report 28. Ministry of Environment and Parks, Recreational Fisheries Branch, Victoria, B.C., 38 p.

Hsu, K.J., *Mountain Building Processes*, London, 1982, pb, 263 pages, 3, $ 55.

Hubbert, M.K. *et al.*, Role of Fluid Pressure in Mechanics of Overthrust Faulting, I. Mechanics of fluid-filled porous solids and its application to overthrust faulting and II. Overthrust Belts in geosynclinal areas of Western Wyoming in light fluid-pressure hypothesis, In: *Bulletin of GSA*, reprint, 1959, volume 70, pages 115-206, 3, $ 10.

Hunt, C.B., *Physiography of the United States*, San Francisco, 1967, cl, 480 pages, 3, $ 25.

Hunt, C.B., *Physiography of the United States*, San Francisco, 1967, cl, 480 pages, 3, slight foxing on edges of pages, $ 20.

Irwin, W.P. *et al., International Atlas of Ophiclites,* 1979, GSA MC-33, Introductory notes, 15 pages, 4 maps: Ophiolitic Belt of the Urals, Ophiolitic Belts of the Himalayan and Tibetan Region, Ophiolitic Belts of the Central Mediterranean, and Ophiolitic Terranes of part of the Western United States, 3, $ 20.

Jacbos, J.A., Russell, R.D., and Wilson, J.T., *Physics and Geology,* 1st Edition, New York, 1959, cl, 424 pages, 3, $ 20.

Jacbos, J.A., Russell, R.D. and Wilson, J.T., *Physics and Geology,* 2nd Edition, New York, 1974, cl, 622 pages, 3, $ 35.

Johnson, H. and Smith, B.L. (Ed.), *The Magatectonics of Continents and Oceans,* New Brunswick, 1970, cl, 282 pages, 3, $ 35.

Jennings, J.N., 1971. *Karst.* M.I.T., London, England, 253 p.

Kasahara, K. (Ed.), *Recent Plate Movements and Deformation,* Geodynamics Series Volume 20, Washington, DC., 1987, cl, 83 pages, 3, $ 45.

Kasbeer, T., *Bibliography of Continental Drift and Plate Tectonics,* Volume 2, GSA SP 164, Boulder, 1975, pb, 151 pages, 3, $ 10.

King, L.C., *Wandering Continents and Spreading Sea Floors on an Explanding Earth,* Chichester, 1983, cl, 232 pages, 3, $ 38.

King, P.B. *et al., Tectonics of the Southern Appalachian,* VPI Department of Geological Sciences Memoir 1, Roanoke, 1964, cl, 114 pages, 8 figs., 2, $ 28.

King, P.B., *Tectonic Map of North America,* 1:5,000,000 scale, 1969, one folded sheet, $ 20.

King, P.B., *An Outline of the Structural Geology of the United States,* 16th International Geological Congress, 1933, Guidebook 28, 57 pages, pb, 1 plate, $ 15.

King, P.B., *The Evolution of North America,* 1st Edition, Princeton, 1959, cl, 189 pages, 3, $ 35.

King, P.B., *The Evolution of North America,* Revised 2nd Edition, Princeton, 1977, pb, 197 pages, 3, $ 20.

King, P.B., The Tectonics of Middle North America: Middle North America East of the Cordilleran System, Princeton, 1951 or 1954 printing or 1951 original, cl, 203 pages, 3, $ 25.

Kröner, A., Hanson, G.N. and Goodwin, A.M., *Archean Geochemistry: The Origin and Evolution of the Archean Continental Crust*, Berlin, 1984, cl, 286 pages, 4, some yellow highlighting, $ 35.

Kröner, A., Hanson, G.N. and Goodwin, A.M., *Archean Geochemistry: The Origin and Evolution of the Archean Continental Crust*, Berlin, 1984, cl, 286 pages, 3, $ 119.

Krumbein, W.C. and L.L. Sloss, 1963. *Stratigraphy and Sedimentation.* W.H. Freeman and Co., San Francisco and London, 660 p.

Kulm, L.D. *et al.*, (Ed.), *Nazca Plate: Crustal Fromation and Andean Convergence,* GSA M 154, New York, 1981, cl, 824 pages, 3, $ 35.

Kulm, L.D., *et al.*, (Ed.), *Nazca Plate: Crustal Formation and Andean Convergence,* GSA M 154, New York, 1981, cl, 824 pages, 1, $ 55.

Lambeck, K., *Geophysical Geodesy: The Slow Deformations of the Earth,* Oxford, 1988, pb, 718 pages, 5, slight creases in back cover, ovg, $ 35.

Larsen, E.S., Jr., *Batholith and Associated Rocks of Corona, Elsinoire, and San Luis Rey Quadrangles,* GSA M 29, Baltimore, 1948, cl, 182 pages, 5, this is a photocopy version bound in cloth, $ 30.

Leggett, J.K., (Ed.), *Trench-Forearc Geology,* Geological Society of London SP 10, Oxford, 1982, cl, 576 pages, 3, $ 75.

Le Mouel, J.L., Smylie, D.E. and Herring, T. (Ed.), *Dynamics of Earth's Deep Interior and Earth Rotation,* Washington, D.C., 1993, cl, 189 pages, 1, $ 40, (Geophysical Monograph 72, IUGG Volume 12).

Leyshon, P.R. and Lisle, R.J., *Stereographic Projection Techniques in Structural Geology,* Oxford, 1996, pb, 104 pages, 7 appendices, computer disc (still sealed) is included, 3, $ 36.

Lobeck, A.K., *Atlas of American Geology,* New York, 1932, cl, 92 atlas sheets with drawings or maps having numerous questions to analyze the contents of each atlas plate, 3, $ 30 (contains excellent cross-section diagram illustrations).

Lobeck, A.K., *Geomorphology,* McGraw Hill, New York, 1939, cl, 731 pages, 5, $ 25.

Lobeck, A.K., *Block Diagrams,* New York, 1924, cl, 206 pages, 3, light spotting on boards, ovg, $ 26.

Lucas, S.B. and St-Onge, M.R., *Geology of the Precambrian Superior and Greenville Provinces and Precambrian Fossils in North America,* GSA DNAG Volume C1, Ottawa, 1998, cl, 387 pages, five maps (in 8 sheets) and one figure (with two parts) in separate slipcase, 1, $ 200.

Luth, R.W. (Ed.), *Short Course Handbook on Experiments at High Pressure and Applications to the Earth's Mantle,* Toronto, 1993, pb, Mineralogical Association of Canada, Volume 21, 485 pages, 3, $ 35.

Machatschek, R., *Das Relief Der Erde,* Berlin, 1955, cl, two volume set, volume with 531 pages, 10 tables, 142 figs., volume two with 594 pages, 10 tables, and 186 figs., 5, $ 145.

McKenzie, G.D., *Gondwana Six: Structure, Tectonics, and Geophysics,* AGU Geophysical Monograph 40, Washington, DC., 1987, cl, 323 pages, 1, $ 90.

McKenzie, G.D., *Gondwana Six: Stratigraphy, Sedimentology, and Paleontology,* AGU Geophysical Monograph 41, Washington, DC., 1987, cl, 250 pages, 1, $ 100.

Malinconico, L.L., Jr. and Lillie, R.J., *Tectonics of the Western Himalayas,* GSA SP 232, Boulder, 1989, pb, 320 pages, 1 pocket plate, 1, $ 45.

Martini, I.P. and Chesworth, W. (Ed.), *Weathering, Soils and Paleosols,* Developments in Earth Surface Processes 2, Amsterdam, 1992, cl, 618 pages 1, $ 355.

Mayer, L. (Ed.), *Extensional Tectonics of the Southwestern U.S.: A Perspective on Processes and Kinematics,* GSA SP 208, Boulder, 1986, pb, 122 pages, 3, $ 25.

McElhinny, M.W., *Palaeomagnetism and Plate Tectonics,* Cambridge, 1973, pb, 358 pages, 3, $ 20.

McKee, E.D. *et al., Paleotectonic Maps: Jurassic System,* USGS Map I-175, 6 pages, 9 plates, Washington, DC., 1956, oversized, folded, $ 35.

McKee, E.G. *et al., Paleotectonic Maps: Triassic System,* USGS Map I-300, 33 pages, 9 plates, Washington, DC., 1959, oversized, folded, $ 35.

Medaris, L.G. *et al.*, (Ed.), *Proterozoic Geology: Selected Paper from An International Symposium*, GSA M 161, Boulder, 1983, cl, 315 pages, 2, $ 75.

Meissner, R., *The Continental Crust: A Geophysical Approach*, 1st Edition, 1986, International Geophysics Series, Volume 34, pb, 426 pages, vg, $ 45.

Menard, H.W., *The Ocean of Truth: A Personal History of Global Tectonics*, Princeton, 1986, cl, 353 pages, 3, $ 20.

Menard, H.W., *The Ocean of Truth: A Personal History of Global Tectonics*, Princeton, 1986, cl, 353 pages, 5, $ 15.

Menke, W. and Abbott, D., *Geophysical Theory*, New York, 1990, cl, 458 pages, 3, $ 70.

Merrill, R.T. and McElhinny, M.W., *The Earth's Magnetic Field: Its Origin and Planetary Perspective*, International Geophysics Series # 32, London, 1983, pb, 401 pages, 3, $ 25.

Meyer, A.W. *et al.*, (Ed.), *Evolution of Mesozoic and Cenozoic Continental Margins*, from a 1990 SEPM Research Symposium, Amsterdam, 1992, cl, 361 pages, 2, $ 100.

Miall, A.D. (Ed.), *Sedimentation and Tectonics in Alluvial Basins*, Geological Association of Canada SP 23, Waterloo, 1981, cl, 272 pages, 1, $ 65.

Miller, D.M., Todd, V.R. and Howard, K.A., *Tectonic and Stratigraphic Studies in the Eastern Great Basin*, GSA M 157, Boulder, 1983, cl, 327 pages, 1, $ 80.

Mittwede, S.K. and Stoddard, E.F., *Ultramafic Rocks of the Appalachian Piedmont*, GSA SP 231, Boulder, 1989, pb, 103 pages, 2, $ 40.

Mitra, G. and Wojtal, S., *Geometries and Mechanisms of Thrusting*, with special reference to the Appalachians, Boulder, 1988, pb, 236 pages, 1, $ 40.

Miyashiro, A., Aki, K. and Sengor, A.M.C., *Orogeny*, Chichester, 1982, pb, 242 pages, 3, $ 50 (English edition of a 1979 Japanese publication).

Morrison, R.B. (Ed.), *Quaternary Nonglacial Geology: Coterminous U.S.*, GSA DNAG Volume K2, Boulder, 1991, cl, 672 pages, 8 plates (in separate slipcase), 3, $ 295.

Murray, G.E., *Geology of the Atlantic and Gulf Coastal Province of North America*, New York, 1961, cl, 692 pages, 3, $ 100.

Neidell, N.S., *Stratigraphic Modelling and Interpretation: Geophysical Principles and Techniques*, Tulsa, 1979, pb, 145 pages, AAPG Educational Course Notes Series # 13, 3, $ 25.

Nesbitt, B.E. (Ed.), *Short Course Handbook: Fluids in Tectonically Active Regimes of the Continental Crust*, Toronto, 1990, pb, 312 pages, Mineralogical Association of Canada Volume 18, 3, $ 30.

Nisbet, E.G., *The Young Earth: An Introduction to Archean Geology*, Boston, 1987, cl, 402 pages, 3, $ 75.

O'Connell, R.J. and Fyfe, W.S. (Ed.), *Evolution of the Earth*, American Geophysical Union Geodynamics Series, Volume 5, Washington, D.C., 1981, cl, 282 pages, 2, $ 30.

Odishaw, H., (Ed.), *Research in Geophysics, Volume 2: Solid Earth and Interface Phenomena, Cambridge*, 1964, cl, 595 pages, 3, $ 30.

Oldow, J.S. and Lallemant, H.G.A. (Ed.), *Architecture of the Central Brooks Range Fold and Thrust Belt, Arctic Alaska*, GSA SP 324, Boulder, 1998, pb, 317 pages, 1, $ 36.

Parasnis, D.S., *Principles of Applied Geophysics*, 1st Edition, New York, 1962, pb, 214 pages, 2, $ 12.

Pakiser, L.C. and Mooney, W.D., *Geophysical Framework of the Continental United States*, Boulder, 1989, cl, 826 pages, GSA Memoir 172, 1, $ 75.

Palmason, G., (Ed.), *Continental and Oceanic Rifts*, Geodynamics Series, Volume 8, Boulder, 1982, cl, 309 pages, 3, $ 55 SOLD.

Parker, R.L., *Geophysical Inverse Theory*, Princeton, 1994, cl, 386 pages, 1, $ 50.

Paterson, M.S. and Weiss, L.E., Symmetry Concepts in the Structural Analysis of Deformed Rocks, In: *GSA Bulletin*, Volume 72, # 6, 1961, pb, pages 841-882, 3, $ 10.

Penck, W., *Morphological Analysis of Land Forms*, Translated into English by H. Czech and K.C. Boswell, New York, 1953, cl, 429 pages, 3, $ 24.

Perry, W.J., Roeder, D.H., and Lageson, D.R., *North American Thrust-Faulted Terranes*, Tulsa, 1984, pb, 3, $ 30 (AAPG Reprint Series # 27; contains articles from the AAPG Bulletin and other journals).

Pettijohn, F.J., 1949. *Sedimentary Rocks*. Harper and Bros., N.Y., New York.

Phillips, F.C., *The Use of Stereographic Projection in Structural Geology*, 2nd Edition, London, 1960, cl, 86 pages, 3, $ 15.

Piper, J.D.A., *Palaeomagnetism and the Continental Crust*, New York, 1987, cl, 434 pages, 3, $ 50.

Poldervaart, A. (Ed.), *Crust of the Earth (A Symposium)*, 1955, cl, 762 pages, GSA SP 62, 3, $ 25.

Price, N.J., *Fault and Joint Development in Brittle and Semi-Brittle Rock*, Oxford, 1966, pb, 176 pages, 3, $ 16.

Price, R.A. and Douglas, R.J.W., *Variations in Tectonics Systems in Canada*, 25th Anniversary Volume of the Geological Association of Canada, Special Paper # 11, Montreal, 1972, cl, 688 pages, 3, $ 30.

Proctor, P.D. *et al.*, *UMR Journal* # 1, containing the following tectonic papers: The Atlantic Continental Margin, Profile of the Folded Appalachians of West Virginia, Tectonic History of Midcontinental U.S., Major Structures of the Rocky Mountains of Colorado and Utah, Tectonic Framework of the Great Basin, Geologic Structure and History of the Sierra Nevada, and the California Coast Range, Rolla, 1968, pb, 156 pages, 3, $ 20.

Quennell, A.M., (Ed.), *Continental Rifts*, New York, 1985, cl, 349 pages, 3, $ 50.

Ragan, D.M., *Structural Geology: An Introduction to Geometrical Techniques*, 2nd Edition, New York, 1973, pb, 208 pages, appendix, 3, $ 15.

Ragan, D.M., *Structural Geology: An Introduction to Geometrical Techniques*, 3rd Edition, New York, 1985, pb, 393 pages, appendix, 3, $ 30.

Ramberg, H., *Gravity, Deformation and the Earth's Crust, In theory, experiments and geological application*, 2nd Edition, London, 1981, cl 452 pages, 3, $ 150.

Rankama, K. (Ed.), *The Precambrian*, Volume 1, The Geologic Systems Series, New York, 1963, cl, 279 pages, 3, $ 30 (Pre-Cambrian of Denmark, Norway, Sweden, and Finland).

Rankama, K. (Ed.), *The Precambrian*, Volume 4, The Geologic System Series, New York, 1970, cl, 288 pages, 3, $ 65 (Pc of South-Eastern and South-Central U.S.; North-Western U.S. Belt Series; Pc of Mexico).

Raymo, C., *The Crust of Our Earth: An Armchair Traveller's Guide to the New Geology,* Englewood Cliffs, 1983, pb, 135 pages, 3, $ 10.

Ramos, V.A. and Keppie, J.D. (Ed.), *Laurentia-Gondwana Connections before Pangea,* GSA SP 336, Boulder, 1999, pb, 276 pages, 1, $ 40.

Raymond, L.A. (Ed.), *Melanges: Their Nature, Origin and Significance,* GSA SP 198, Boulder, 1984, pb, 170 pages, 1, $ 50.

Raymond, L.A. (Ed.), *Melanges: Their Nature, Origin and Significance,* GSA SP 198, Boulder, 1984, pb, 170 pages, 3, $ 40.

Ramos, V.A. and Keppie, J.D. (Ed.), *Laurentia-Gondwana Connections Before Pangea,* GSA SP 336, Boulder, 1999, pb, 276 pages, 1, $ 45.

Redfern, R., *The Making of A Continent,* New York, 1983, cl, 242, pages, 5, $ 15.

Redfern, R., *The Making of A Continent,* New York, 1983, cl, 242, pages, 3, $ 20.

Redfern, R., *The Making of A Continent,* New York, 1983, pb, cl, 242, pages, 3, $ 10.

Reed, J.C., Jr. *et al.* (Ed.), *Precambrian: Coterminous U.S.,* DNAG Volume C2, Boulder, 1993, cl, 657 pages, 1 microfiche, 7 plates in separate slipcase, 1, $ 255.

Reiche, P., *A Survey of Weathering Processes and Products,* Albuquerque, 1950, pb, 95 pages, 3, $ 7.

Reiche, R.E. (Ed.), *Rio Grande Rift: Tectonics and Magnatism,* Washington, DC., 1979, cl, 438 pages, 1, $ 55.

Ringwood, A.E., *Composition and Petrology of the Earth's Mantle,* New York, 1975, cl, 618 pages, 3, $ 25.

Ritsema, A.R. (Ed.), *The Upper Mantle, Developments in Geotectonics 4,* Amsterdam, 1972, cl, 644 pages, 3, $ 45.

Ritter, D.F., *Process Geomorphology,* 1st Edition, Dubuque, 1978, cl, 603 pages, 3, $ 20.

Ritter, D.F., *Process Geomorphology,* 2nd Edition, Dubuque, 1986, cl, 579 pages, 3, $ 65.

Robertson, E.C., *The Nature of the Solid Earth,* Cambridge, 1972, cl, 677 pages, 3, $ 35.

Roddick, J.A. (Ed.), *Circum-Pacific Plutonic Terranes*, GSA Memoir 159, Boulder, 1983, cl, 316 pages, 2, $ 55.

Roddick, J.A. (Ed.), *Circum-Pacific Plutonic Terranes*, GSA Memoir 159, Boulder, 1983, cl, 316 pages, 3, $ 45.

Rodgers, J., *The Tectonics of the Appalachians*, Regional Geology Series, New York, 1970, cl, 271 pages, tectonic plate of Appalachians (in two parts) and abbreviation key (in pocket), 3, $ 65.

Rowland, S.M. and Duebendorfer, E.M., *Structural Analysis and Synthesis: A Laboratory Course in Structural Geology*, 2nd Edition, Oxford, 1994, pb, 279 pages, 1, $ 25.

Runcon, S.K. (Ed.), *Continental Drift*, International Geophysics Series, Volume 3, New York, 1962, cl, 338 pages, 3, $ 25.

Ryder, J.M. and D.E. Howes, 1984. *Terrain Information: A User's Guide to Terrain Maps in British Columbia*. British Columbia, Ministry of Environment, Surveys and Resource Mapping Branch, 16 p.

Schaer, J.P. and Rodgers, J. (Ed.), *The Anatomy of Mountain Ranges*, Princeton, 1987, cl, 298 pages, 1, $ 70.

Schinger, C.M., *The Magnetic Petrology of the Deep Crust and The Interpretation of Regional Magnetic Anomalies*, University Microfilms International, 1983, pb, 241 pages, 3, $ 20.

Schmidt, C.J. and Perry, W.J. Jr., *Interaction of the Rocky Mountain Foreland and the Cordilleran Thrust Belt*, Boulder, 1988, cl, 582 pages, 3 plates in pocket, 1 mf, 1, $ 75.

Schumm, S.A., *The Fluvial System*, New York, 1977, cl, 338 pages, 3, $ 20.

Scovel, J.L., *et al.*, *Atlas of Landforms*, U.S., Military Academy, West Point, 1965, spiral bound inside a stiff plastic binder, 155 pages, + bibliography + index, 3, $ 55.

Scrutton, R.A. (Ed.), *Dynamics of Passive Margins*, Boulder, 1982, cl, 200 pages, American Geophysical Union Geodynamics Series, Volume 6, 3, $ 60.

Secor, D.T., Jr., (Ed.), *Anatomy of the Alleghanian Orogeny as Seen from the Piedmont of South Carolina and Georgia*, Carolina Geological Society, 50th Anniversary Meeting, Columbia, 1987, pb, 97 pages, 1 plate, 3, $ 25.

Selley, R.C., *Concepts and Methods of Subsurface Facies Analysis*, AAPG Continuing Course Note Series # 9, Tulsa, 1978, pb, 82 pages, 3, $ 20.

Seyfert, C.K., *The Encyclopedia of Structural Geology and Plate Tectonics*, New York, 1987, cl, 876 pages, 3, $ 75.

Sharma, P.V., *Geophysical Methods in Geology*, New York, 1982, pb, 428 pages, 3, slight crease on cover and spine, ovg, $ 15.

Sheridan, R.E. and Grow, J.A., *The Atlantic Continental Margin: U.S.*, Boulder, 1988, cl, 610 pages, 8 plates (total of 16 sheets) in separate slipcase, 1, $ 30.

Shrock, R.R., *Sequence in Layered Rocks: A Study of Features and Structures Useful for Determining Top and Bottom or Order of Succession in Bedded and Tabular Rock Bodies*, New York, 1948, cl, 507 pages, 3, $ 40.

Simkin, T. *et al.*, *This Dynamic Planet: World Map of Volcanoes, Earthquakes and Plate Tectonics*, 1 large sheet, folded, 1:30,000,000 scale, 3, $ 10.

Sinha, A.K., Whalen, J.B. and Hogan, J.P., *The Nature of Magmatism in the Appalachian Orogen*, GSA M 191, Boulder, 1997, cl, 429 pages, 1, $ 45.

Slemmons, D.B., *Determination of Volcanism and Plutonic Plagioclases Using a Three or Four-Axis Universal Stage*, GSA SP 69, New York, 1962, pb, 64 pages, 5 figs., 12 plates, 3 tables, 3, $ 20.

Slemmons, D.B. *et al.*. *Neotectonics of North America*, GSA DNAG Tectonic Volume, Boulder, 1991, cl, 498 pages, 1, $ 75.

Smith, P.J., *Topics in Geophysics*, 1st Edition, Cambridge, 1973, cl, 246 pages, 3, $ 25.

Smith, R.B. and Eaton, G.P. (Ed.), *Cenozoic Tectonics and Regional Geophysics of the Western Cordillera*, GSA M 152, Boulder, 1978, cl, 388 pages, 9 plates in slipcase, 3, $ 45.

Stacey, F.D., *Physics of the Earth*, 1st Edition, New York, 1969, cl, 324 pages, 3, $ 25.

St. Julien, P. and Beland, J., *Major Structural Zones and Faults of the Northern Appalachians*, Waterloo, 1982, cl, 280 pages, 3, $ 55.

Stowell, H.H. and McClelland, W.C. (Ed.), *Tectonics of the Coastal Mountains, Southeastern Alaska and British Columbia*, GSA SP 343, Boulder, 2000, pb, 289 pages, 1, $ 45.

Strahler, A.N., *Principles of Physical Geology,* 1st Edition, New York, 1977, cl, 419 pages, $ 25.

Strangway, D.W., *History of the Earth's Magnetic Field,* New York, 1970, cl, 168 pages, 3, $ 10.

Sullivan, W., *Landprints on the Magnificent American Landscape,* New York, cl, 384 pages, 3, $ 25.

Sullivan, W., *Continents in Motion: The New Earth Debate,* 2nd Edition, New York, 1991, pb, 430 pages, 3, $ 25.

Sugimura, A. and Uyeda, S., *Island Arcs: Japan and Its Environs, Developments in Geotectonics Series # 3,* Amsterdam, 1973, cl, 247 pages, 3, $ 80.

Sutton G.H., *et al.,* (Ed.), *The Geophysics of the Pacific Ocean Basin and Its Margin,* AGU Geophysical Monograph 19, Washington, DC., 1976, cl, 480 pages, 3, $ 40.

Swanston, D.N., 1974. *The Forest Ecosystem of Southeast Alaska.* 5. Soil mass movement. U.S.D.A. Forest Service General Technical Report PNW-17, 22 p.

Swanston, D.N. and F.J. Swanson, 1976. Timber Harvesting, Mass Erosion, and Steepland Forest Geomorphology in the Pacific Northwest. In: *Geomorphology and Engineering.* Edited by D.R. Coates. Dowden, Hutchinson, and Ross. Inc., Stroudsburg, Pa.: 199-221.

Takeuchi, H., Uyeda, S., and Kanamori, H., *Debate about the Earth: Approach to Geophysics through Analysis of Continental Drift,* San Francisco, 1969, cl, 253 pages, 3, $ 25.

Takeuchi, H., Uyeda, S., and Kanamori, H., *Debate about the Earth: Approach to Geophysics through Analysis of Continental Drift,* Revised Edition, San Francisco, 1970, cl, 253 pages, 3, $ 10.

Tarling, D.H., *Palaeomagnetism: Principles and Applications in Geology, Geophysics, and Archaeology,* London, 1983, pb, 379 pages, 3, $ 45.

Taylor, S.R. and McLennan S.M., *The Continental Crust: Its Composition and Evolution,* Oxford, 1985, pb, 312 pages, 2, $ 65.

Telford, W.M. *et al., Applied Geophysics,* Cambridge, 1976, cl, 860 pages, 5, $ 55.

Telford, W.M. *et al., Applied Geophysics,* Cambridge, 1976, pb, 860 pages, 5, $ 25.

The Asphalt Institute, 1978. *Soils Manual for the Design of Pavement "U" Structures "U"*. Manual Series No. 10 (MS-10). The Asphalt Institute: 51-69.

Thornes, J.B., and Brunsden, D., *Geomorphology and Time*, New York, 1977, cl, 208 pags, 5, $ 25.

Timofeev, P.P. (Ed.), *Neotectonics and the Recent Geodynamics of Mobil Belts*, Academy of Sciences of the USSR, Transaction 427, in Russian, Mockba, 1988, cl, 365 pages, table of contents in English (pages 364-365), 3, $ 75.

Tobi, A.C. and Touret, J.L.R. (Ed.), *The Deep Proterozoic Crust in the North Atlantic Provinces*, Nato ASI Series, C, Volume 158, Dordrecht, 1985, cl, 603 pages, 3, $ 235.

Tornbury, W.D., *Regional Geomorphology of the United States*, 1st Edition, New York, 1965, cl, 609 pages, 3, $ 75.

Tozer, E.T. *et al.*, *Caledonian-Appalachian Orogen of the North Atlantic Region*, Geological Survey of Canada, Paper 78-13, Project 27, Caledonide Orogen, Ottawa, 1978, pb, 3, $ 40.

Trettin, H.P., *Geology of the Innuitian Orogen and Arctic Platform of Canada and Greenland*, GSA DNAG Volume E, Ottawa, cl, 569 pages, 9 figures in separate slipcase, 1, $ 125.

Tucker, R.H. *et al.*, *Global Geophysics*, 1st Edition, London, 1970, cl, 199 pages, 3, $ 25.

Tull, J.F., Bearce, D.N. and Guthrie, G.M., *Early Evolution of the Appalachian Miogeocline: Upper Precambrian—Lower Paleozoic Stratigraphy of the Talladega Slate Belt*, Guidebook for the 22nd Annual Trip, Alabama Geological Society, 1985, spiral bound, 92 pages, 2, $ 20.

Underwood, M.B., (Ed.), *Thermal Evolution of the Tertiary Shimanto Belt, Southwest Japan: An Example of Ridge-Trench Interaction*, GSA SP 273, Boulder, 1993, pb, 172 pages, 1, $ 38.

Uyeda, S., *The New View of the Earth-Moving Continents and Moving Oceans*, San Francisco, 1978, pb, 217 pages, 2, $ 10.

Valentino, D.W. and Gates, A.E. (Ed.), *The Mid-Atlantic Piedmont: The Missing Link of the Appalachians*, GSA SP 330, Boulder, 1999, pb, 139 pages, 1, $ 38.

Van Andel, T.H., *New Views of an Old Planet: Continental Drift and the History of the Earth*, 1st Edition, New York, 1985, cl, 324 pages, some underlining in text, ovg, 3, $ 15.

Vacquier, V., Steenland, C., Henderson, R.G. and Zietz, I., *Interpretation of Aeromagnetic Maps*, GSA Memoir 47, New York, 1951, cl, 151 pages, 3, $ 25.

Varnes, D.J., 1978. Slope Movement Types and Processes. In: *Landslides Analysis and Control*. Edited by R.L. Schuster and R.J. Krizek. National Academy of Sciences, Special Report 176, Washington, DC.: 11-33.

Veevers, J.J. and Powell, C. McA., (Ed.), *Permian-Triassic Pangean Basins and Foldbelts Along the Pathalassan Margin of Gondwanaland*, GSA M, 184, Boulder, 1994 cl, 368 pages, 3, $ 55.

Veevers, J.J. and Tewari, R.C., *Gondwana Master Basin of Peninsular India: Between Tethys and the Interior of the Gondwanaland Province of Pangea*, GSA M 187, Boulder, 1995, cl, 72 pages, 1, $ 26.

von Engeln, O.D., *Geomorhology: Systematic and Regional*, New York, 1942, cl, 655 pages, 3, $ 25.

von Höfer, H., *Anleitung Zum Geologischen Beobachten, Kartieren Und Profilieren*, Braunschweig, 1915, cl, 82 seite, 3, $ 30.

Wegener, A., *The Origin of Continents and Oceans*, Translated by J. Biram, New York, 1996, pb, 246 pages, 3, $ 5.

Wessel, G.R., *The Geology of Plate Tectonics*, Boulder, 2005, wall chart 39″ × 53″, folded, 1, $ 16 (Now back in stock in a larger version you can actually read without a magnifying glass).

Weyman, D., *Tectonic Processes*, London, 1981, pb, 102 pages, 3, $ 15.

White, S.E., 1981. *Alpine Mass Movement Forms (Noncatastrophic): Classification, Description, and Significance*. Arctic and Alpine Research, 13: 127-137.

Whitten, E.H.T., *Structural Geology of Folded Rocks*, Chicago, 1966, cl, 663 pages, index, 3, $ 75.

Williams, H., *Geology of the Appalachian-Caledonian Orogen in Canada and Greenland*, GSA DNAG Volume F1, Ottawa, 1995, cl, 944 pages, 1 plate in pocket, 1, $ 125.

Willis, B. and Willis, R., *Geologic Structures*, 2nd Edition, Revised and Enlarged, New York, 1929, cl, 518 pages, 3, $ 35.

Windley, B.F., *The Evolving Continents,* 1st Edition, New York, 1977, cl, 385 pages, 3, $ 25.

Windley, B.F., *The Evolving Continents,* 1st Edition, New York, 1977, pb, 385 pages, 3, $ 20.

Windley, B.F., *The Evolving Continents,* 2nd Edition, Chichester, 1984, pb, 399 pages, 3, $ 30.

Winkler-Hermaden, A., *Geologisches Kraftespiel Und Landformung,* Wein, 1957, cl, 822 seite, 120 textabbilden, 5 tafeln, 3, $ 75.

Wones, D.R. (Ed.), *The Caledonides in the USA,* Proceedings of the IGCP Project 27: Caledonide Orogen, 1979 Meeting, Blacksburg, 1980, pb, 329 pages, 19 page appendix, 3, $ 35.

Wycoff, J., *Rock, Time, and Landforms,* New York, 1963, cl, 372 pages, 3, $ 25.

Wyllie, P.J., *The Way the Earth Works: An Introduction to the New Global Geology and Its Revolutionary Development.* New York, 1976, cl, 296 pages, 3, $ 20.

Zen, E-An White, W.S. and Hadley, J.B., *Studies of Appalachian Geology: Northern and Maritime,* New York, 1968, cl, 475 pages, 3, $ 95.

INDEX